AF699267

Shu-fang Xu

An Introduction to Inverse Algebraic Eigenvalue Problems

Shu-fang Xu

An Introduction to Inverse Algebraic Eigenvalue Problems

Peking University Press

Vieweg Publishing

Professor Shu-fang Xu
School of Mathematical Sciences
Peking University
People's Republic of China

Mathematical Subject Classification: 15A18, 65F15, 93B60

All rights reserved
© Peking University Press, Beijing, 1998, and
© Friedr. Vieweg & Sohn Verlagsgesellschaft mbH, Braunschweig/Wiesbaden, 1998

Vieweg is a subsidiary company of the Bertelsmann Professional Information.

No part of this publication may be reproduced, stored in a retrieval system or transmitted, mechanical, photocopying or otherwise, without prior permission of the copyright holder.

http://www.vieweg.de

Typesetting: Peking University Press
Printing and binding: C & C OFFSET PRINTING CO., LTD. (H. K.)
Printed in Hong Kong

ISBN 7-301-03602-7/O · 405
ISBN 3-528-06684-9

Preface

Inverse algebraic eigenvalue problem, roughly speaking, is how to determine the elements of a matrix from its spectrum data. This kind of problem has been of great value for many applications, including control theory, structure mechanics, geology, molecular spectroscopy, and so on. Therefore, in recent years, many authors have been devoted to the study of this kind of problem and some significant progress has been made. One of the purposes of this book is to provide the reader with the recent developments in this field.

Although a given inverse algebraic eigenvalue problem is just a matrix computational problem, it is often unable to be solved with the traditional skills from matrix computations. In fact, its participants need skills from various areas. The other purpose of this book is to introduce the reader to these skills as they pertain to inverse algebraic eigenvalue problems.

Since the mid-1950's, many different kinds of inverse eigenvalue problems have appeared from varied applications. Therefore, it would be neither practical nor sensible to cover all aspects of this area in a single book. Instead, this book is confined to four relatively well-established problems: Jacobi matrix inverse eigenvalue problems (Chapter 2), pole assignment problems (Chapter 3), additive and multiplicative inverse eigenvalue problems (Chapter 4), and inverse eigenvalue problems for nonnegative matrices (Chapter 5).

Chapter 1 presents some useful concepts and facts which are used repeatedly throughout this book. It is intended that this material be used to serve the reader as a short review prior to beginning the book or as a convenient reference when necessary. For each problem in Chapters 2–4, the contents follow the logical order of intellectual activity, progressing from problem introduction, through solvability and sensitivity analysis, to numerical methods. In addition to the rigorously mathematical theory, it is those efficient and stable algorithms that are emphasized. The last chapter is mainly devoted to the smallest realizable spectral radius. In my opinion, perhaps this is one of the most important developments in the field of the inverse eigenvalue problems for nonnegative matrices in recent years. At the end of each chapter some notes and references are given, which emphasize original sources and historical development. They also point out some untouched topics to the reader.

This book is written for graduate students, computational scientists and engineers, numerical analysts, and mathematicians seeking an introduction to inverse algebraic eigenvalue problems. The text is largely self-contained and is based on classical analysis and modern matrix computation techniques.

I wish to express my sincere appreciation to my Ph.D. thesis adviser, Professor Ji-guang Sun, for leading me to the interesting area of inverse algebraic eigenvalue problems and for his long-standing and continuing friendship, help, and encouragement. I dedicate this book to him.

I wish to thank my colleague, Dr. Xing-zhi Zhan, for reading and commenting on the manuscript. Parts of this book were written while I was visiting the University of California, Santa Barbara, U.S.A. I thank the University of California for providing facilities, and my host, Professor A.J. Laub, and his colleague, Professor C.S. Kenney, for helping to make my stay both pleasant and scientifically profitable. I also thank Professor G.H. Golub and Professor G.S. Miminis for providing me with several useful references.

I owe most to my wife, Hui-min Lian, who typeset all the manuscript using LaTeX 2_ε, and who has been a source of steady support and encouragement.

Finally I would like to thank the editors, Shu-qing Qiu and Yong Liu, who have contributed to the publication of this book.

Beijing, Spring, 1997 *Shu-fang Xu*

Contents

Chapter 1

Preliminaries

This preliminary chapter contains a short exposition of the basic notation and of the useful concepts and facts which are used repeatedly throughout this book. It begins with a summary of the basic notation and definitions, continues with a review of norms, eigenvalues and eigenvectors, singular value decomposition, symmetric matrices, Vandermonde matrices, and the Kronecker product. The chapter ends with a brief discussion of the implicit function theorem and the differentiability of eigenvalues and eigenvectors of a real matrix dependent on several real parameters.

It is intended that this material be used to serve the reader as a short review prior to beginning the book or as a convenient reference when necessary. Some of it will be familiar to the reader and some of it will probably be new. We suggest that the reader read the chapter through "lightly" at first; and then refer back to it for details as needed.

1.1 Notation and Definitions

Throughout this book we will use the following notational conventions. Scalar integers will be denoted by lower case Roman letters such as i, j, k, l, m, and n. Non-integer scalars (real or complex numbers) will be denoted either by lower case Greek letters such as α, β, ε, λ, and σ, or by lower case Roman letters other than i, j, k, l, m, or n. Column vectors will be denoted by lower case Roman letters such as x and y. Matrices are denoted by upper case Roman letters such as A, T, and X except for diagonal matrices whose diagonal entries are eigenvalues or singular value. Such diagonal matrices will be denoted by upper case Greek letters such as Λ or Σ. Diagonal matrices may also be denoted by $\operatorname{diag}(\alpha_1, \dots, \alpha_n)$.

The symbol 0 will stand for anything that is zero (a number, vector, or matrix), according to context. The symbol I will denote the identity matrix of size implied by the context. Sometimes we also use I_p to denote the identity matrix of order p.

The symbol e_i will denote the ith column of I. That is, e_i is the ith unit vector, whose component are zero except for the ith components that is one. The symbol δ_{ij} will denote the Kronecker delta, that is,

$$\delta_{ij} = \begin{cases} 0, & \text{if } i \neq j, \\ 1, & \text{if } i = j. \end{cases}$$

The set of all real numbers will be denoted by $\mathbf{R}$. The set of all n-dimensional column vectors with components in $\mathbf{R}$ will be denoted by $\mathbf{R}^n$. The set of all $m \times n$ matrices with elements in $\mathbf{R}$ will be denoted by $\mathbf{R}^{m\times n}$. The corresponding complex sets will be denoted by $\mathbf{C}$, $\mathbf{C}^n$, and $\mathbf{C}^{m\times n}$.

Let A be an $m \times n$ matrix. The notation A^T, A^*, and $A^\dagger$ will denote the transpose, the conjugate transpose, and the Moore-Penrose inverse of A, respectively. If A is a square matrix, that is, $m = n$, the notation $\operatorname{tr} A$, $\det A$, and $\operatorname{rank} A$ will denote the trace, the determinant, and the rank of A, respectively. A square matrix A is said to be **nonsingular** if $\det A \neq 0$. If A is nonsingular, the notation A^{-1} will denote the inverse of A. The notation $\operatorname{adj} A$ will denote the **adjoint matrix** of a square matrix. It is a fact that

$$A(\operatorname{adj} A) = (\operatorname{adj} A)A = (\det A)I$$

holds for any square matrix A. In particular, in the case of *a* nonsingular matrix A we have $A^{-1} = \dfrac{1}{\det A}\operatorname{adj} A$.

Let A be an $m\times n$ complex matrix. The notation $\overline{A}$, ReA, and ImA will denote the complex conjugate, the real part, and the imaginary part of the elements of A. For example, if

$$A = [0,\ 3 + 2i,\ i],$$

then

$$\overline{A} = [0,\ 3 - 2i,\ -i], \quad \operatorname{Re} A = [0,\ 3,\ 0], \quad \operatorname{Im} A = [0,\ 2,\ 1].$$

The notation $A = [\alpha_{ij}] \in \mathbf{R}^{m\times n}$ means that the $(i,\ j)$ element of A is α_{ij}. In some case we also use $A(i,\ j)$ to denote the $(i,\ j)$ element of a matrix A. The notation $x = (\xi_1, \dots, \xi_n)^T \in \mathbf{R}^n$ means that the ith component of x is ξ_i. Sometimes we also use $x(i)$ to denote the ith component of a vector x.

The notation $k = i : l : j$ means that k takes all the values starting from i until j with step l; if $l = 1$ the above may be written as $k = i : j$ for simplicity. Given a matrix A, the MATLAB-like notation $A(r_1 : r_2, c_1 : c_2)$ will denote the submatrix of A that consists of rows r_1 through r_2 and columns c_1 through c_2. If all rows (or columns) of A are included, the above submatrix may be written as $A(:, c_1 : c_2)$ (or $A(r_1 : r_2, :)$). For example, if

$$A = \begin{bmatrix} 1 & 2 & 3 & 4 \\ 5 & 6 & 7 & 8 \\ 9 & 10 & 11 & 12 \\ 13 & 14 & 15 & 16 \end{bmatrix},$$

then

$$A(2:3,2:4) = \begin{bmatrix} 6 & 7 & 8 \\ 10 & 11 & 12 \end{bmatrix},$$

$$A(:,1:1) = \begin{bmatrix} 1 \\ 5 \\ 9 \\ 13 \end{bmatrix}, \quad A(2:4,:) = \begin{bmatrix} 5 & 6 & 7 & 8 \\ 9 & 10 & 11 & 12 \\ 13 & 14 & 15 & 16 \end{bmatrix}.$$

Let A be an $m \times n$ matrix. In general we can partition both rows and columns of A to obtain

$$A = \begin{bmatrix} A_{11} & A_{12} & \cdots & A_{1q} \\ A_{21} & A_{22} & \cdots & A_{2q} \\ \vdots & \vdots & & \vdots \\ A_{p1} & A_{p2} & \cdots & A_{pq} \end{bmatrix} \begin{matrix} m_1 \\ m_2 \\ \\ m_p \end{matrix}, \qquad \begin{matrix} n_1 & n_2 & & n_q \end{matrix}$$

where $m_1 + \cdots + m_p = m$ and $n_1 + \cdots + n_q = n$. That is, A_{ij} is the $m_i \times n_j$ submatrix of A: $A_{ij} = A(r_i+1 : r_i+m_i, s_j+1 : s_j+n_j)$, where $r_i = m_1+\cdots+m_{i-1}$, $s_j = n_1 + \cdots + n_{j-1}$, and $n_0 = m_0 = 0$. We say that $A = [A_{ij}]$ is a $p \times q$ **block matrix**. Partitioning of matrices is often a convenient device for perception of useful structure.

In general, for an $m \times n$ matrix A, we frequently write

$$A = [a_1, a_2, \ldots, a_n].$$

This means that a_j is the jth column of A.

Let $x_1, \ldots, x_m \in \mathbf{R}^n$. If $\alpha_1 x_1 + \cdots + \alpha_m x_m = 0$ implies $\alpha_1 = \alpha_2 = \cdots = \alpha_m = 0$, then these vectors are said to be **linearly independent**. Otherwise, if $\alpha_1 x_1 + \cdots + \alpha_m x_m = 0$ with $\alpha_i \neq 0$ for at least one i, then the vectors are said to be **linearly dependent**. The set of all linear combinations of $x_1, \ldots, x_n$ is referred to as the **span** of $x_1, \ldots, x_m$ and is denoted by span $\{x_1, \ldots, x_m\}$. That is,

$$\operatorname{span}\{x_1, \ldots, x_m\} = \left\{ \sum_{j=1}^{m} \alpha_j x_j \ : \ \alpha_j \in \mathbf{R} \right\}.$$

Let $\mathcal{S}$ be a subset of $\mathbf{R}^n$, i.e., $\mathcal{S} \subset \mathbf{R}^n$. If $\mathcal{S}$ is also a vector space, then $\mathcal{S}$ is said to be a **subspace** of $\mathbf{R}^n$. Obviously, span$\{x_1, \ldots, x_m\}$ is a subspace. For any subspace $\mathcal{S}$, it is possible to find linearly independent vectors $x_1, \ldots, x_m \in \mathcal{S}$ such that $\mathcal{S} = \text{span}\ \{x_1, \ldots, x_m\}$. A set of such vectors $x_1, \ldots, x_m$ is said to be a **basis** for the subspace. All bases for a subspace $\mathcal{S}$ have the same number of elements. This number is referred to as the **dimension** of $\mathcal{S}$ and is denoted by dim($\mathcal{S}$).

Given any subset $\mathcal{S} \subset \mathbf{R}^n$, the orthogonal complement of $\mathcal{S}$ is the set:

$$\mathcal{S}^{\perp} = \left\{ x \in \mathbf{R}^n \ : \ x^T y = 0 \text{ for all } y \in \mathcal{S} \right\}.$$

Even if $\mathcal{S}$ is not a subspace, $\mathcal{S}^{\perp}$ is always a subspace.

There are two important subspaces associated with an $m \times n$ matrix A. The **range** of A is defined by

$$\mathcal{R}(A) = \{ y \in \mathbf{R}^m \ : \ y = Ax \text{ for some } x \in \mathbf{R}^n \}$$

and the **null space** of A is defined by

$$\mathcal{N}(A) = \{ x \in \mathbf{R}^n \ : \ Ax = 0 \}.$$

1.2 Vector and Matrix Norms

A **vector norm** on $\mathbf{R}^n$ is a function that assign to each $x \in \mathbf{R}^n$ a real number $||x||$, called the norm of x, such that the following three properties are satisfied for all x, $y \in \mathbf{R}^n$ and all $\alpha \in \mathbf{R}$:

(1) **Positive definite property**: $||x|| > 0$ if $x \neq 0$ and $||0|| = 0$;
(2) **Absolute homogeneity**: $||\alpha x|| = |\alpha|\, ||x||$;
(3) **Triangular inequality**: $||x + y|| \leq ||x|| + ||y||$.

A useful class of vector norms are the ***p*-norms** defined by

$$||x||_p = \left(\sum_{i=1}^{n} |x(i)|^p \right)^{\frac{1}{p}}.$$

Of these the following three norms are the ones most commonly used in practice:

$$||x||_1 = \sum_{i=1}^{n} |x(i)|,$$

$$||x||_2 = \left(\sum_{i=1}^{n} |x(i)|^2 \right)^{\frac{1}{2}},$$

$$||x||_\infty = \max_{1 \leq i \leq n} |x(i)|.$$

The 2-norm is also called the **Euclidean norm**. Unless otherwise stated, a unit vector x means that $||x||_2 = 1$.

A very important result concerning 2-norm is the **Cauchy-Schwarz inequality**:

$$|x^T y| \leq ||x||_2 \, ||y||_2, \quad x, \, y \in \mathbf{R}^n.$$

A very important property of vector norms on $\mathbf{R}^n$ is that all vector norms on $\mathbf{R}^n$ are equivalent, i.e., if $||\cdot||_\alpha$ and $||\cdot||_\beta$ are two norms on $\mathbf{R}^n$, then there exist positive constants c_1 and c_2 such that

$$c_1||x||_\alpha \leq ||x||_\beta \leq c_2||x||_\alpha$$

for all $x \in \mathbf{R}^n$. For example, if $x \in \mathbf{R}^n$, then

$$||x||_2 \leq ||x||_1 \leq \sqrt{n}||x||_2,$$
$$||x||_\infty \leq ||x||_2 \leq \sqrt{n}||x||_\infty,$$
$$||x||_\infty \leq ||x||_1 \leq n||x||_\infty.$$

Let us now turn our attention to matrix norms. A **matrix norm** is a function that assigns to each $A \in \mathbf{R}^{n\times n}$ a real number $||A||$, called the norm of A, such that the following four properties are satisfied for all $A, \, B \in \mathbf{R}^{n\times n}$ and all $\alpha \in \mathbf{R}$:

(1) **Positive definite property**: $||A|| > 0$ if $A \neq 0$ and $||0|| = 0$;
(2) **Absolute homogeneity**: $||\alpha A|| = |\alpha| \, ||A||$;
(3) **Triangular inequality**: $||A + B|| \leq ||A|| + ||B||$;
(4) **Consistency**: $||AB|| \leq ||A|| \, ||B||$.

One of the most frequently used matrix norms is the **Frobenius norm** defined by

$$||A||_F = \left(\sum_{j=1}^{n} \sum_{i=1}^{n} |A(i,j)|^2 \right)^{\frac{1}{2}}.$$

Because it is the same as the Euclidean norm on $\mathbf{R}^n$, we already know that it satisfies properties 1, 2, and 3. The consistency property 4 can be deduced from the Cauchy-Schwarz inequality and its proof is left as an exercise.

Every vector norm on $\mathbf{R}^n$ can be used to define a matrix norm on $\mathbf{R}^{n\times n}$ in a nature way. Given the vector norm $||\cdot||_v$, the matrix norm induced by $||\cdot||_v$ is defined by

$$||A||_M = \max_{x\neq 0} \frac{||Ax||_v}{||x||_v}.$$

Some of the most important matrix norms are induced by p-norms. For $1 \leq p \leq \infty$, the norm induced by the p-norm is called the **matrix p-norm:**

$$||A||_p = \max_{x \neq 0} \frac{||Ax||_p}{||x||_p}.$$

The matrix 2-norm is also known as the **spectral norm**. This norm has great theoretical importance. Its drawback is that it is hard to compute; while it is easy for the Frobenius norm. The other important cases are $p = 1$ and $p = \infty$. These norms can be computed easily:

$$||A||_1 = \max_{1 \leq j \leq n} \sum_{i=1}^{n} |A(i,j)|,$$

$$||A||_\infty = \max_{1 \leq i \leq n} \sum_{j=1}^{n} |A(i,j)|.$$

Thus $||A||_1$ is found by summing the absolute values of the elements in each column of A and then taking the largest of these column sums. Therefore the matrix 1-norm is sometimes called the **column sum norm**. Similarly, the matrix ∞-norm is sometimes called the **row sum norm.**

1.3 Eigenvalues and Eigenvectors

In this section we survey the basic definitions and properties of eigenvalues and eigenvectors which is the corner stone of this book. In the mathematical analysis of the eigenproblem it is convenient to work in the complex field.

Let $A \in \mathbf{C}^{n \times n}$. Define

$$p_A(z) = \det(zI - A), \quad z \in C.$$

Then $p_A(z)$ is a monic polynomial of degree n and is referred to as the **characteristic polynomial** of A. By the fundamental theorem of algebra it follows that $p_A(z)$ has exactly n roots, counting multiplicities. These roots are called the **eigenvalues** of A. The set of these roots is called the **spectrum** of A and is denoted by $\lambda(A)$. The nonnegative real number defined by

$$\rho(A) = \max\{|\lambda| \ : \ \lambda \in \lambda(A)\}$$

is called the **spectral radius** of A. This is just the radius of the smallest disc centered at the origin in the complex plane that includes all the eigenvalues of A.

Let $\lambda \in \lambda(A)$. Then $\lambda I - A$ is singular. The nonzero vector $x \in \mathbf{C}^n$ is called a **right eigenvector** of A associated with λ if $Ax = \lambda x$ and a **left eigenvector** if $x^*A = \lambda x^*$. In general, these two concepts are different.

For given $\lambda \in \lambda(A)$, the set of all vectors $x \in \mathbf{C}^n$ satisfying $Ax = \lambda x$ is called the **eigenspace** of A corresponding to the eigenvalue λ and is denoted by $\mathcal{L}_\lambda$, i.e.,

$$\mathcal{L}_\lambda = \{x \in \mathbf{C}^n \ : \ Ax = \lambda x\}.$$

Note that every nonzero element of $\mathcal{L}_\lambda$ is an eigenvector of A corresponding to λ. The dimension of $\mathcal{L}_\lambda$ is called the **geometric multiplicity** of the eigenvalue λ. The multiplicity of λ as a root of the characteristic polynomial $p_A(z)$ is called the **algebraic multiplicity** of the eigenvalue λ. It is easy to show that the geometric multiplicity of an eigenvalue λ is never more, and can be less, than its algebraic multiplicity.

Throughout this book we will follow the following convention. Unless otherwise stated, "eigenvector" means "right eigenvector", and "multiplicity" means "algebraic multiplicity".

Let $\lambda \in \lambda(A)$, and let x be an eigenvector of A associated with λ. Sometimes the pair $(\lambda,\ x)$ is called an **eigenpair** of A.

Let $\lambda \in \lambda(A)$. If the multiplicity of λ is 1, then λ is referred to as a **simple eigenvalue** of A. Otherwise, if the multiplicity of λ is greater than 1, then λ is referred to as a **multiple eigenvalue** of A.

Let $x_1, x_2, \ldots, x_k$ be eigenvectors of $A \in \mathbf{C}^{n\times n}$ associated with distinct eigenvalues $\lambda_1, \lambda_2, \ldots, \lambda_m$. Then $x_1, \ldots, x_m$ are linearly independent.

If $A \in \mathbf{R}^{n\times n}$, then the characteristic polynomial $p_A(z)$ of A has real coefficients. Therefore, if a real matrix has complex eigenvalues, they must occur in complex conjugate pairs; that is, if $\lambda = \alpha + i\beta$ $(\alpha,\ \beta \in \mathbf{R})$ is an eigenvalue of a real matrix A, then $\lambda = \alpha - i\beta$ is also an eigenvalue of A. Moreover, every real eigenvalue of a real matrix has a real eigenvector associated with it.

A matrix $B \in \mathbf{C}^{n\times n}$ is said to be **similar** to a matrix $A \in \mathbf{C}^{n\times n}$ if there exists a nonsingular matrix X such that

$$B = X^{-1}AX$$

The transformation $A \longrightarrow X^{-1}AX$ is called a **similarity transformation** by the **similarity matrix** X. The relation "B is similar to A" is sometimes abbreviated $B \sim A$

Let $A, B \in \mathbf{C}^{n\times n}$. It is easy to show that if A is similar to B, then the characteristic polynomial of B is the same as that of A. Hence, $A \sim B$ implies $\lambda(A) = \lambda(B)$. Furthermore, if $B = X^{-1}AX$, then v is an eigenvector of A corresponding to an eigenvalue λ if and only if $X^{-1}v$ is an eigenvector of B corresponding to λ. Therefore, a similarity transformation preserves the algebraic and geometric multiplicity of each eigenvalue.

Let $A \in \mathbf{C}^{n\times n}$. If A is similar to a diagonal matrix, then A is said to be **diagonalizable**. It is easy to verify that A is diagonalizable if and only if A has n linearly independent eigenvectors. In particular, if A has n distinct eigenvalues, then A is diagonalizable. In view of this fact, sometimes a diagonalizable matrix is said to be a **nondefective matrix.**

A major theme of matrix theory is the reduction of a matrix to a simple form by similarity transformations. Perhaps the most fundamentally useful facts of elementary matrix theory are Schur's unitary triangularization theorem and the Jordan decomposition theorem. We now survey the two important results, beginning with some elementary concepts.

A matrix $U \in \mathbf{C}^{n\times n}$ is said to be a **unitary matrix** if $U^*U = I$, that is, $U^* = U^{-1}$. A real unitary matrix is called an **orthogonal matrix**; that is, $Q \in \mathbf{R}^{n\times n}$ is orthogonal if and only if $Q^T = Q^{-1}$. A matrix $T = [t_{ij}] \in \mathbf{C}^{n\times n}$ is said to be an **upper-triangular matrix** if $t_{ij} = 0$ whenever $i > j$.

Two matrices A, $B \in \mathbf{C}^{n\times n}$ are said to be **unitarily similar** if there exists a unitary matrix $U \in \mathbf{C}^{n\times n}$ such that $B = U^*AU$. If A, B, and U are real, then U is orthogonal and A and B are said to be **orthogonally similar.** Unitary similarity transformations have some nice properties not possessed by similarity transformations in general.

Theorem 1.3.1 (**Schur Decomposition Theorem**) *Given $A \in \mathbf{C}^{n\times n}$, with eigenvalues $\lambda_1, \dots, \lambda_n$ in any prescribed order, then there exists a unitary matrix $U \in \mathbf{C}^{n\times n}$ such that $U^*AU = T = [t_{ij}]$ is upper triangular, with diagonal elements $t_{ii} = \lambda_i$, $i = 1, 2, \dots, n$. That is, every square matrix A is unitarily similar to a triangular matrix T, whose diagonal elements are the eigenvalues of A in a prescribed order.*

The proof of Schur's theorem is not difficult. But the result is one of the most important tools in theoretical and computational linear algebra. It is well known that the extensively used and numerically reliable QR algorithm is based on this result.

If $A \in \mathbf{R}^{n\times n}$, using only orthogonal similarity transformations we can reduce A into a block upper triangular form with either 1×1 or 2×2 diagonal blocks.

Theorem 1.3.2 (**Real Schur Decomposition Theorem**) *Let $A \in \mathbf{R}^{n\times n}$. Then there exists an orthogonal matrix $Q \in \mathbf{R}^{n\times n}$ such that*

$$Q^TAQ = \begin{bmatrix} T_{11} & T_{12} & \cdots & T_{1m} \\ 0 & T_{22} & \cdots & T_{2m} \\ \vdots & \vdots & & \vdots \\ 0 & 0 & \cdots & T_{mm} \end{bmatrix}$$

is a block triangular matrix, where each main diagonal block T_{ii} is in either $\mathbf{R}$ *or* $\mathbf{R}^{2\times 2}$. *Each 1×1 block is a real eigenvalue of A. Each 2×2 block has its eigenvalues as a complex conjugate pair of eigenvalues of A.*

Notice that the Schur upper triangular form is the simplest form achievable under unitary similarity transformations. If we wish to get much simpler form, we must use the nonunitary similarity transformations. In fact, by using the nonunitary similarity transformations, it is possible to reduce a given matrix into the "almost diagonal" form — the Jordan canonical form.

Theorem 1.3.3 (**Jordan Decomposition Theorem**) *If $A \in \mathbf{C}^{n\times n}$, then there exists a nonsingular matrix $X \in \mathbf{C}^{n\times n}$ such that*

$$X^{-1}AX = J \equiv \operatorname{diag}(J_1, J_2, \ldots, J_k),$$

called the **Jordan canonical form** *of A, where*

$$J_i = \begin{bmatrix} \lambda_i & 1 & 0 & \cdots & 0 \\ 0 & \lambda_i & 1 & \ddots & \vdots \\ \vdots & \ddots & \ddots & \ddots & 0 \\ \vdots & & \ddots & \ddots & 1 \\ 0 & \cdots & \cdots & 0 & \lambda_i \end{bmatrix} \in \mathbf{R}^{n_i \times n_i},$$

referred to as **Jordan blocks**, *and $n_1 + \cdots + n_k = n$. The Jordan canonical form of A is unique up to permutation of the diagonal Jordan blocks. The eigenvalues λ_i, $i = 1, 2, \ldots, k$, are not necessarily distinct. If A is a real matrix with only real eigenvalues, then the similarity matrix X can be taken to be real.*

Once one knows the Jordan canonical form of a matrix, all the linear algebraic information about the given matrix is known at a glance. For example, it is easy to see that:

- The number k of Jordan blocks (counting multiple occurrences of the same block) is the number of linearly independent eigenvectors of J.
- The matrix J is diagonalizable if and only if $k = n$.
- The number of Jordan blocks corresponding to a given eigenvalues is the geometric multiplicity of the eigenvalue. The sum of the orders of all the Jordan blocks corresponding to a given eigenvalue is the algebraic multiplicity of the eigenvalue.

We conclude this section by stating the well-known Cayley-Hamilton theorem, which will be used in the sequel.

Theorem 1.3.4 (Cayley-Hamilton Theorem) *Let $A \in \mathbf{C}^{n\times n}$, and let $p_A(z)$ be the characteristic polynomial of A. Then $p_A(A) = 0$.*

1.4 The Singular Value Decomposition

In this section we survey the extremely useful singular value decomposition. We begin with a few definitions.

A set of vectors $\{x_1, \ldots, x_m\}$ in $\mathbf{R}^m$ is **orthogonal** if $x_i^T x_j = 0$ whenever $i \neq j$ and **orthonormal** if $x_i^T x_j = \delta_{ij}$. The vectors $v_1, \ldots, v_k$ form a **orthonormal basis** for a subspace $\mathcal{S} \subset \mathbf{R}^n$ if they are orthonormal and $\text{span}\{v_1, \ldots, v_k\} = \mathcal{S}$. If $Q = [q_1, q_2, \ldots, q_n] \in \mathbf{R}^n$ is orthogonal, then the q_i form an orthonormal basis for $\mathbf{R}^n$.

Theorem 1.4.1 (The Singular Value Decomposition Theorem) *Let $A \in \mathbf{R}^{m\times n}$ have rank r. Then there exist real numbers $\sigma_1 \geq \sigma_2 \geq \cdots \geq \sigma_r > 0$ and orthogonal matrices*

$$U = [u_1, \ldots, u_m] \in \mathbf{R}^{m\times m} \quad \textit{and} \quad V = [v_1, \ldots, v_n] \in \mathbf{R}^{n\times n}$$

such that

$$U^T A V = \begin{bmatrix} \Sigma_r & 0 \\ 0 & 0 \end{bmatrix}, \tag{1.4.1}$$

where $\Sigma_r = \text{diag}\,(\sigma_1, \sigma_2, \ldots, \sigma_r)$.

The decomposition (1.4.1) is called the **singular value decomposition** of A. The numbers $\sigma_1, \ldots, \sigma_r$ are called **singular values** of A. Let $k = \min\{n, m\}$. If $r < k$, it is usual to adjoin $k - r$ zero singular values $\sigma_{r+1} = \ldots = \sigma_k = 0$. The vectors u_i and v_i are called the ith **left singular vector** and the ith **right singular vector** respectively. It follows from (1.4.1) that

$$\begin{aligned} &Av_i = \sigma_i u_i, \quad A^T u_i = \sigma_i v_i, \quad i = 1, 2, \ldots, r, \\ &Av_i = 0, \quad i = r+1, \ldots, n, \\ &A^T u_i = 0, \quad i = r+1, \ldots, m. \end{aligned} \tag{1.4.2}$$

From (1.4.2) it is easy to see that the singular value decomposition displays the orthonormal bases for the four spaces $\mathcal{R}(A)$, $\mathcal{N}(A)$, $\mathcal{R}(A^T)$, and $\mathcal{N}(A^T)$. It is clear that

$$\mathcal{R}(A^T) = \text{span}\,\{v_1, \ldots, v_r\}, \quad \mathcal{R}(A) = \text{span}\,\{u_1, \ldots, u_r\},$$
$$\mathcal{N}(A) = \text{span}\,\{v_{r+1}, \ldots, v_n\}, \quad \mathcal{N}(A^T) = \text{span}\,\{u_{r+1}, \ldots, u_m\}.$$

Let $U_r = [u_1, \dots, u_r]$ and $V_r = [v_1, \dots, v_r]$. Then the singular value decomposition (1.4.1) can be written in the form

$$A = U_r \Sigma_r V_r^T,$$

which is called the **singular value factorization** of A.

It is convenient to have the following notation for designating singular values: $\sigma_i(A)$ will denote the ith largest singular value of A; $\sigma_{\max}(A)$ will denote the largest singular value of A; and $\sigma_{\min}(A)$ will denote the smallest singular value of A.

There are many applications of the singular value decomposition. Here we give several elementary consequences.

Let $A \in \mathbf{R}^{n\times n}$, with singular values $\sigma_1 \geq \cdots \geq \sigma_n \geq 0$, and let the singular value decomposition of A be given by (1.4.1). Firstly, both the spectral norm and the Frobenius norm are neatly characterized in terms of the singular values:

$$||A||_2 = \sigma_1, \quad ||A||_F = \left(\sum_{j=i}^{n} \sigma_j^2\right)^{\frac{1}{2}}.$$

Secondly, for any $1 \leq k \leq n-1$, if $\sigma_k > 0$, then we have

$$\sigma_{k+1} = \min\{||A - B||_2 \ : \ B \in \mathbf{R}^{n\times n} \text{ with } \operatorname{rank}(B) = k\} = ||A - A_k||_2,$$

where

$$A_k = U_k \Sigma_k V_k^T,$$

with

$$U_k = [u, \dots, u_k], \quad V_k = [v_1, \dots, v_k],$$
$$\Sigma_k = \operatorname{diag}(\sigma_1, \dots, \sigma_k).$$

Finally, assume A is nonsingular, and suppose that A_s be the singular matrix that is closest to A, in the sense that $||A - A_s||_2$ is as small as possible. Then

$$||A - A_s||_2 = \sigma_n = \sigma_{\min}(A).$$

1.5 Symmetric Matrices

A matrix $A \in \mathbf{R}^{n\times n}$ is said to be a **symmetric matrix** if $A^T = A$. A real symmetric matrix has many elegant properties, and we will present only several most important classical results: the spectral theorem for symmetric matrices, Cauchy's interlacing theorem, Courant-Fischer minimax theorem, and Weyl's theorem.

Theorem 1.5.1 (The Spectral Theorem for Symmetric Matrices) *Let $A \in \mathbf{R}^{n\times n}$ be given. Then A is symmetric if and only if there is an orthogonal matrix $Q \in \mathbf{R}^{n\times n}$ and a real diagonal matrix $\Lambda \in \mathbf{R}^{n\times n}$ such that $A = Q\Lambda Q^T$.*

It is easy to see from this theorem that if an $n \times n$ real matrix A is symmetric, then

(1) All eigenvalues of A are real;
(2) A is nondefective; and
(3) A has n orthonormal real eigenvectors, which form an orthonormal basis for $\mathbf{R}^n$.

Since the eigenvalues of a symmetric matrix $A \in \mathbf{R}^{n\times n}$ are real, we will adopt the convention that they are labeled according to increasing size:

$$\lambda_{\min}(A) = \lambda_1(A) \le \lambda_2(A) \le \cdots \le \lambda_n(A) = \lambda_{\max}(A),$$

where the $\lambda_i(A)$ denote the eigenvalues of A.

Theorem 1.5.2 (Cauchy's Interlacing Theorem) *Let $A \in \mathbf{R}^{n\times n}$ be a symmetric matrix with eigenvalues*

$$\lambda_1 \le \lambda_2 \le \cdots \le \lambda_n,$$

and let

$$\mu_1 \le \mu_2 \le \cdots \le \mu_{n-1}$$

be the eigenvalues of a principal submatrix of A of order $n-1$. Then

$$\lambda_1 \le \mu_1 \le \lambda_2 \le \mu_2 \le \cdots \le \mu_{n-1} \le \lambda_n.$$

Theorem 1.5.3 (Courant-Fischer Minimax Theorem) *Let $A \in \mathbf{R}^{n\times n}$ be a given symmetric matrix with eigenvalues*

$$\lambda_1 \le \lambda_2 \le \cdots \le \lambda_n,$$

and let k be a given integer with $1 \le k \le n$. Then

$$\begin{aligned}\lambda_k &= \min_{\dim \mathcal{X}=k}\ \max_{0\ne x\in\mathcal{X}} \frac{x^TAx}{x^Tx} \\ &= \max_{\dim \mathcal{X}=n-k+1}\ \min_{0\ne x\in\mathcal{X}} \frac{x^TAx}{x^Tx}.\end{aligned}$$

In particular, for the smallest and the largest eigenvalues we have

$$\lambda_1 = \min_{x\ne 0} \frac{x^TAx}{x^Tx} \quad \text{and} \quad \lambda_n = \max_{x\ne 0} \frac{x^TAx}{x^Tx}.$$

This theorem shows that the eigenvalues of a symmetric matrix can be characterized as the solutions of a series of optimization problems. This result has many important applications. One of the simplest is to the problem of comparing the eigenvalues of $A+E$ with those of A.

Theorem 1.5.4 (Weyl's Theorem) *Let A, $E \in \mathbf{R}^{n\times n}$ be symmetric and let the eigenvalues $\lambda_i(A)$, $\lambda_i(E)$, and $\lambda_i(A+E)$ be arranged in increasing order. Then for each $k = 1, 2, \ldots, n$ we have*

$$\lambda_k(A) + \lambda_1(E) \le \lambda_k(A+E) \le \lambda_k(A) + \lambda_n(E).$$

The proof of this theorem is not difficult and is left as an exercise.

Weyl's theorem gives two-sided bounds for the eigenvalues of $A+E$ for any symmetric matrices A and E. Notice that for any symmetric matrix $A \in \mathbf{R}^{n\times n}$ we have

$$\|A\|_2 = \max\{|\lambda_1(A)|, |\lambda_n(A)|\}.$$

We can easily get a more conventional form of Weyl's theorem as follows.

Corollary 1.5.5 *Under the hypotheses of Weyl's Theorem, we have*

$$|\lambda_k(A) - \lambda_k(A+E)| \le \|E\|_2$$

for each $k = 1, 2, \ldots, n$.

1.6 Vandermonde Matrices

A **Vandermonde matrix** of order n is a matrix of the form

$$V = V(\lambda_1, \lambda_2, \cdots, \lambda_n) = \begin{bmatrix} 1 & 1 & \cdots & 1 \\ \lambda_1 & \lambda_2 & \cdots & \lambda_n \\ \vdots & \vdots & & \vdots \\ \lambda_1^{n-1} & \lambda_2^{n-1} & \cdots & \lambda_n^{n-1} \end{bmatrix} \quad (n > 1), \qquad (1.6.1)$$

where λ_i are real or complex numbers, that is, $V = [v_{ij}]$ with $v_{ij} = \lambda_j^{i-1}$. It is a fact

$$\det V = \prod_{1\le i<j\le n} (\lambda_i - \lambda_j), \qquad (1.6.2)$$

so a Vandermonde matrix is nonsingular if and only if the n parameters $\lambda_1, \ldots, \lambda_n$ are distinct. In this section, we present a norm estimate for inverse of Vandermonde matrices. Such a estimate will be useful in the subsequent chapters.

In what follows we assume that the n parameters $\lambda_1, \ldots, \lambda_n$ are distinct. We denote by S_k the kth **elementary symmetric function** of the n numbers $\lambda_1, \ldots, \lambda_n$

$$S_k = S_k(\lambda_1, \ldots, \lambda_n) = \sum_{1 \le i_1 < \cdots < i_k \le n} \prod_{j=1}^{k} \lambda_{i_j}, \tag{1.6.3}$$

where $1 \le k \le n$. For example, $S_1(\lambda_1, \ldots, \lambda_n) = \lambda_1 + \cdots + \lambda_n$ is the sum of the λ_i and $S_n(\lambda_1, \ldots, \lambda_n) = \lambda_1 \cdots \lambda_n$ is the product of the λ_i. Let $p(t) = \prod_{j=1}^{n}(t - \lambda_j)$. Then it is easy to verify that

$$p(t) = \sum_{k=0}^{n} (-1)^k S_k t^{n-k}. \tag{1.6.4}$$

Let $V^{-1} = [\xi_{ij}]$. Then $V^{-1}V = I$, that is,

$$\xi_{i1} + \lambda_j \xi_{i2} + \cdots + \lambda_j^{n-1} \xi_{in}, = \delta_{ij}, \quad i, j = 1, 2, \ldots, n. \tag{1.6.5}$$

On the other hand, if we define

$$L_i(t) = \prod_{\substack{k=1 \\ k \ne i}}^{n} \frac{t - \lambda_k}{\lambda_i - \lambda_k}, \quad i = 1, 2, \ldots, n, \tag{1.6.6}$$

referred to as **Lagrange elementary interpolating polynomials**, then it is easy to see that

$$L_i(\lambda_j) = \delta_{ij}, \quad i = 1, 2, \ldots, n.$$

Note that each polynomial $L_i(t)$ has degree $n - 1$ and can be expressed as

$$L_i(t) = \sum_{k=0}^{n-1} (-1)^k S_k^{(i)} t^{n-k-1} \Bigg/ \prod_{\substack{k=1 \\ k \ne i}}^{n} (\lambda_1 - \lambda_k),$$

where $S_k^{(i)} = S_k(\lambda_1, \ldots, \lambda_{i-1}, \lambda_{i+1}, \ldots, \lambda_n)$, that is, $S_k^{(i)}$ denotes the kth elementary symmetric function of the $n - 1$ numbers $\lambda_1, \ldots, \lambda_{i-1}, \lambda_{i+1}, \ldots, \lambda_n$ with λ_i missing. Hence we have

$$\frac{\displaystyle\sum_{k=0}^{n-1} (-1)^k S_k^{(i)} \lambda_j^{n-k-1}}{\displaystyle\prod_{\substack{k=1 \\ k \ne i}}^{n} (\lambda_i - \lambda_k)} = \delta_{ij}, \quad i, j = 1, 2, \ldots, n. \tag{1.6.7}$$

Comparing (1.6.5) and (1.6.7) gives

$$\xi_{ij} = \frac{(-1)^{n-j} S_{n-j}^{(i)}}{\prod_{\substack{k=1 \\ k \neq i}}^{n} (\lambda_i - \lambda_k)}, \quad i, j = 1, 2, \ldots, n. \tag{1.6.8}$$

Therefore,

$$\|V^{-1}\|_\infty = \max_{1 \leq i \leq n} \sum_{j=1}^{n} |\xi_{ij}| = \max_{1 \leq i \leq n} \frac{\sum_{j=1}^{n} |S_{n-j}^{(i)}|}{\prod_{\substack{k=1 \\ k \neq i}}^{n} |\lambda_i - \lambda_k|}. \tag{1.6.9}$$

Thus in order to estimate an upper bound of $\|V^{-1}\|_\infty$ we need the following lemma.

Lemma 1.6.1 *Let $\lambda_1, \ldots, \lambda_n$ be n complex numbers. Then*

$$1 + |S_1| + \cdots + |S_n| \leq \prod_{k=1}^{n} (1 + |\lambda_k|). \tag{1.6.10}$$

Proof From (1.6.4) we have

$$p(-1) = \sum_{k=0}^{n} (-1)^k S_k (-1)^{n-k} = (-1)^n \sum_{k=0}^{n} S_k,$$

that is,

$$\prod_{j=1}^{n} (1 + \lambda_j) = \sum_{k=0}^{n} S_k.$$

Thus, if all $\lambda_j \geq 0$, then all $S_k \geq 0$, by definition, and so

$$\sum_{k=0}^{n} |S_k| = \sum_{k=0}^{n} S_k = \prod_{j=1}^{n} (1 + |\lambda_j|). \tag{1.6.11}$$

On the other hand, by the definition of S_k we easily know that

$$|S_k(\lambda_1, \ldots, \lambda_n)| \leq S_k(|\lambda_1|, |\lambda_2|, \ldots, |\lambda_n|), \; k \geq 1.$$

Hence, using (1.6.11), we find

$$\sum_{k=0}^{n} |S_k(\lambda_1, \cdots, \lambda_n)| \leq \sum_{k=0}^{n} S_k(|\lambda_1|, |\lambda_2|, \ldots, |\lambda_n|) = \prod_{k=1}^{n} (1 + |\lambda_k|),$$

which is the desired result. □

Using (1.6.9) and the Lemma, we obtain the following theorem immediately.

Theorem 1.6.2 *Let $\lambda_1, \dots, \lambda_n$ be n distinct numbers. Then we have*

$$\|V(\lambda_1, \dots, \lambda_n)^{-1}\|_\infty \le \max_{1 \le i \le n} \prod_{\substack{k=1 \\ k \ne i}}^{n} \frac{1 + |\lambda_k|}{|\lambda_i - \lambda_k|}.$$

1.7 The Kronecker Product

Let $A = [\alpha_{ij}] \in \mathbf{C}^{m\times n}$, and let $B \in \mathbf{C}^{p\times q}$. The $mp \times nq$ matrix

$$A \otimes B = \begin{bmatrix} \alpha_{11}B & \alpha_{12}B & \cdots & \alpha_{1n}B \\ \alpha_{21}B & \alpha_{22}B & \cdots & \alpha_{3n}B \\ \vdots & \vdots & & \vdots \\ \alpha_{m1}B & \alpha_{m2}B & \cdots & \alpha_{mn}B \end{bmatrix}$$

is called the **Kronecker product** of A and B.

The basic properties of the Kronecker product can be summarized in the following theorem.

Theorem 1.7.1 *Let $A \in \mathbf{C}^{m\times n}$, and let $B \in \mathbf{C}^{p\times q}$. Then we have*

(1) $\operatorname{rank}(A \otimes B) = \operatorname{rank} A \cdot \operatorname{rank} B$;
(2) $(A \otimes B)^T = A^T \otimes B^T$;
(3) $(A \otimes B)(C \otimes D) = (AC) \otimes (BD)$, *where* $C \in \mathbf{C}^{n\times k}$, $D \in \mathbf{C}^{q\times r}$;
(4) *If* $m = n$ *and* $p = q$, *then* $\det(A \otimes B) = (\det A)^p (\det B)^m$;
(5) *If both A and B are nonsingular, then $A \otimes B$ is also nonsingular and*

$$(A \otimes B)^{-1} = A^{-1} \otimes B^{-1}.$$

The proof of these results is not difficult and is left as an exercise.

The **vec operator** is a linear operator which maps matrices into vectors by stacking the columns. For instance,

$$\operatorname{vec}(G) = (g_1^T, g_2^T, \dots, g_m^T)^T \in \mathbf{C}^{nm},$$

for any $G = [g_1, \dots, g_m] \in \mathbf{C}^{n\times m}$.

Lemma 1.7.2 *Let $G \in \mathbf{C}^{m\times n}, V \in \mathbf{C}^{n\times l}, H \in \mathbf{C}^{l\times k}$. Then*

$$\operatorname{vec}(GVH) = (H^T \otimes G)\operatorname{vec}(V).$$

Proof Let $V = [v_{ij}] = \sum_{i,j} v_{ij} e_j e_i^T, G = [g_1, \cdots, g_n]$, and $H^T = [h_1, \cdots, h_l]$. Then, we have

$$\begin{aligned}\text{vec}\,(GVH) &= \text{vec}\left(\sum_{i,j} v_{ij} G e_i e_j^T H\right) = \sum_{i,j} v_{ij} \text{vec}\,(G e_i e_j^T H) \\ &= \sum_{i,j} v_{ij} \text{vec}\,(g_i h_j^T) = \sum_{i,j} v_{ij} (h_j \otimes g_i) = (H^T \otimes G) \text{vec}\,(V),\end{aligned}$$

which is the desired result. □

As an application of Lemma 1.7.2, we consider the following linear matrix equation

$$AX - XB = C, \tag{1.7.1}$$

where $A \in \mathbf{C}^{m\times m}, B \in \mathbf{C}^{n\times n}$, and $C \in \mathbf{C}^{m\times n}$ are given, and $X \in \mathbf{C}^{m\times n}$ is to be found. Applying Lemma 1.7.2, in terms of the vec operator and the Kronecker product we can rewrite (1.7.1) as

$$(I_n \otimes A - B^T \otimes I_m) \text{vec}\,(X) = \text{vec}\,(C). \tag{1.7.2}$$

Theorem 1.7.3 *Let $A \in \mathbf{C}^{m\times m}$ with $\lambda(A) = \{\lambda_1, \ldots, \lambda_m\}$, and let $B \in \mathbf{C}^{n\times n}$ with $\lambda(B) = \{\mu_1, \ldots, \mu_n\}$. Then*

$$\lambda(I_n \otimes A - B^T \otimes I_m) = \left\{ \lambda_i - \mu_j \ : \ \begin{array}{c} i = 1, 2, \ldots, m \\ j = 1, 2, \ldots, n \end{array} \right\}.$$

Proof Let the Schur's decompositions of A and B^T are respectively

$$A = QT_AQ^* \text{ and } B^T = PT_BP^*,$$

where $Q \in \mathbf{C}^{m\times m}$ and $P \in \mathbf{C}^{n\times n}$ are unitary, and $T_A \in \mathbf{C}^{m\times m}$ and $T_B \in \mathbf{R}^{n\times n}$ are upper triangular. Applying Theorem 1.7.1 and Lemma 1.7.2, we have

$$\begin{aligned}I_n \otimes A &- B^T \otimes I_m \\ &= PI_nP^* \otimes QT_AQ^* - PT_BP^* \otimes QI_mQ^* \\ &= (P \otimes Q)(I_n \otimes T_A)(P \otimes Q)^* - (P \otimes Q)(T_B \otimes I_m)(P \otimes Q)^* \\ &= (P \otimes Q)(I_n \otimes T_A - T_B \otimes I_m)(P \otimes Q)^*.\end{aligned} \tag{1.7.3}$$

Noting that $P \otimes Q$ is unitary, $I_n \otimes T_A - T_B \otimes I_m$ is upper triangular, and the diagonal elements of T_A and T_B are $\lambda_1, \ldots, \lambda_m$ and $\mu_1, \ldots, \mu_n$, respectively, it follows from (1.7.3) that the eigenvalues of $I_n \otimes A - B^T \otimes I_m$ are $\lambda_i - \mu_j$, $i = 1, 2, \ldots, m$, $j = 1, 2, \ldots, n$. □

With Theorem 1.7.3 and Equation (1.7.2), we can easily establish the unique-existence theorem for the solution of Equation (1.7.1) as follows.

Corollary 1.7.4 *For any* $C \in \mathbf{R}^{m\times n}$ *the linear matrix equation* (1.7.1) *has a unique solution if and only if* $\lambda(A) \cap \lambda(B) = \emptyset$.

1.8 The Implicit Function Theorem

In this section we will review some basic concepts and facts from the function theory and give a generalization of the implicit function theorem, which will be used in the subsequent chapters.

Suppose that the **vector-valued function**

$$f\ :\ \mathcal{D} \subset \mathbf{R}^n \longrightarrow \mathbf{R}^m$$

with

$$f(x) = \big(f_1(x), f_2(x), \dots, f_m(x)\big), \quad x = (\xi_1, \dots, \xi_n)^T,$$

is defined on an open set $\mathcal{D}$ of $\mathbf{R}^n$, and let $x_0 \in \mathcal{D}$. If there exists a matrix $A \in \mathbf{R}^{m\times n}$ such that

$$\lim_{\|h\|_2 \longrightarrow 0} \frac{\|f(x_0+h) - f(x_0) - Ah\|_2}{\|h\|_2} = 0, \quad h \in \mathcal{D}.$$

then f is said to be **differentiable** at x_0, the matrix A is called the **differential** of f at x_0 and is designated $f'(x_0)$; if f is differentiable at each point in $\mathcal{D}$, then f is said to be differentiable on $\mathcal{D}$.

We gather together in the next theorem the familiar facts of differentiation.

Theorem 1.8.1 *Let f be the vector-valued function defined above. Then we have*

(1) *If f is differentiable at x_0, then the $f'(x_0)$ is uniquely determined* ;

(2) *f is differentiable at x_0 if and only if its component functions f_i, $i = 1, 2, \dots, m$, are differentiable at x_0. Moreover, if f is differentiable at x_0, then*

$$f'(x_0) = \begin{bmatrix} \frac{\partial f_1(x)}{\partial \xi_1} & \frac{\partial f_1(x)}{\partial \xi_2} & \cdots & \frac{\partial f_1(x)}{\partial \xi_n} \\ \vdots & \vdots & & \vdots \\ \frac{\partial f_m(x)}{\partial \xi_1} & \frac{\partial f_m(x)}{\partial \xi_2} & \cdots & \frac{\partial f_m(x)}{\partial \xi_n} \end{bmatrix},$$

which is called the **Jacobian matrix** *of f at x_0*;

(3) (**Chain Rule**) *If $f : \mathcal{D} \longrightarrow \mathbf{R}^m$ is differentiable at x_0, and $g : \mathbf{R}^m \longrightarrow \mathbf{R}^l$ is differentiable at $y_0 = f(x_0)$, the composite-function $g \circ f\ :\ \mathcal{D} \longrightarrow \mathbf{R}^l$ is differentiable at x_0, and*

$$(g \circ f)'(x_0) = g'\big(f(x_0)\big) f'(x_0)\,;$$

(4) (**The Mean-Value Formula**) *If f is differentiable at each point in $\mathcal{L} \equiv \{x_0 + th : 0 \le t \le 1\} \subset \mathcal{D}$, where $x_0 \in \mathcal{D}$ and $h \in \mathbf{R}^n$ are given, then there is a $\tau \in [0,1]$ such that*

$$\|f(x_0+h) - f(x_0)\|_2 \le \|f'(x_0+\tau h)h\|_2.$$

Some of the most frequently used vector-valued functions are the functions from some open sets of $\mathbf{R}^l$ to $\mathbf{R}^{m\times n}$, i.e., matrices dependent on several parameters. Here we call such a function a **matrix-valued function**. Let $A : \mathcal{D} \subset \mathbf{R}^l \longrightarrow \mathbf{R}^{m\times n}$ with

$$A(x) = \left[a_{ij}(x)\right], \quad x = (\xi_1, \cdots, \xi_l)^T.$$

be differentiable on $\mathcal{D}$. It is often convenient to adopt the following notation

$$\frac{\partial A(x)}{\partial \xi_k} = \left[\frac{\partial a_{ij}(x)}{\partial \xi_k}\right], \quad k = 1, 2, \ldots, l.$$

In particular, when $l = 1$, it is frequently written as

$$\dot{A}(x) = \left[\frac{\mathrm{d}a_{ij}(x)}{\mathrm{d}x}\right], \quad x \in \mathcal{D} \subset \mathbf{R}.$$

Suppose that $A(x) \in \mathbf{R}^{p\times n}$ and $B(x) \in \mathbf{R}^{n\times m}$ are two differentiable matrix-valued functions defined on an open subset $\mathcal{D}$ of $\mathbf{R}^l$. Using the above notation it is easy to verify that

$$\frac{\partial(A(x)B(x))}{\partial \xi_k} = \frac{\partial A(x)}{\partial \xi_k}B(x) + A(x)\frac{\partial A(x)}{\partial \xi_k}$$

for each $k = 1, 2, \ldots, l$, which are frequently used in the subsequent chapters.

A function $f : \mathcal{D} \subset \mathbf{R}^l \longrightarrow \mathbf{R}$ is **analytic** in some neighborhood of $x_0 \in \mathcal{D}$ if f can be expanded in a power series about the point x_0, that is,

$$f(x) = f(x_0) + \sum_{r=1}^{\infty}\sum_{|t|=r} a_{t_1,t_2,\ldots,t_l}\left(x(1) - x_0(1)\right)^{t_1}\cdots\left(x(l) - x_0(l)\right)^{t_l}$$

holds in the neighborhood of x_0, where $|t| = t_1 + \cdots + t_l$, the t_i are nonnegative integers. A vector-valued function is **analytic** in some neighborhood of the origin if all its component functions are analytic in the neighborhood of the origin.

Let $f : \mathcal{D} \subset \mathbf{R}^n$ be differentiable. We define the **Jacobian** $\frac{\partial(f_1,\cdots,f_n)}{\partial(\xi_1,\cdots,\xi_n)}(x_0)$ of f at $x_0 \in \mathcal{D}$ by

$$\frac{\partial(f_1,\cdots,f_n)}{\partial(\xi_1,\cdots,\xi_n)}(x_0) = \det f'(x_0).$$

We now state the implicit theorem, which is one of the most useful tools in function theory.

Theorem 1.8.2 (The Implicit Function Theorem) *Let $\mathcal{D}$ be an open subset of $\mathbf{R}^{k+l}$ with $0 \in \mathcal{D}$. If the vector-valued function $f : \mathcal{D} \longrightarrow \mathbf{R}^k$ with*

$$f(x;y) = \big(f_1(x;y), \dots, f_k(x;y)\big)^T, \quad x = (\xi_1, \dots, \xi_k)^T, \quad y = (\eta_1, \dots, \eta_l)^T$$

is analytic in some neighborhood of the origin of $\mathbf{R}^{k+l}$, if $f(0;0) = 0$, and if the Jacobian

$$\frac{\partial(f_1, \dots, f_k)}{\partial(\xi_1, \dots, \xi_k)}(0) \neq 0,$$

then there exists a neighborhood $\mathcal{B}$ of the origin of $\mathbf{R}^l$ and a unique vector-valued function $g : \mathcal{B} \longrightarrow \mathbf{R}^k$ such that

(1) *g is analytic in $\mathcal{B}$,*
(2) *$g(0) = 0$,*
(3) *$f(g(y);y) = 0$ for each $y \in \mathcal{B}$.*

Finally, we apply Theorem 1.8.2 to prove a generalization of the implicit function theorem. We need the following notation. Let $f : \mathbf{R}^{k+l} \longrightarrow \mathbf{R}^m$ be differentiable. Define

$$f'_x(x_0;y_0) = \begin{bmatrix} \frac{\partial f_1(x_0;y_0)}{\partial \xi_1} & \cdots & \frac{\partial f_1(x_0;y_0)}{\partial \xi_k} \\ \vdots & & \vdots \\ \frac{\partial f_m(x_0;y_0)}{\partial \xi_1} & \cdots & \frac{\partial f_m(x_0;y_0)}{\partial \xi_k} \end{bmatrix}$$

and

$$f'_y(x_0;y_0) = \begin{bmatrix} \frac{\partial f_1(x_0;y_0)}{\partial \eta_1} & \cdots & \frac{\partial f_1(x_0;y_0)}{\partial \eta_l} \\ \vdots & & \vdots \\ \frac{\partial f_m(x_0;y_0)}{\partial \eta_1} & \cdots & \frac{\partial f_m(x_0;y_0)}{\partial \eta_l} \end{bmatrix},$$

where

$$x = (\xi_1, \dots, \xi_k)^T \in \mathbf{R}^k, \quad y = (\eta_1, \dots, \eta_l)^T \in \mathbf{R}^l,$$

and

$$f(x;y) = \big(f_1(x;y), \dots, f_m(x;y)\big)^T.$$

Theorem 1.8.3 (A Generalization of the Implicit Function Theorem) *Let $\mathcal{D}$ be an open subset of $\mathbf{R}^{k+l}$. Suppose that the vector-valued function $f : \mathcal{D} \longrightarrow \mathbf{R}^m$ with*

$$f(x;y) = \big(f_1(x;y), \dots, f_m(x;y)\big)^T, \quad x = (\xi_1, \dots, \xi_k)^T, \quad y = (\eta_1, \dots, \eta_l)^T,$$

is analytic in some neighborhood of the origin of $\mathbf{R}^{k+l}$, *where* $m < k$. *If*

$$f(0;0) = 0 \quad \textit{and} \quad \operatorname{rank} f'_x(0;0) = m, \tag{1.8.1}$$

then for any $W = [\omega_{ij}] \in \mathbf{R}^{k\times l}$ *that satisfies*

$$f'_x(0;0)W + f'_y(0;0) = 0, \tag{1.8.2}$$

there exists a neighborhood $\mathcal{B}_y$ *of the origin of* $\mathbf{R}^l$, *and a vector-valued function* $g : \mathcal{B}_y \longrightarrow \mathbf{R}^k$ *such that*

(1) g *is analytic in* $\mathcal{B}_y$,
(2) $g(0) = 0$ *and* $g'(0) = W$,
(3) $f(g(y); y) = 0$ *for each* $y \in \mathcal{B}_y$.

Remark 1.8.1 Condition (1.8.1) and $m < k$ imply that the matrix equation (1.8.2) has infinite many solutions W. Hence in practice we can use such freedom to make a suitable choice of W to meet the practical requirements.

Proof of Theorem 1.8.3 Without loss of generality we may assume that

$$\frac{\partial(f_1, \dots, f_m)}{\partial(\xi_1, \dots, \xi_m)}(0) \neq 0.$$

Let

$$x_1 = (\xi_1, \dots, \xi_m)^T, \quad x_2 = (\xi_{m+1}, \dots, \xi_k)^T.$$

By the implicit function theorem, there exists a neighborhood $\mathcal{B}$ of $\mathbf{R}^{(k-m)+l}$ and a unique analytic vector-valued function $h : \mathcal{B} \longrightarrow \mathbf{R}^m$ such that $h(0;0) = 0$ and

$$f\big(h(x_2; y), x_2; y\big) = 0$$

for each $(x_2^T, y^T)^T \in \mathcal{B}$.

We now take

$$g_i(y) = \omega_{i1}\eta_1 + \omega_{i2}\eta_2 + \cdots + \omega_{il}\eta_l, \quad i = m+1, \dots, k.$$

Then there must have a neighborhood $\mathcal{B}_y$ of the origin of $\mathcal{R}^l$ such that

$$\big(g_{m+1}(y), \dots, g_k(y), y^T\big)^T \in \mathcal{B}$$

for $y \in \mathcal{B}_y$. Define

$$g_i(y) = h_i\big(g_{m+1}(y), \dots, g_k(y); y\big), \quad y \in \mathcal{B}_y,$$

for $i = 1, 2, \ldots, m$. Then it is easy to verify that $g(y) = \big(g_1(y), \ldots, g_k(y)\big)^T$ is analytic in $\mathcal{B}_y$ and

$$f\big(g(y); y\big) = 0 \quad \text{for each} \quad y \in \mathcal{B}_y. \tag{1.8.3}$$

Next we prove that $g'(0) = W$. Obviously,

$$\frac{\partial g_i(0)}{\partial \eta_j} = \omega_{ij}, \quad i = m+1, \ldots, k, \; j = 1, 2, \ldots, l. \tag{1.8.4}$$

Differentiating (1.8.3) at $y = 0$ yields

$$f'_x(0;0)g'(0) + f'_y(0;0) = 0. \tag{1.8.5}$$

Subtracting (1.8.5) from (1.8.2) and using (1.8.4), we have

$$f'_{x_1}(0;0;0)\left(\begin{bmatrix} \omega_{11} & \cdots & \omega_{1l} \\ \vdots & & \vdots \\ \omega_{m1} & \cdots & \omega_{ml} \end{bmatrix} - \begin{bmatrix} \frac{\partial g_1(0)}{\partial \eta_1} & \cdots & \frac{\partial g_1(0)}{\partial \eta_l} \\ \vdots & & \vdots \\ \frac{\partial g_m(0)}{\partial \eta_1} & \cdots & \frac{\partial g_m(0)}{\partial \eta_l} \end{bmatrix}\right) = 0,$$

and so

$$\begin{bmatrix} \frac{\partial g_1(0)}{\partial \eta_1} & \cdots & \frac{\partial g_1(0)}{\partial \eta_l} \\ \vdots & & \vdots \\ \frac{\partial g_m(0)}{\partial \eta_1} & \cdots & \frac{\partial g_m(0)}{\partial \eta_l} \end{bmatrix} = \begin{bmatrix} \omega_{11} & \cdots & \omega_{1l} \\ \vdots & & \vdots \\ \omega_{m1} & \cdots & \omega_{ml} \end{bmatrix}$$

since $\det f'_{x_1}(0;0;0) = \frac{\partial(f_1, \ldots, f_m)}{\partial(\xi_1 \cdots, \xi_m)}(0) \neq 0$. This, together with (1.8.4), proves that $g'(0) = W$. □

1.9 The Derivative of a Simple Eigenvalue

Let $A(x) \in \mathbf{R}^{n \times n}$ be an analytic function in some neighborhood $\mathcal{B}$ of the origin of $\mathbf{R}^m$. In this section we will establish the differentiability theorem of the eigenvalues and eigenvectors of $A(x)$.

Lemma 1.9.1 *Assume that*

$$A(x) = \begin{bmatrix} A_{11}(x) & A_{12}(x) \\ A_{21}(x) & A_{22}(x) \end{bmatrix}$$

with $A_{11}(x) \in \mathbf{R}^{r \times r}$ *and* $A_{22}(x) \in \mathbf{R}^{(n-r)\times(n-r)}$ *satisfying that*

$$A_{21}(0) = 0, \quad A_{12}(0) = 0, \quad \lambda(A_{11}(0)) \cap \lambda(A_{22}(0)) = \emptyset.$$

Then there exist two analytic functions $U(x), V(x) \in \mathbf{R}^{n\times n}$ in some neighborhood $\mathcal{B}_0 \subset \mathcal{B}$ of the origin of $\mathbf{R}^m$ satisfying $U(0) = V(0) = I$ and $V(x)^T U(x) = I$ such that

$$V(x)^T A(x) U(x) = \begin{bmatrix} \widetilde{A}_{11}(x) & 0 \\ 0 & \widetilde{A}_{22}(x) \end{bmatrix}, \quad x \in \mathcal{B}_0,$$

where $\widetilde{A}_{11}(x) \in \mathbf{R}^{r\times r}$ with $\widetilde{A}_{11}(0) = A_{11}(0)$ and $\widetilde{A}_{22}(x) \in \mathbf{R}^{(n-r)\times(n-r)}$ with $\widetilde{A}_{22}(0) = A_{22}(0)$.

Proof Define a vector-value function f by

$$f(x;y) = \operatorname{vec}\big(A_{21}(x) - YA_{11}(x) + A_{22}(x)Y - YA_{12}(x)Y\big), \tag{1.9.1}$$

where $x \in \mathcal{B}$, $Y \in \mathbf{R}^{(n-r)\times r}$, $y = \operatorname{vec}(Y)$, and the vec denotes the vec operator defined in Section 1.7. From the definition of f it is easy to see that $f(x;y)$ is analytic for $y \in \mathbf{R}^{r(n-r)}$ and $x \in \mathcal{B}$, $f(0;0) = 0$, and

$$\det f'_y(0;0) = \det(I_r \otimes A_{22}(0) - A_{11}(0)^T \otimes I_{n-r}) \neq 0,$$

by Theorem 1.7.3 and the hypothesis that $\lambda(A_{11}(0)) \cap \lambda(A_{22}(0)) = \emptyset$. Hence by the implicit function theorem there exists a neighborhood $\mathcal{B}_0 \subset \mathcal{B}$ of the origin of $\mathbf{R}^m$ and a unique analytic function $y = y(x)$ with $y(0) = 0$ such that

$$f\big(x; y(x)\big) = 0 \quad \text{for each } x \in \mathcal{B}_0. \tag{1.9.2}$$

Now let $Y(x) \in \mathbf{R}^{(n-r)\times r}$ satisfy that $\operatorname{vec}(Y(x)) = y(x)$ in $\mathcal{B}_0$. It follows from (1.9.1) and (1.9.2) that

$$\begin{bmatrix} I_r & 0 \\ -Y(x) & I_{n-r} \end{bmatrix} \begin{bmatrix} A_{11}(x) & A_{12}(x) \\ A_{21}(x) & A_{22}(x) \end{bmatrix} \begin{bmatrix} I_r & 0 \\ Y(x) & I_{n-r} \end{bmatrix} = \begin{bmatrix} \widetilde{A}_{11}(x) & \widetilde{A}_{12}(x) \\ 0 & \widetilde{A}_{22}(x) \end{bmatrix}$$

for each $x \in \mathcal{B}_0$, where

$$\widetilde{A}_{11}(x) = A_{11}(x) + A_{12}(x)Y(x), \quad \widetilde{A}_{12}(x) = A_{12}(x),$$
$$\widetilde{A}_{22}(x) = A_{22}(x) - Y(x)A_{12}(x).$$

Thus, $\widetilde{A}_{11}(x)$, $\widetilde{A}_{22}(x)$, and $\widetilde{A}_{12}(x)$ are analytic in $\mathcal{B}_0$, and $\widetilde{A}_{11}(0) = A_{11}(0)$, $\widetilde{A}_{22}(0) = A_{22}(0)$. Hence, without loss of generality we may assume that $\mathcal{B}_0$ is sufficiently small so that

$$\lambda(\widetilde{A}_{11}(x)) \cap \lambda(\widetilde{A}_{22}(x)) = \emptyset \quad \text{for each } x \in \mathcal{B}_0. \tag{1.9.3}$$

By Corollary 1.7.4, (1.9.3) implies the matrix equation

$$\widetilde{A}_{11}(x)Z - Z\widetilde{A}_{22}(x) + \widetilde{A}_{12}(x) = 0$$

has a unique solution $Z = Z(x) \in \mathbf{R}^{r\times(n-r}$ that is analytic in $\mathcal{B}_0$ and $Z(0) = 0$. It is easy to see that such a $Z(x)$ satisfies

$$\begin{bmatrix} I_r & -Z(x) \\ 0 & I_{n-r} \end{bmatrix} \begin{bmatrix} \widetilde{A}_{11}(x) & \widetilde{A}_{12}(x) \\ 0 & \widetilde{A}_{22}(x) \end{bmatrix} \begin{bmatrix} I_r & Z(x) \\ 0 & I_{n-r} \end{bmatrix} = \begin{bmatrix} \widetilde{A}_{11}(x) & 0 \\ 0 & \widetilde{A}_{22}(x) \end{bmatrix}$$

for each $x \in \mathcal{B}_0$.

Hence, if we set

$$U(x) = \begin{bmatrix} I_r & 0 \\ Y(x) & I_{n-r} \end{bmatrix} \begin{bmatrix} I_r & Z(x) \\ 0 & I_{n-r} \end{bmatrix}$$

and

$$V(x)^T = \begin{bmatrix} I_r & -Z(x) \\ 0 & I_{n-r} \end{bmatrix} \begin{bmatrix} I_r & 0 \\ -Y(x) & I_{n-r} \end{bmatrix},$$

then $U(x), V(x) \in \mathbf{R}^{n\times n}$ are analytic in $\mathcal{B}_0$ and have the desired properties. □

As a natural consequence of Lemma 1.9.1, we have

Theorem 1.9.2 *Suppose that λ is a real simple eigenvalue of $A(0)$ and $u, v \in \mathbf{R}^n$ are associated right and left eigenvectors with $v^T u = 1$, that is, there exist $U = [u, U_2], V = [v, V_2] \in \mathbf{R}^{n\times n}$ such that*

$$V^T A(0) U = \begin{bmatrix} \lambda & 0 \\ 0 & A_2 \end{bmatrix}, \quad V^T U = I, \quad \lambda \notin \lambda(A_2).$$

Then there exists a neighborhood $\mathcal{B}_0 \subset \mathcal{B}$ of the origin of $\mathbf{R}^m$ and real analytic functions $\lambda(x) \in \mathbf{R}$ and $u(x), v(x) \in \mathbf{R}^n$, with $\lambda(0) = \lambda$, $u(0) = u$, $v(0) = v$, and $v^T \frac{\partial u(0)}{\partial \xi_i} = 0$, $i = 1, \ldots, m$, such that

$$A(x)u(x) = \lambda(x)u(x),$$

$$v(x)^T A(x) = \lambda(x)v(x)^T, \quad v(x)^T u(x) = 1,$$

for each $x \in \mathcal{B}_0$, and we have

$$\frac{\partial \lambda(0)}{\partial \xi_i} = v^T \frac{\partial A(0)}{\partial \xi_i} u,$$

$$\frac{\partial u(0)}{\partial \xi_i} = U_2(\lambda I - A_2)^{-1} V_2^T \frac{\partial A(0)}{\partial \xi_i} u,$$

$$\frac{\partial v(0)^T}{\partial \xi_i} = v_1^T \frac{\partial A(0)}{\partial \xi_i} U_2(\lambda I - A_2)^{-1} V_2^T,$$

where $i = 1, 2, \ldots, m$, $x = (\xi_1, \ldots, \xi_m)^T$.

Proof Apply Lemma 1.9.1 and differentiate the equations

$$A(x)u(x) = \lambda(x)u(x) \quad \text{and} \quad v(x)^T A(x) = \lambda(x)v(x)^T$$

at $x = 0$. The details are left as an exercise (see also the proof of Theorem 1.9.5 below). □

Noting that the singular values and the singular vectors of a real matrix A are closely related to the eigenvalues and the eigenvectors of the symmetric matrices $A^T A$ and AA^T, we immediately obtain the following theorem from Theorem 1.9.2.

Theorem 1.9.3 *Suppose $B(x) \in \mathbf{R}^{m\times n} (m \geq n)$ is a real analytic matrix-valued function of $x = (\xi_1, \dots, \xi_l)^T$ in some neighborhood $\mathcal{B}$ of the origin of $\mathbf{R}^l$. If σ is a simple non-zero singular value of $B(0)$, and if $v \in \mathbf{R}^n$ and $u \in \mathbf{R}^m$ are associated unit right and unit left singular vectors, respectively, then there exists a neighborhood $\mathcal{B}_0 \subset \mathcal{B}$ of the origin and real analytic functions $\sigma(x) \in \mathbf{R}, u(x) \in \mathbf{R}^m$, and $v(x) \in \mathbf{R}^n$, with $\sigma(0) = \sigma$, $u(0) = u$, and $v(0) = v$, such that for any $x \in \mathcal{B}_0, \sigma(x)$ is a simple singular value of $B(x)$, and $v(x)$ and $u(x)$ are associated unit right and unit left singular vectors. Moreover, assume further that the singular value decomposition of $B(0)$ is*

$$U^T B(0) V = \begin{bmatrix} \sigma & 0 \\ 0 & \Sigma_2 \end{bmatrix}, \quad \Sigma_2 = \begin{bmatrix} \operatorname{diag}(\sigma_2, \dots, \sigma_n) \\ 0 \end{bmatrix} \in \mathbf{R}^{(m-1)\times(n-1)},$$

where $U = [u, U_2] \in \mathbf{R}^{m\times m}$ and $V = [v, V_2] \in \mathbf{R}^{n\times n}$ are orthogonal and $\sigma_j \geq 0$ with $\sigma \neq \sigma_j$, $j = 2, \dots, n$. Then we have

$$\frac{\partial \sigma(0)}{\partial \xi_j} = u^T \frac{\partial B(0)}{\partial \xi_j} v,$$

$$\frac{\partial v(0)}{\partial \xi_j} = V_2 \Lambda V_2^T \frac{\partial B(0)^T}{\partial \xi_j} u + V_2 \Omega^T U_2^T \frac{\partial B(0)}{\partial \xi_j} v,$$

$$\frac{\partial u(0)}{\partial \xi_j} = U_2 \Omega V_2^T \frac{\partial B(0)^T}{\partial \xi_j} u + U_2 \Sigma U_2^T \frac{\partial B(0)}{\partial \xi_j} v,$$

where

$$\Lambda = \sigma(\sigma^2 I - \Sigma_2^T \Sigma_2)^{-1}, \quad \Omega = \Sigma_2(\sigma^2 I - \Sigma_2^T \Sigma_2)^{-1},$$
$$\Sigma = \sigma(\sigma^2 I - \Sigma_2 \Sigma_2^T)^{-1}.$$

Now we consider the differentiability of the complex eigenvalues of $A(x)$, beginning with the case of $n = 2$

Lemma 1.9.4 *Let*

$$T(x) = \begin{bmatrix} a_{11}(x) & a_{12}(x) \\ a_{21}(x) & a_{22}(x) \end{bmatrix} \in \mathbf{R}^{2\times 2}$$

be analytic in some neighborhood $\mathcal{B}$ of the origin of $\mathbf{R}^m$, with $\Delta(0) < 0$, where

$$\Delta(x) = \big(a_{11}(x) - a_{22}(x)\big)^2 + 4a_{12}(x)a_{21}(x).$$

Then there exists a neighborhood $\mathcal{B}_0 \subset \mathcal{B}$ of the origin such that $\Delta(x) < 0$ for all $x \in \mathcal{B}_0$, and so $T(x)$ has a pair of complex conjugate eigenvalues :

$$\lambda_1(x) = \frac{1}{2}\big(a_{11}(x) + a_{22}(x)\big) + \frac{i}{2}\sqrt{-\Delta(x)},$$
$$\lambda_2(x) = \frac{1}{2}\big(a_{11}(x) + a_{22}(x)\big) - \frac{i}{2}\sqrt{-\Delta(x)},$$

for each $x \in \mathcal{B}$, and the corresponding right and left eigenvectors can be respectively taken as

$$u_1(x) = \begin{bmatrix} 1 \\ \dfrac{\lambda_1(x) - a_{11}(x)}{a_{12}(x)} \end{bmatrix}, \qquad v_1(x) = \begin{bmatrix} \dfrac{\lambda_1(x) - a_{22}(x)}{i\sqrt{-\Delta(x)}} \\ \dfrac{a_{12}(x)}{i\sqrt{-\Delta(x)}} \end{bmatrix},$$

$$u_2(x) = \begin{bmatrix} 1 \\ \dfrac{\lambda_2(x) - a_{11}(x)}{a_{12}(x)} \end{bmatrix}, \qquad v_2(x) = \begin{bmatrix} -\dfrac{\lambda_2(x) - a_{22}(x)}{i\sqrt{-\Delta(x)}} \\ -\dfrac{a_{12}(x)}{i\sqrt{-\Delta(x)}} \end{bmatrix}.$$

Obviously, $\lambda_1(x), \lambda_2(x), u_1(x), v_1(x), u_2(x)$, and $v_2(x)$ all are analytic in $\mathcal{B}_0$, and $v_1(x)^T u_1(x) = v_2(x)^T u_2(x) = 1$ for every $x \in \mathcal{B}_0$.

The proof is straightforward and is left as an exercise.

By using Lemmas 1.9.1 and 1.9.4, it is easy to establish the following theorem.

Theorem 1.9.5 *Suppose that λ is a simple complex eigenvalue of $A(0)$, that is, there exist two $n \times n$ matrices $U = [u_1, u_2, U_3]$ and $V = [v_1, v_2, V_3]$, with $u_1, u_2, v_1, v_2 \in \mathbf{C}^n$, and $U_3, V_3 \in \mathbf{R}^{n\times(n-2)}$, such that*

$$V^T A(0) U = \begin{bmatrix} \lambda & 0 & 0 \\ 0 & \overline{\lambda} & 0 \\ 0 & 0 & A_3 \end{bmatrix} \quad \text{and} \quad V^T U = I, \tag{1.9.4}$$

where

$$A_3 \in \mathbf{R}^{(n-2)\times(n-2)}, \quad \lambda = \nu + i\mu \notin \lambda(A_3), \quad \nu, \mu \in R, \quad \mu \neq 0,$$
$$\overline{u}_2 = u_1 = w_1 + iw_2, \quad w_1, w_2 \in \mathbf{R}^n,$$
$$\overline{v}_2 = v_1 = y_1 + iy_2, \quad y_1, y_2 \in \mathbf{R}^n.$$

Then there exists a neighborhood $\mathcal{B}_0 \subset \mathcal{B}$ of the origin and real analytic functions $\nu(x), \mu(x) \in \mathbf{R}$ and $w_1(x), w_2(x), y_1(x), y_2(x) \in \mathbf{R}^n$, with $\nu(0) = \nu$, $\mu(0) = \mu$, $w_1(0) = w_1$, $w_2(0) = w_2, y_1(0) = y_1$, and $y_2(0) = y_2$, such that

$$A(x)u(x) = \lambda(x)u(x), \quad v(x)^T u(x) = 1,$$
$$v(x)^T A(x) = \lambda(x)v(x)^T, \quad x \in \mathcal{B}_0,$$
$$v_1^T \frac{\partial u(0)}{\partial \xi_j} = 0, \quad j = 1, \ldots, m,$$

where $\lambda(x) = \nu(x) + i\mu(x)$, $u(x) = w_1(x) + iw_2(x)$, and $v(x) = y_1(x) + iy_2(x)$. Furthermore, we have

$$\frac{\partial \nu(0)}{\partial \xi_j} = y_1^T \frac{\partial A(0)}{\partial \xi_j} w_1 - y_2^T \frac{\partial A(0)}{\partial \xi_j} w_2, \tag{1.9.5}$$
$$\frac{\partial \mu(0)}{\partial \xi_j} = y_1^T \frac{\partial A(0)}{\partial \xi_j} w_2 + y_2^T \frac{\partial A(0)}{\partial \xi_j} w_1, \tag{1.9.6}$$
$$\frac{\partial w_1(0)}{\partial \xi_j} = G \frac{\partial A(0)}{\partial \xi_j} w_1 + H \frac{\partial A(0)}{\partial \xi_j} w_2, \tag{1.9.7}$$
$$\frac{\partial w_2(0)}{\partial \xi_j} = G \frac{\partial A(0)}{\partial \xi_j} w_2 - H \frac{\partial A(0)}{\partial \xi_j} w_1, \tag{1.9.8}$$
$$\frac{\partial y_1(0)^T}{\partial \xi_j} = y_1^T \frac{\partial A(0)}{\partial \xi_j} G + y_2^T \frac{\partial A(0)}{\partial \xi_j} H, \tag{1.9.9}$$
$$\frac{\partial y_2(0)^T}{\partial \xi_j} = y_2^T \frac{\partial A(0)}{\partial \xi_j} G - y_1^T \frac{\partial A(0)}{\partial \xi_j} H, \tag{1.9.10}$$

for each $j = 1, 2, \ldots, m$. Where

$$G = U_3(\nu I - A_3)\left[\mu^2 I + (\nu I - A_3)^2\right]^{-1} V_3^T - \frac{1}{2\mu}(w_1 y_2^T + w_2 y_1^T),$$
$$H = \mu U_3 \left[\mu^2 I + (\nu I - A_3)^2\right]^{-1} V_3^T + \frac{1}{2\mu}(w_1 y_1^T - w_2 y_2^T).$$

Proof Let

$$\widetilde{U} = U\mathrm{diag}\,(P, I_{n-2}), \quad \widetilde{V} = V\mathrm{diag}\,(Q, I_{n-2}), \quad \widetilde{A}(x) = \widetilde{V}^T A(x)\widetilde{U},$$

where

$$P = \frac{1}{2}\begin{bmatrix} 1 & -i \\ 1 & i \end{bmatrix}, \quad Q = \begin{bmatrix} 1 & i \\ 1 & -i \end{bmatrix}.$$

Then it is easy to verify that $\widetilde{U} = [w_1, w_2, U_3]$, $\widetilde{V} = [2y_1, -2y_2, V_3] \in \mathbf{R}^{n\times n}$, $\widetilde{V}^T\widetilde{U} = I$, and

$$\widetilde{A}(0) = \widetilde{V}^T A(0)\widetilde{U} = \begin{bmatrix} T & 0 \\ 0 & A_3 \end{bmatrix},$$

where

$$T = \frac{1}{2}\begin{bmatrix} 1 & i \\ 1 & -i \end{bmatrix}^T \begin{bmatrix} \lambda & 0 \\ 0 & \overline{\lambda} \end{bmatrix} \begin{bmatrix} 1 & -i \\ 1 & i \end{bmatrix} = \begin{bmatrix} \nu & \mu \\ -\mu & \nu \end{bmatrix}.$$

Now, applying Lemma 1.9.1 to $\widetilde{A}(x)$, we know that there exists two analytic functions $U(x), V(x) \in \mathbf{R}^{n\times n}$ in some neighborhood $\mathcal{B}_1 \subset \mathcal{B}$ of the origin of $\mathbf{R}^m$ satisfying $U(0) = V(0) = I$ and $V(x)^T U(x) = I$, such that

$$V(x)^T \widetilde{A}(x) U(x) = \begin{bmatrix} T(x) & 0 \\ 0 & A_3(x) \end{bmatrix}$$

where $T(x) \in \mathbf{R}^{2\times 2}$ with $T(0) = T$ and $A_3(x) \in \mathbf{R}^{(n-2)\times(n-2)}$ with $A_3(0) = A_3$. By Lemma 1.9.4 it follows that there exist real analytic functions $\nu(x), \mu(x) \in \mathbf{R}$ and $\widetilde{w}_1(x), \widetilde{w}_2(x), \widetilde{y}_1(x), \widetilde{y}_2(x) \in \mathbf{R}^2$ in some neighborhood $\mathcal{B}_0 \subset \mathcal{B}_1$ of the origin of $\mathbf{R}^n$ with $\nu(0) = \nu$, $\mu(0) = \mu$, and

$$\widetilde{w}_1(0) = \begin{bmatrix} 1 \\ 0 \end{bmatrix}, \quad \widetilde{w}_2(0) = \begin{bmatrix} 0 \\ 1 \end{bmatrix}, \quad \widetilde{y}_1(0) = \begin{bmatrix} \frac{1}{2} \\ 0 \end{bmatrix}, \quad \widetilde{y}_2(0) = \begin{bmatrix} 0 \\ -\frac{1}{2} \end{bmatrix},$$

such that

$$T(x)\widetilde{u}(x) = \lambda(x)\widetilde{u}(x), \quad \widetilde{v}(x)^T\widetilde{u}(x) = 1,$$
$$\widetilde{v}(x)^T T(x) = \lambda(x)\widetilde{v}(x)^T, \quad x \in \mathcal{B}_0,$$

where $\lambda(x) = \nu(x) + i\mu(x)$, $\widetilde{u}(x) = \widetilde{w}_1(x) + i\widetilde{w}_2(x)$, and $\widetilde{v}(x) = \widetilde{y}_1(x) + i\widetilde{y}_2(x)$.

Let

$$\alpha(x) = e_1^T \operatorname{diag}(P, I_{n-2}) U(x) \begin{bmatrix} \widetilde{u}(x) \\ 0 \end{bmatrix}.$$

Then it follows that $\alpha(x)$ is analytic in $\mathcal{B}_0$ and satisfies that

$$\alpha(0) = e_1^T \operatorname{diag}(P, I_{n-2}) \begin{bmatrix} \widetilde{u}(0) \\ 0 \end{bmatrix} = 1.$$

Therefore, without loss of generality we may assume that $\alpha(x) \neq 0$ for all $x \in \mathcal{B}_0$. Now let

$$u(x) = \frac{1}{\alpha(x)}\widetilde{U}U(x)\begin{bmatrix} \widetilde{u}(x) \\ 0 \end{bmatrix} \quad \text{and} \quad v(x) = \alpha(x)\widetilde{V}V(x)\begin{bmatrix} \widetilde{v}(x) \\ 0 \end{bmatrix}.$$

Clearly, $w_1(x) = \operatorname{Re}\big(u(x)\big)$, $w_2(x) = \operatorname{Im}\big(u(x)\big)$, $y_1(x) = \operatorname{Re}\big(v(x)\big)$, $y_2(x) = \operatorname{Im}\big(v(x)\big)$ are real analytic functions in $\mathcal{B}_0$ and satisfy that $w_i(0) = w_i$ and $y_i(0) = y_i$, $i = 1, 2$. Moreover, we have

$$A(x)u(x) = \lambda(x)u(x), \quad v(x)^T u(x) = 1,$$
$$v(x)^T A(x) = \lambda(x)v(x)^T, \quad x \in \mathcal{B}_0.$$

Differentiating the equality

$$A(x)u(x) = \lambda(x)u(x),$$

we obtain

$$\frac{\partial A(x)}{\partial \xi_j}u(x) + A(x)\frac{\partial u(x)}{\partial \xi_j} = \frac{\partial \lambda(x)}{\partial \xi_j}u(x) + \lambda(x)\frac{\partial u(x)}{\partial \xi_j},$$

that is,

$$\big(A(x) - \lambda(x)I\big)\frac{\partial u(x)}{\partial \xi_j} = \frac{\partial \lambda(x)}{\partial \xi_j}u(x) - \frac{\partial A(x)}{\partial \xi_j}u(x).$$

Letting $x = 0$ in the above equation gives

$$\big(\lambda I - A(0)\big)\frac{\partial u(0)}{\partial \xi_j} = \frac{\partial A(0)}{\partial \xi_j}u_1 - \frac{\partial \lambda(0)}{\partial \xi_j}u_1. \tag{1.9.11}$$

Substituting (1.9.4) into (1.9.11), we have

$$U\begin{bmatrix} 0 & 0 & 0 \\ 0 & 2i\mu & 0 \\ 0 & 0 & \lambda I_{n-2} - A_3 \end{bmatrix}V^T\frac{\partial u(0)}{\partial \xi_j} = \frac{\partial A(0)}{\partial \xi_j}u_1 - \frac{\partial \lambda(0)}{\partial \xi_j}u_1.$$

Multiplying it on the left by V^T yields

$$\begin{bmatrix} 0 & 0 & 0 \\ 0 & 2i\mu & 0 \\ 0 & 0 & \lambda I_{n-1} - A_3 \end{bmatrix}V^T\frac{\partial u(0)}{\partial \xi_j} = V^T\frac{\partial A(0)}{\partial \xi_j}u_1 - \frac{\partial \lambda(0)}{\partial \xi_j}e_1,$$

and so

$$0 = v_1^T\frac{\partial A(0)}{\partial \xi_j}u_1 - \frac{\partial \lambda(0)}{\partial \xi_j} \tag{1.9.12}$$

and

$$\begin{bmatrix} 2i\mu & 0 \\ 0 & \lambda I_{n-2} - A_3 \end{bmatrix} [v_2,\, V_3]^T \frac{\partial u(0)}{\partial \xi_j} = [v_2,\, V_3]^T \frac{\partial A(0)}{\partial \xi_j} u_1. \tag{1.9.13}$$

Thus, (1.9.5) and (1.9.6) immediately follow from (1.9.12) by equating its real and imaginary parts. Moreover, since $\lambda \notin \lambda(A_3)$, it follows from (1.9.13) that

$$[v_2,\, V_3]^T \frac{\partial u(0)}{\partial \xi_j} = \begin{bmatrix} -\frac{i}{2\mu} & 0 \\ 0 & [\lambda I_{n-2} - A_3]^{-1} \end{bmatrix} [v_2,\, V_3]^T \frac{\partial A(0)}{\partial \xi_j} u_1. \tag{1.9.14}$$

This says that $\frac{\partial u(0)}{\partial \xi_j}$ is a solution of an underdetermined system. Notice that

$$[v_2,\, V_3]^T [u_2,\, U_3] = I_{n-1} \quad \text{and} \quad [v_2,\, V_3]^T u_1 = 0.$$

By the standard theory of linear system, it follows from (1.9.14) that

$$\frac{\partial u(0)}{\partial \xi_j} = [u_2,\, U_3] \begin{bmatrix} -\frac{i}{2\mu} & 0 \\ 0 & (\lambda I_{n-2} - A_3)^{-1} \end{bmatrix} [v_2,\, V_3]^T \frac{\partial A(0)}{\partial \xi_j} u_1 + \alpha u_1 \tag{1.9.15}$$

for some $\alpha \in \mathbf{C}$. By the definition of $u(x)$ it is easy to see that $u(x)$ has the form

$$u(x) = [u_1, u_2, U_3] \begin{bmatrix} 1 \\ z(x) \end{bmatrix}, \quad z(x) \in \mathbf{C}^{n-1}.$$

This, together with $V^T U = I$, gives

$$v_1^T \frac{\partial u(0)}{\partial \xi_j} = v_1^T [u_2,\, U_3]^T \frac{\partial z(0)}{\partial \xi_j} = 0, \tag{1.9.16}$$

and so, multiplying (1.9.15) on the left by v_1^T gives

$$0 = v_1^T \frac{\partial u(0)}{\partial \xi_j} = \alpha v_1^T u_1 = \alpha.$$

Hence, we have

$$\begin{aligned} \frac{\partial u(0)}{\partial \xi_j} &= [u_2,\, U_3] \begin{bmatrix} -\frac{i}{2\mu} & 0 \\ 0 & (\lambda I_{n-2} - A_3)^{-1} \end{bmatrix} [v_2,\, V_3]^T \frac{\partial A(0)}{\partial \xi_j} u_1 \\ &= \left[-\frac{i}{2\mu} u_2 v_2^T + U_3 (\lambda I_{n-2} - A_3)^{-1} V_3^T \right] \frac{\partial A(0)}{\partial \xi_j} u_1. \end{aligned} \tag{1.9.17}$$

Let

$$(\lambda I_{n-2} - A_3)^{-1} = X + iY, \quad X, Y \in \mathbf{R}^{(n-2)\times(n-2)}.$$

Then

$$(i\mu I_{n-2} + \nu I_{n-2} - A_3)(X + iY) = I_{n-2}.$$

Comparing the real and imaginary parts gives

$$\begin{cases} (\nu I_{n-2} - A_3)X - \mu Y = I_{n-2}, \\ \mu X + (\nu I_{n-2} - A_3)Y = 0. \end{cases}$$

Solving this system, we obtain

$$X = (\nu I_{n-2} - A_3)\big(\mu^2 I_{n-2} + (\nu I_{n-2} - A_3)^2\big)^{-1}, \tag{1.9.18}$$

$$Y = -\mu\big(\mu^2 I_{n-2} + (\nu I_{n-2} - A_3)^2\big)^{-1}. \tag{1.9.19}$$

Thus, equating real and imaginary parts of (1.9.17) and using (1.9.18) and (1.9.19) we immediately obtain (1.9.7) and (1.9.8). Similarly, we can prove that

$$\frac{\partial v^T(0)}{\partial \xi_j} = v_1^T \frac{\partial A(0)}{\partial \xi_j}\left(-\frac{i}{2\mu}u_2 v_2^T + U_3 X V_3^T + iU_3 Y V_3^T\right) + \beta v_1^T \tag{1.9.20}$$

for some $\beta \in \mathbf{C}$, where X and Y are defined by (1.9.18) and (1.9.19), respectively. Notice that $v(x)^T u(x) = 1$ for $x \in \mathcal{B}_0$. We have

$$\frac{\partial v(0)^T}{\partial \xi_j} u_1 + v_1^T \frac{\partial u(0)}{\partial \xi_j} = 0.$$

This, together with (1.9.16), implies that

$$\frac{\partial v(0)^T}{\partial \xi_j} u_1 = 0.$$

Thus, it must have $\beta = 0$, and so (1.9.9) and (1.9.10) immediately follow from (1.9.20) by comparing its real and imaginary parts. □

So far our discussion has been on the differentiability of a simple eigenvalue and its corresponding right and left eigenvectors of a real matrix analytically dependent on several real parameters. What about the differentiability of a multiple eigenvalue? This is a very complicated problem, and the rigorous discussion about it is too long to give here. However, we present here a classical result.

Theorem 1.9.6 (Rellech's Theorem) *Let $A(t)$ be an $n \times n$ real symmetric matrix-valued analytic function in some neighborhood $\mathcal{B}$ of the origin of $\mathbf{R}$, and let λ be a multiple eigenvalue of $A(0)$ with multiplicity $r > 1$. Then there exists a neighborhood $\mathcal{B}_0 \subset \mathcal{B}$ of the origin and real analytic functions $\lambda_1(t), \dots, \lambda_r(t)$ and*

$x_1(t), \ldots, x_r(t)$, *in which* $\lambda_j(t) \in \mathbf{R}$ *and* $x_j(t) \in \mathbf{R}^n$ *for each* $j = 1, 2, \ldots, r$, *such that*

$$A(t)x_j(t) = \lambda_j(t)x_j(t), \quad \lambda_j(0) = \lambda_j,$$
$$x_j(t)^T x_j(t) = 1, \quad t \in \mathcal{B}_0, \quad j = 1, \ldots, r.$$

Furthermore, we have

$$\frac{\mathrm{d}\lambda_j(0)}{\mathrm{d}t} = x_j(0)^T \dot{A}(0) x_j(0), \quad j = 1, \ldots, r.$$

It is worth while to point out that Theorem 1.9.6 does not state that $\lambda_1(t) \leq \cdots \leq \lambda_r(t)$. In fact, in general such a requirement cannot be satisfied. For example, let $A(t) = \begin{bmatrix} t & 0 \\ 0 & -t \end{bmatrix}$. If we wanted $\lambda_1(t) \leq \lambda_2(t)$ we would need to define $\lambda_1(t) = -|t|$, $\lambda_2(t) = |t|$, but these are not analytic.

Notes and References

In this book, we assume the reader has a knowledge of basic matrix theory. Those readers unfamiliar with the fundamentals of matrix theory, or those wishing a review of the subject, are referred to the books by Gantmacher [77], Bellman [11], or Horn and Johnson [118]. This book also presupposes a knowledge of numerical linear algebra, for which there are many good introductory texts. Among them, we recommend the books by Varga [216], Wilkinson [221], Forsythe [69], Stewart [193], Parlett [174], Golub and Van Loan [90], Watkins [219], and Xu [235].

The notation adopted in this book is the one that is commonly used by numerical analysts. For the specification of algorithms the MATLAB-like language is used in this book. Matlab is an elegant interactive system idea for matrix computation work. For the formal MATLAB language consult [162] and [146].

Stewart and Sun [195] have an excellent discussion of vector and matrix norms; see also [90], [118], [120], and [193].

The mathematical properties of eigenvalues and eigenvectors are elegantly covered in [221] and [193]. The first proof of the Schur decomposition theorem is given by Schur [186]. The real Schur decomposition was originally presented in [165]. The famous QR algorithm for computing the real Schur decomposition is elegantly described in [221], [193], and [90]. The proofs of Jordan decomposition theorem and Cayley-Hamilton theorem can be found in many elementary linear algebra texts. For those who need further review we recommend [118]. For the theory and computations of the symmetric eigenvalue problem we highly recommend the book by Parlett [174]; see also [38], [39], [90], and [193].

Golub and Van Loan [90] offer an excellent proof of singular value decomposition. Some further references include [120] and [69]. For the detailed proofs of Theorems 1.5.1–1.5.4 the reader is referred to [118].

Theorem 1.6.2 is due to Gautschi [78] and [79]. For the Kronecker product consult Marcus [145].

The vector-valued function calculus and the implicit function theorem we introduced are detailed in Chapter 3 of [141]; see also [16]. Theorem 1.8.3 is taken from [191], but here some significant improvement has been made. The material of Section 1.9 is mainly based on [197] and [205], but here we confine our discussion to the case of a real matrix dependent on several real parameters. Theorem 1.9.6 is due to Rellech [180].

Chapter 2

Jacobi Matrix Inverse Eigenvalue Problems

2.1 Introduction

An $n \times n$ matrix J is called a **Jacobi matrix** if it is of the following form

$$J = \begin{bmatrix} \alpha_1 & \beta_2 & & & & \\ \beta_2 & \alpha_2 & \beta_3 & & 0 & \\ & \ddots & \ddots & \ddots & & \\ & & \ddots & \ddots & \ddots & \\ & 0 & & \beta_{n-1} & \alpha_{n-1} & \beta_n \\ & & & & \beta_n & \alpha_n \end{bmatrix}, \tag{2.1.1}$$

where α_i and β_i are real, and all β_i are positive. For a given Jacobi matrix, throughout this chapter we will use the symbol $J_{\hat{i}}$ to denote the $(n-1) \times (n-1)$ principal submatrix of J obtained from J by deleting the ith row and column. Especially, the symbol $J_{\hat{1}}$ denotes the $(n-1) \times (n-1)$ lower principal submatrix of J, which is obtained from J by deleting the first row and column.

A Jacobi matrix inverse eigenvalue problem, roughly speaking, is how to construct a Jacobi matrix from its spectral data. One of the most important problems of this type is the following inverse eigenvalue problem:

Problem 2.1 *Given two sets of real numbers $\{\lambda_i\}_1^n$ and $\{\mu_i\}_1^{n-1}$, satisfying the strictly* **interlacing property**

$$\lambda_1 < \mu_1 < \lambda_2 < \cdots < \mu_{n-1} < \lambda_n, \tag{2.1.2}$$

find an $n \times n$ Jacobi matrix J such that the given values $\{\lambda_i\}_1^n$ and $\{\mu_i\}_1^{n-1}$ are the eigenvalues of J and $J_{\hat{1}}$, respectively.

In this chapter we will mainly discuss this problem. It is well known that the eigenvalues of $J_{\hat{1}}$ strictly separate those of J, so that the condition (2.1.2) is necessary for the existence of a solution.

Much of motivation for this problem came about from an interest in the inverse Sturm-Liouville problem. Frequently, the matrix approximation to the Sturm-Liouville operator is a tridiagonal matrix. The inverse Sturm-Liouville problem arose in numerous physical investigations. For example, in quantum mechanics one encounters the Schrödinger equation

$$\psi'' + \big(E - g(x)\big)\psi = 0,$$

where $g(x)$ is a potential function. There it is desirable to deduce $g(x)$ from the knowledge of the spectrum of the operator.

Another typical example is a composite pendulum. Suppose that a pendulum is suspended from a fixed point and consists of n beads of masses $m_i > 0$ connected by weightless infinite flexible strings of length l_i. Here m_1 is the mass of the lowest bead. Assume that the vibrations of the pendulum are small and that the strings do not stretch under tension. Let u_i denote the deviation of the ith bead from its position at rest, let $\sigma_i = g\sum_{j=1}^{i} m_i$, which is the tension of the ith string, where g is the gravity, and let $\alpha_i = \sigma_i / l_i$. It is well known that if ω is an eigenfrequency of the pendulum, then $\lambda = \omega^2$ is an eigenvalue of the following eigenvalue problem:

$$Au = \lambda M u,$$

where

$$A = \begin{bmatrix} \alpha_1 & -\alpha_1 & & & \\ -\alpha_1 & \alpha_1+\alpha_2 & -\alpha_2 & & \\ & \ddots & \ddots & \ddots & \\ & & -\alpha_{n-2} & \alpha_{n-2}+\alpha_{n-1} & -\alpha_{n-1} \\ & & & -\alpha_{n-1} & \alpha_{n-1}+\alpha_n \end{bmatrix},$$

$$M = \operatorname{diag}(m_1, \ldots, m_n), \quad u = (u_1, \ldots, u_n)^T.$$

Let now the uppermost bead, with mass m_n, be fixed. What remains is a smaller composite pendulum which consists of $n-1$ beads and $n-1$ strings. It will be called the truncated pendulum. The problem now is: can we determine the lengths of the strings $(l_1, \ldots, l_n)$ and the masses of the beads $(m_1, \ldots, m_n)$ from the given eigenfrequencies of the composite pendulum and the given eigenfrequencies of the truncated pendulum. This problem can be reduced to Problem 2.1 provided m_1 is known.

Great progress has been made in the study of Problem 2.1 since the mid-1960's. Now, the basic theory of existence and uniqueness of the solution for Problem 2.1

has been established and the several fast and stable numerical algorithms for solving this problem have been developed. In the subsequent sections of this chapter we will introduce these results.

In Section 2.2 some basic properties of Jacobi matrices will be reviewed, which will be required in the subsequent sections. The theory of the existence and uniqueness of the solution for Problem 2.1 will be established in Section 2.3. A sensitivity analysis of Problem 2.1 will be done in Section 2.4. In Section 2.5 several efficient and reliable algorithms for solving Problem 2.1 will be developed. In the last five sections we will show how the methods developed in the previous sections can be used to solve some other inverse problems.

2.2 Preparation Theorems

In order to present the theory of Jacobi matrix inverse eigenvalue problems in a concise way, it is useful to give some preliminary results which play a fundamental role throughout the rest of this chapter.

Let J be an $n \times n$ Jacobi matrix defined as in (2.1.1) and for any $1 \le k \le n$ define

$$\widetilde{J}_k = \begin{bmatrix} \alpha_k & \beta_{k+1} & & & \\ \beta_{k+1} & \alpha_{k+1} & \beta_{k+2} & & \\ & \ddots & \ddots & \ddots & \\ & & \beta_{n-1} & \alpha_{n-1} & \beta_n \\ & & & \beta_n & \alpha_n \end{bmatrix}.$$

We denote the characteristic polynomials of the submatrices $\widetilde{J}_k$ by

$$\widetilde{p}_k(t) = \det(tI - \widetilde{J}_k), \quad k = 1, 2, \dots, n.$$

Then $\widetilde{p}_k$ is a monic polynomial of degree $n - k + 1$, and by expanding the determinant it is easy to verify that the sequence $\{\widetilde{p}_k\}$ satisfies the **three-term recurrence**

$$\widetilde{p}_k(t) = (t - \alpha_k)\widetilde{p}_{k+1}(t) - \beta_{k+1}^2 \widetilde{p}_{k+2}(t), \quad k = 1, 2, \dots, n, \tag{2.2.1}$$

where we define $\widetilde{p}_{n+1}(t) \equiv 1$, $\widetilde{p}_{n+2}(t) \equiv 0$, and $\beta_{n+1} = 0$.

Using the three-term recurrence and some properties of a real symmetric matrix we can easily prove the following theorem.

Theorem 2.2.1 *Let J be an $n \times n$ Jacobi matrix. Then all eigenvalues of J are real and distinct. Moreover, if the eigenvalues of J and $J_{\hat{1}}$ are respectively*

$$\lambda_1 < \lambda_2 < \cdots < \lambda_n,$$

and

$$\mu_1 < \mu_2 < \cdots < \mu_{n-1},$$

then

$$\lambda_1 < \mu_1 < \lambda_2 < \cdots < \mu_{n-1} < \lambda_n,$$

that is, the eigenvalues of $J_{\hat{1}}$ *strictly separate those of* J.

Proof Left to the reader. □

Theorem 2.2.2 *Let* J *be an* $n \times n$ *Jacobi matrix. Then there exists an orthogonal matrix* $Q = [q_{ij}] \in \mathbf{R}^{n\times n}$ *with* $q_{1j}q_{nj} \neq 0$, *for all* j, *such that*

$$Q^T JQ = \operatorname{diag}(\lambda_1, \ldots, \lambda_n),$$

where $\lambda_1, \ldots, \lambda_n$ *are the eigenvalues of* J.

Proof Let $q_j = (q_{1j}, q_{2j}, \ldots, q_{nj})^T$ be a unit eigenvector associated with λ_j, i.e.,

$$Jq_j = \lambda_j q_j \quad \text{with } \|q_j\|_2 = 1 \tag{2.2.2}$$

for $j = 1, 2, \ldots, n$. Then it follows that

$$(\lambda_j - \lambda_i)q_i^T q_j = \lambda_j q_j^T q_i - \lambda_i q_i^T q_j = q_j^T Jq_i - q_i^T Jq_j = 0,$$

and so $q_i^T q_j = 0$ if $i \neq j$ since $\lambda_j \neq \lambda_i$ by Theorem 2.2.1.

Thus, if we set

$$Q = [q_1, q_2, \ldots, q_n],$$

then Q is orthogonal and (2.2.2) can be rewritten in matrix form as

$$JQ = Q\operatorname{diag}(\lambda_1, \lambda_2, \ldots, \lambda_n),$$

that is,

$$Q^T JQ = \operatorname{diag}(\lambda_1, \ldots, \lambda_n).$$

On the other hand, equating the ith element on each side of (2.2.2) and rearranging terms we have

$$\beta_{i+1}q_{i+1,j} = (\lambda_j - \alpha_i)q_{ij} - \beta_i q_{i-1,j}, \quad i = 1, 2, \ldots, n, \tag{2.2.3}$$

where we define $\beta_1 = \beta_{n+1} = 0$ and $q_{n+1,j} = q_{0j} = 0$. If for some j, q_{1j} or q_{nj} were zero, it would follow from the above three- term recurrence that $q_{ij} = 0$ for all i contradicting the hypothesis that $\|q_j\|_2 = 1$. □

Theorem 2.2.3 **(Paige's Formula)** *Let J be an $n \times n$ Jacobi matrix, and let its Schur decomposition be $J = Q\Lambda Q^T$, where $\Lambda = \operatorname{diag}(\lambda_1, \dots, \lambda_n)$ with $\lambda_1 < \lambda_2 < \cdots < \lambda_n$, and $Q = [q_{ij}]$ is an $n \times n$ orthogonal matrix. Then*

$$\operatorname{adj}(\lambda_j I - J) = \prod_{\substack{k=1\\k\neq j}}^{n} (\lambda_j - \lambda_k) Q e_j e_j^T Q^T. \tag{2.2.4}$$

Especially, we have

$$\prod_{k=2}^{n} \beta_k = q_{1j} q_{nj} \prod_{\substack{k=1\\k\neq j}}^{n} (\lambda_j - \lambda_k) \tag{2.2.5}$$

and

$$q_{ij}^2 = \prod_{l=1}^{j-1} \frac{\lambda_j - \mu_l}{\lambda_j - \lambda_l} \prod_{l=j+1}^{n} \frac{\lambda_j - \mu_{l-1}}{\lambda_j - \lambda_l} \tag{2.2.6}$$

for all i and j, where $\mu_1 \le \mu_2 \le \cdots \le \mu_{n-1}$ are the eigenvalues of $J_{\hat{i}}$, which is obtained from J by deleting the ith row and column.

Proof Let

$$p(t) = \det(tI - J) \quad \text{and} \quad q(t) = \det(tI - J_{\hat{i}}).$$

By the hypotheses, it follows that

$$p(t) = \prod_{i=1}^{n} (t - \lambda_i) \quad \text{and} \quad q(t) = \prod_{i=1}^{n-1} (t - \mu_i).$$

Recall from linear algebra that for any $t \notin \lambda(J)$ the matrix $(tI - J)$ is invertible and

$$(tI - J)^{-1} = \frac{1}{p(t)} \operatorname{adj}(tI - J), \tag{2.2.7}$$

where $\operatorname{adj}(tI - J)$ denotes the adjoint matrix of $(tI - J)$.

On the other hand, since $J = Q\Lambda Q^T$ the inverse can be also expressed as

$$(tI - J)^{-1} = Q \operatorname{diag}\left(\frac{1}{t - \lambda_1}, \frac{1}{t - \lambda_2}, \cdots, \frac{1}{t - \lambda_n} \right) Q^T. \tag{2.2.8}$$

Combining (2.2.7) and (2.2.8) yields

$$\operatorname{adj}(tI - J) = Q \operatorname{diag}\left(\frac{p(t)}{t - \lambda_1}, \frac{p(t)}{t - \lambda_2}, \cdots, \frac{p(t)}{t - \lambda_n} \right) Q^T.$$

Since the λ_i are distinct and all the elements of the adjoint matrix adj$(tI - J)$ are continuous functions of t, letting $t \to \lambda_j$ we obtain

$$\operatorname{adj}(\lambda_j I - J) = p'(\lambda_j) Q e_j e_j^T Q^T, \tag{2.2.9}$$

which is the desired result. Especially, equating the (i,i)-element on each side of (2.2.9) we have

$$q(\lambda_j) = p'(\lambda_j) q_{ij}^2,$$

from which it immediately follows that (2.2.6) is true. Similarly, comparing the $(1, n)$-element on each side of (2.2.9) gives the formula (2.2.5). □

Theorem 2.2.4 *Let $Q^T A Q = J$, where A is an $n \times n$ real symmetric matrix, J is an $n \times n$ Jacobi matrix, and $Q = [q_1, q_2, \ldots, q_n]$ is an $n \times n$ orthogonal matrix. Then J and Q are uniquely determined by A and q_1.*

Proof The hypothesis can be rewritten as

$$QJ = AQ. \tag{2.2.10}$$

Now equating the jth column on each side of (2.2.10) and rearranging terms, we obtain an important relation

$$\beta_{j+1} q_{j+1} = A q_j - \alpha_j q_j - \beta_j q_{j-1}. \tag{2.2.11}$$

This holds for $j = 1, 2, \ldots, n$ if we define $\beta_1 = \beta_{n+1} = 0$ and $q_0 = q_{n+1} = 0$. Using the orthogonality of Q, from (2.2.11) it is easy to derive that

$$\alpha_j = q_j^T A q_j, \qquad \beta_{j+1} = \|A q_j - \alpha_j q_j - \beta_j q_{j-1}\|_2,$$
$$q_{j+1} = \left(A q_j - \alpha_j q_j - \beta_j q_{j-1}\right) / \beta_{j+1}.$$

The last equation holds since $\beta_{j+1} > 0$ by hypothesis. From these relations it is easy to see that q_1 alone determines α_1, β_2, and q_2, in turn, α_1, β_2, q_1, and q_2 determine α_2, β_3, and q_3, and so on. Hence, by induction, q_1 determines uniquely all the elements of J and Q. □

Lemma 2.2.5 *Let $\Lambda = \operatorname{diag}(\lambda_1, \ldots, \lambda_n)$ with $\lambda_i \neq \lambda_j$ if $i \neq j$, and let $x = (\xi_1, \ldots, \xi_n)^T$ with $\xi_i \neq 0$ for all i. Then the n vectors $x, \Lambda x, \Lambda^2 x, \ldots, \Lambda^{n-1} x$ are linearly independent.*

Proof This is true because

$$\det[x, \Lambda x, \Lambda^2 x, \ldots, \Lambda^{n-1} x] = \det(V(\lambda_1, \ldots, \lambda_n)) \prod_{j=1}^{n} \xi_j \neq 0,$$

where $V(\lambda_1, \dots, \lambda_n)$ denotes the Vandermonde matrix determined by $\lambda_1, \dots, \lambda_n$. □

Theorem 2.2.6 *Let*

$$Q\Lambda Q^T = J = \begin{bmatrix} \alpha_1 & \beta_2 & & & \\ \beta_2 & \alpha_2 & \beta_3 & & \\ & \ddots & \ddots & \ddots & \\ & & \beta_{n-1} & \alpha_{n-1} & \beta_n \\ & & & \beta_n & \alpha_n \end{bmatrix} \tag{2.2.12}$$

with $\beta_i \geq 0$ for all i, where $\Lambda = \operatorname{diag}(\lambda_1, \dots, \lambda_n)$ with $\lambda_i \neq \lambda_j$ if $i \neq j$, and Q is orthogonal with all the elements of its first row nonzero. Then J is a Jacobi matrix, that is, $\beta_i > 0$ for all i.

Proof If $\beta_{i_0} = 0$ for some $1 \leq i_0 \leq n$, then it would follow from (2.2.12) that

$$\Lambda[q_1, \dots, q_{i_0-1}] = [q_1, \dots, q_{i_0-1}] \begin{bmatrix} \alpha_1 & \beta_2 & & & \\ \beta_2 & \alpha_2 & \beta_3 & & \\ & \ddots & \ddots & \ddots & \\ & & \beta_{i_0-2} & \alpha_{i_0-2} & \beta_{i_0-1} \\ & & & \beta_{i_0-1} & \alpha_{i_0-1} \end{bmatrix}, \tag{2.2.13}$$

where q_i denotes the ith row of Q, that is, $Q^T = [q_1, q_2, \dots, q_n]$. Consequently, the n vectors $q_1, \Lambda q_1, \dots, \Lambda^{n-1} q_1$ would be linearly dependent, contradicting the assumptions that $\lambda_i \neq \lambda_j$ if $i \neq j$ and all components of q_1 are nonzero by Lemma 2.2.5. □

2.3 Existence and Uniqueness

In this section, we will establish the theorem of the existence and uniqueness of the solution for Problem 2.1 in Section 2.1. We begin with a fundamental theorem.

Theorem 2.3.1 *Let J be an $n \times n$ Jacobi matrix. Then J is a solution of Problem 2.1 in Section 2.1 if and only if there exists an $n \times n$ orthogonal matrix $Q = [q_{ij}]$ such that*

$$J = Q\Lambda Q^T \tag{2.3.1}$$

and

$$q_{1j}^2 = \prod_{i=1}^{n-1}(\lambda_j - \mu_i) \Bigg/ \prod_{\substack{i=1\\i\neq j}}^{n}(\lambda_j - \lambda_i) \tag{2.3.2}$$

for $j = 1, 2, \dots, n$. *Where* $\Lambda = \mathrm{diag}(\lambda_1, \lambda_2, \dots, \lambda_n)$.

Proof We first prove the necessity. Assume that J is a solution of Problem 2.1. Then the eigenvalues of J and $J_{\hat{1}}$ are $\lambda_1, \lambda_2, \dots, \lambda_n$ and $\mu_1, \mu_2, \dots, \mu_{n-1}$, respectively. By Theorem 2.2.3 , it follows that the $n \times n$ orthogonal matrix $Q = [q_{ij}]$ which consists of the eigenvectors of J satisfies (2.3.1) and (2.3.2).

Now we prove the sufficiency. Suppose that Jacobi matrix J and orthogonal matrix $Q = [q_{ij}]$ satisfy (2.3.1) and (2.3.2). Let

$$p_n(\lambda) = \det(\lambda I - J) \quad \text{and} \quad p_{n-1}(\lambda) = \det(\lambda I - J_{\hat{1}}). \tag{2.3.3}$$

Applying Theorem 2.2.3 again, we have

$$q_{1j}^2 = p_{n-1}(\lambda_j)\,/p_n'(\lambda_j) \quad \text{for } j = 1, \dots, n. \tag{2.3.4}$$

Notice that (2.3.1) implies that $p_n(\lambda) = \prod_{i=1}^{n}(\lambda - \lambda_i)$. It follows from (2.3.2) and (2.3.4) that

$$p_{n-1}(\lambda_j) = \prod_{i=1}^{n-1}(\lambda_j - \mu_i) \quad \text{for } j = 1, 2, \dots, n, \tag{2.3.5}$$

which implies that

$$p_{n-1}(\lambda) = \prod_{i=1}^{n-1}(\lambda - \mu_i)$$

because the only polynomial of degree at most $n-1$ which is zero at n distinct points is the zero polynomial. That is, $J_{\hat{1}}$ has eigenvalues $\mu_1, \cdots, \mu_{n-1}$. Hence, J is a solution of Problem 2.1. □

To establish the existence, we will use the tridiagonalization theorem in matrix computation, which is useful in its own right. Its proof is not difficult and is left as an exercise(see also Section 2.5).

Lemma 2.3.2 (Tridiagonalization Theorem) *Let* A *be an* $n \times n$ *symmetric* *matrix. Then there exists an $n \times n$ orthogonal matrix Q with $Qe_1 = e_1$ such that*

$$QAQ^T = \begin{bmatrix} \alpha_1 & \beta_2 & & & & \\ \beta_2 & \alpha_2 & \beta_3 & & 0 & \\ & \ddots & \ddots & \ddots & & \\ & & \ddots & \ddots & \ddots & \\ & 0 & & \beta_{n-1} & \alpha_{n-1} & \beta_n \\ & & & & \beta_n & \alpha_n \end{bmatrix},$$

with $\beta_i \geq 0$ for all i.

We are now in a position to state the fundamental **unique-existence theorem** of this chapter.

Theorem 2.3.3 *Let $\{\lambda_i\}_1^n$ and $\{\mu_i\}_1^{n-1}$ be given and satisfy the strictly interlacing property*

$$\lambda_1 < \mu_1 < \lambda_2 < \cdots < \mu_{n-1} < \lambda_n. \tag{2.3.6}$$

Then there exists a unique $n \times n$ Jacobi matrix J such that the given values $\{\lambda_i\}_1^n$ and $\{\mu_i\}_1^{n-1}$ are the eigenvalues of J and $J_{\hat{1}}$, respectively. That is, Problem 2.1 has a unique solution.

Proof Existence. We first use the $2n-1$ given numbers to construct an $(n+1) \times (n+1)$ bordered diagonal matrix

$$A = \begin{bmatrix} 0 & q^T \\ q & \Lambda \end{bmatrix},$$

where $\Lambda = \operatorname{diag}(\lambda_1, \cdots, \lambda_n)$, and $q = (\omega_1, \cdots, \omega_n)^T$ with

$$\omega_j = \left(\prod_{i=1}^{n-1} (\lambda_j - \mu_i) \Big/ \prod_{\substack{i=1 \\ i \neq j}}^{n} (\lambda_j - \lambda_i) \right)^{\frac{1}{2}} > 0 \tag{2.3.7}$$

for $j = 1, 2, \ldots, n$. The positive condition (2.3.7) is guaranteed by the strictly interlacing property (2.3.6). By Lemma 2.3.2, we know that there exists an $n \times n$ orthogonal matrix Q such that

$$\begin{bmatrix} 1 & 0 \\ 0 & Q \end{bmatrix} \begin{bmatrix} 0 & q^T \\ q & \Lambda \end{bmatrix} \begin{bmatrix} 1 & 0 \\ 0 & Q \end{bmatrix}^T = \begin{bmatrix} 0 & \beta e_1^T \\ \beta e_1 & J \end{bmatrix}, \tag{2.3.8}$$

where $\beta = \|q\|_2$ and

$$J = \begin{bmatrix} \alpha_1 & \beta_2 & & & & \\ \beta_2 & \alpha_2 & \beta_3 & & 0 & \\ & \ddots & \ddots & \ddots & & \\ & & \ddots & \ddots & \ddots & \\ & 0 & & \beta_{n-1} & \alpha_{n-1} & \beta_n \\ & & & & \beta_n & \alpha_n \end{bmatrix}$$

with $\beta_i \geq 0$ for all i. Comparing two sides of (2.3.8), we have

$$Q\Lambda Q^T = J \quad \text{and} \quad Qq = \beta e_1. \tag{2.3.9}$$

This implies that J is a Jacobi matrix by Theorem 2.2.6. Moreover, similar to the proof of the sufficiency of Theorem 2.3.1, we from (2.3.9) and (2.3.7) can prove that

$$\det(\lambda I - J_{\hat{1}}) = \beta^{-2} \prod_{i=1}^{n-1} (\lambda - \mu_i),$$

which implies that $\beta = 1$ and the eigenvalues of $J_{\hat{1}}$ are the given μ_i. Thus we have proved that the Jacobi matrix J obtained above is a solution of Problem 2.1.

Uniqueness. Assume that both J and $\widetilde{J}$ are solutions of Problem 2.1. Then it follows from Theorem 2.3.1 that there exist two orthogonal matrices $Q = [q_{ij}]$ and $\widetilde{Q} = [\widetilde{q}_{ij}]$ such that

$$J = Q\Lambda Q^T \quad \text{and} \quad \widetilde{J} = \widetilde{Q}\Lambda\widetilde{Q}^T, \tag{2.3.10}$$

and the first row elements of both Q and $\widetilde{Q}$ satisfy (2.3.2). Moreover, without loss of generality we can assume that

$$\widetilde{q}_{1j} > 0 \quad \text{and} \quad q_{1j} > 0$$

for $j = 1, 2, \cdots n$. Thus, it follows that

$$\widetilde{q}_{1j} = q_{1j}, \quad \text{for } j = 1, 2, \cdots n. \tag{2.3.11}$$

By Theorem 2.2.4, we have $Q = \widetilde{Q}$ and $J = \widetilde{J}$. That is, Problem 2.1 has only one solution. □

As an immediate consequence of the proof of Theorem 2.3.3, we have

Corollary 2.3.4 *Let $\{\lambda_i\}_1^n$ and $\{\mu_i\}_1^{n-1}$ be given and satisfy*

$$\lambda_1 < \mu_1 < \lambda_2 < \cdots < \mu_{n-1} < \lambda_n.$$

Then

$$\sum_{j=1}^{n}\left(\prod_{i=1}^{n-1}(\lambda_j-\mu_i)\Big/\prod_{\substack{i=1\\i\neq j}}^{n}(\lambda_j-\lambda_i)\right)=1.$$

In addition, the proof of Theorem 2.3.3 also implies the following important result.

Theorem 2.3.5 *Let $\lambda_1,\dots,\lambda_n$ be n distinct real numbers, with $\lambda_1<\lambda_2<\cdots<\lambda_n$, and let $\omega_1,\omega_2,\dots,\omega_n$ be n positive numbers, with $\sum_{i=1}^{n}\omega_i^2=1$. Then there exists a unique $n\times n$ Jacobi matrix J_n such that its eigenvalues are the given values λ_i and the first components of its normalized eigenvectors are exactly the ω_i.*

2.4 Sensitivity

In this section we will do a sensitivity analysis of Problem 2.1, that is, we will study how the solution of Problem 2.1 is affected by the small changes in the given numbers. More precisely, suppose that $\{\lambda_i\}_1^n$, $\{\widetilde{\lambda}_i\}_1^n$, $\{\mu_i\}_1^{n-1}$, and $\{\widetilde{\mu}_i\}_1^{n-1}$ are given, satisfying

$$\lambda_1<\mu_1<\lambda_2<\cdots<\mu_{n-1}<\lambda_n, \tag{2.4.1}$$

$$\widetilde{\lambda}_1<\widetilde{\mu}_1<\widetilde{\lambda}_2<\cdots<\widetilde{\mu}_{n-1}<\widetilde{\lambda}_n, \tag{2.4.2}$$

and suppose that J and $\widetilde{J}$ are the solutions of Problem 2.1 for the given data $(\{\lambda_i\}_1^n,\{\mu_i\}_1^{n-1})$ and $(\{\widetilde{\lambda}_i\}_1^n,\{\widetilde{\mu}_i\}_1^{n-1})$, respectively. Our aim is to develop an upper bound on $\|J-\widetilde{J}\|_F$, which is reasonably sharp. For this, it will be helpful to define some notation. Define

$$\beta=\max_{1\le i\le n}|\lambda_i|,$$

$$\theta_j=\sum_{i=1}^{n-1}\frac{1}{|\lambda_j-\mu_i|}+\sum_{\substack{i=1\\i\neq j}}^{n}\frac{1}{|\lambda_j-\lambda_i|}+\left|\sum_{i=1}^{n-1}\frac{1}{\lambda_j-\mu_i}-\sum_{\substack{i=1\\i\neq j}}^{n}\frac{1}{\lambda_j-\lambda_i}\right|,$$

$$\omega_j=\left(\prod_{i=1}^{n-1}(\lambda_j-\mu_i)\Big/\prod_{\substack{i=1\\i\neq j}}^{n}(\lambda_j-\lambda_i)\right)^{\frac12}, \tag{2.4.3}$$

$$\widetilde{\omega}_j = \left(\prod_{i=1}^{n-1} (\widetilde{\lambda}_j - \widetilde{\mu}_i) \Big/ \prod_{\substack{i=1 \\ i \neq j}}^{n} (\widetilde{\lambda}_j - \widetilde{\lambda}_i) \right)^{\frac{1}{2}},$$

$$\eta = \sum_{i=1}^{n} \omega_i^2 \left(\frac{1}{4}\theta_j + \sum_{j=2}^{n} \lambda_i^{2(j-2)} \left(j - 1 + \frac{1}{2}|\lambda_i|\theta_i \right)^2 \right),$$

$$\varepsilon = \left(\sum_{i=1}^{n} |\lambda_i - \widetilde{\lambda}_i|^2 + \sum_{i=1}^{n-1} |\mu_i - \widetilde{\mu}_i|^2 \right)^{\frac{1}{2}},$$

$$A = \begin{bmatrix} \omega_1 & \lambda_1\omega_1 & \cdots & \lambda_1^{n-1}\omega_1 \\ \omega_2 & \lambda_2\omega_2 & \cdots & \lambda_2^{n-1}\omega_2 \\ \vdots & \vdots & & \vdots \\ \omega_n & \lambda_n\omega_n & \cdots & \lambda_n^{n-1}\omega_n \end{bmatrix}.$$

We are now in a position to state the main result of this section.

Theorem 2.4.1 *If ε is sufficiently small, then*

$$\|J - \widetilde{J}\|_F \le \left(1 + 2\beta\sqrt{2\eta}\|A^{-1}\|_2\right)\varepsilon + O(\varepsilon^2), \tag{2.4.4}$$

where ε, β, η, and A are defined as in (2.4.3).

Remark 2.4.1 Notice that

$$\|J\|_F = \left(\sum_{i=1}^{n} \lambda_i^2 \right)^{\frac{1}{2}}$$

and that condition (2.4.1) implies that

$$\left(\sum_{i=1}^{n} \lambda_i^2 + \sum_{i=1}^{n-1} \mu_i^2 \right)^{\frac{1}{2}} \le \sqrt{2} \left(\sum_{i=1}^{n} \lambda_i^2 \right)^{\frac{1}{2}}.$$

From (2.4.4) we get

$$\frac{\|J - \widetilde{J}\|_F}{\|J\|_F} \le \sqrt{2}\left(1 + 2\beta\sqrt{2\eta}\|A^{-1}\|_2\right)\delta + O(\delta^2), \tag{2.4.5}$$

where

$$\begin{aligned} \delta &= \varepsilon \left(\sum_{i=1}^{n} \lambda_i^2 + \sum_{i=1}^{n-1} \mu_i^2 \right)^{-\frac{1}{2}} \\ &= \left(\frac{\sum_{i=1}^{n} (\lambda_i - \widetilde{\lambda}_i)^2 + \sum_{i=1}^{n-1} (\mu_i - \widetilde{\mu}_i)^2}{\sum_{i=1}^{n} \lambda_i^2 + \sum_{i=1}^{n-1} \mu_i^2} \right)^{\frac{1}{2}}. \end{aligned}$$

Hence, by Rice's theory of condition [181], the scalar $\kappa(\{\lambda_i\}_1^n, \{\mu_i\}_1^{n-1})$ defined by

$$\kappa(\{\lambda_i\}_1^n, \{\mu_i\}_1^{n-1}) = \sqrt{2}\left(1 + 2\beta\sqrt{2\eta}\|A^{-1}\|_2\right) \tag{2.4.6}$$

can be called the **condition number** of Problem 2.1 for the given data $\{\lambda_i\}_1^n$ and $\{\mu_i\}_1^{n-1}$. Generally speaking, if the condition number $\kappa(\{\lambda_i\}_1^n, \{\mu_i\}_1^{n-1})$ is very large, we will call Problem 2.1 with this given data **ill-conditioned.**

Presumably, this theorem shows that there is some intrinsic relation between the sensitivity of the solution to Problem 2.1 and the separation of the given eigenvalues. See Example 2.4.1.

In order to give the proof of Theorem 2.4.1, we will need a perturbation theorem for QR factors. Recall that the **QR factorization** of a matrix $A \in \mathbf{R}^{n\times n}$ with $\text{rank}A = n$ is a decomposition of the form $A = QR$, in which $R \in \mathbf{R}^{n\times n}$ is an upper triangular matrix with positive diagonal elements, and $Q \in \mathbf{R}^{n\times n}$ is orthogonal. The matrices Q and R are referred to as the **QR factors** of A. It is well known that the QR factorization of a full rank matrix is unique.

Lemma 2.4.2 *Let $A = QR$ be the QR factorization of $A \in \mathbf{R}^{n\times n}$ with* $\text{rank}\,(A) = n$. *Let $E \in \mathbf{R}^{n\times n}$ satisfy*

$$\|A^{-1}\|_2\,\|E\|_2 < 1.$$

Then $A + E$ has a unique QR factorization

$$A + E = (Q + W)(R + F)$$

where

$$\|W\|_F \le \sqrt{2}\|A^{-1}\|_2\,\|E\|_F + O(\|E\|_F^2).$$

The proof of this lemma is too long to be given here. See [208] or [31] for the detailed proof.

With the forgoing preparations we may now give the proof of Theorem 2.4.1.

Proof of Theorem 2.4.1 By Theorem 2.3.1, we have

$$J = Q\Lambda Q^T, \qquad \widetilde{J} = \widetilde{Q}\widetilde{\Lambda}\widetilde{Q}^T, \tag{2.4.7}$$

where $\Lambda = \text{diag}\,(\lambda_1, \ldots \lambda_n)$, $\widetilde{\Lambda} = \text{diag}\,(\widetilde{\lambda}_1, \ldots \widetilde{\lambda}_n)$, and Q and $\widetilde{Q}$ are orthogonal matrices with

$$Q^T e_1 = q = (\omega_1, \ldots, \omega_n)^T \quad \text{and} \quad \widetilde{Q}^T e_1 = \widetilde{q} = (\widetilde{\omega}_1, \ldots, \widetilde{\omega}_n)^T.$$

Here the ω_i and $\widetilde{\omega}_i$ are defined as in (2.4.3).

Let now

$$W = \widetilde{Q} - Q, \qquad \Sigma = \widetilde{\Lambda} - \Lambda.$$

Then

$$\widetilde{J} - J = \widetilde{Q}\widetilde{\Lambda}\widetilde{Q}^T - Q\Lambda Q^T = \widetilde{Q}\Sigma\widetilde{Q}^T + W\Lambda\widetilde{Q}^T + Q\Lambda W^T.$$

Hence, we have

$$\|\widetilde{J} - J\|_F \le \|\Sigma\|_F + 2\|\Lambda\|_2\,\|W\|_F. \tag{2.4.8}$$

Thus, in order to give an upper bound on $\|J - \widetilde{J}\|_F$ we need to develop an upper bound on $\|W\|_F$. Notice that the matrix A defined in (2.4.3) can be written as

$$A = [\lambda_i^{j-1}\omega_i] = [q, \Lambda q, \dots, \Lambda^{n-1}q]. \tag{2.4.9}$$

Thus, using (2.4.7) we can easily prove that

$$Q^T A = [e_1, Je_1, \dots, J^{n-1}e_1],$$

where Q is the orthogonal matrix in (2.4.7). Let

$$R = [e_1, Je_1, \dots, J^{n-1}e_1].$$

Then R is an upper triangular matrix with positive diagonal elements, and $A = QR$ is the QR factorization of A.

Similarly, we can get the QR factorization of $\widetilde{A} = [\widetilde{q}, \widetilde{\Lambda}\widetilde{q}, \dots, \widetilde{\Lambda}^{n-1}\widetilde{q}]$ as follows:

$$\widetilde{A} = \widetilde{Q}\widetilde{R},$$

where $\widetilde{Q}$ is the orthogonal matrix in (2.4.7), and $\widetilde{R} = [e_1, \widetilde{J}e_1, \dots, \widetilde{J}^{n-1}e_1]$ is an upper triangular matrix with positive diagonal elements.

Without loss of generality we may assume that $\varepsilon \neq 0$. Now, let

$$A(t) = \begin{bmatrix} \omega_1(t) & \lambda_1(t)\omega_1(t) & \cdots & \lambda_1(t)^{n-1}\omega_1(t) \\ \omega_2(t) & \lambda_2(t)\omega_2(t) & \cdots & \lambda_2(t)^{n-1}\omega_2(t) \\ \vdots & \vdots & & \vdots \\ \omega_n(t) & \lambda_n(t)\omega_n(t) & \cdots & \lambda_n(t)^{n-1}\omega_n(t) \end{bmatrix}, \qquad t \in [0,\, 1],$$

where

$$\lambda_j(t) = \lambda_j + t\varepsilon^{-1}(\widetilde{\lambda}_j - \lambda_j), \tag{2.4.10}$$

$$\omega_j(t) = \left(\frac{\prod_{i=1}^{n-1}[\lambda_j - \mu_i + t\varepsilon^{-1}(\widetilde{\lambda}_j - \lambda_j + \mu_i - \widetilde{\mu}_i)]}{\prod_{i=1, i\neq j}^{n}[\lambda_j - \lambda_i + t\varepsilon^{-1}(\widetilde{\lambda}_j - \widetilde{\lambda}_i + \lambda_i - \lambda_j)]} \right)^{\frac{1}{2}}, \tag{2.4.11}$$

for $j = 1, 2, \ldots, n$. Then $A(t)$ is analytic and the Taylor expansion for $A(t)$ about $t = 0$ gives at $t = \varepsilon$

$$\widetilde{A} = A(\varepsilon) = A(0) + \dot{A}(0)\varepsilon + O(\varepsilon^2) = A + \dot{A}(0)\varepsilon + O(\varepsilon^2). \tag{2.4.12}$$

Hence, it follows from (2.4.12) that

$$\|A^{-1}\|_2 \, \|A - \widetilde{A}\|_2 < 1$$

when ε is sufficiently small, and so, by Lemma 2.4.2, we have

$$\|W\|_F \leq \sqrt{2}\, \|A^{-1}\|_2 \, \|E\|_F + O(\|E\|_F^2) \tag{2.4.13}$$

when ε is sufficiently small, where $E = \widetilde{A} - A$.

Differentiating (2.4.10) and (2.4.11) at $t = 0$, we have

$$\begin{aligned}
\dot{\lambda}_j(0) &= \varepsilon^{-1}(\widetilde{\lambda}_j - \lambda_j), \\
\dot{\omega}_j(0) &= \frac{1}{2}\omega_j \left(\sum_{i=1}^{n-1} \frac{\widetilde{\lambda}_j - \lambda_j + \mu_i - \widetilde{\mu}_i}{\varepsilon(\lambda_j - \mu_i)} - \sum_{\substack{i=1 \\ i \neq j}}^{n} \frac{\widetilde{\lambda}_j - \widetilde{\lambda}_i + \lambda_i - \lambda_j}{\varepsilon(\lambda_j - \lambda_i)} \right) \\
&= \frac{1}{2}\omega_j \left(\left[\sum_{i=1}^{n-1} \frac{1}{\lambda_j - \mu_i} - \sum_{\substack{i=1 \\ i \neq j}}^{n} \frac{1}{\lambda_j - \lambda_i} \right] \frac{\widetilde{\lambda}_j - \lambda_j}{\varepsilon} \right. \\
&\qquad \left. + \sum_{i=1}^{n-1} \frac{\mu_i - \widetilde{\mu}_i}{\varepsilon(\lambda_j - \mu_i)} - \sum_{\substack{i=1 \\ i \neq j}}^{n} \frac{\lambda_i - \widetilde{\lambda}_i}{\varepsilon(\lambda_j - \lambda_i)} \right).
\end{aligned}$$

From this it follows directly that

$$|\dot{\lambda}_j(0)| = \varepsilon^{-1}|\widetilde{\lambda}_j - \lambda_j| \leq 1 \quad \text{and} \quad |\dot{\omega}_j(0)| \leq \frac{1}{2}\omega_j\theta_j,$$

where θ_j is defined as in (2.4.3), and so

$$\begin{aligned}
\|\dot{A}(0)\|_F^2 &= \sum_{i=1}^{n} \left(\dot{\omega}_i(0)^2 + \sum_{j=2}^{n} \left((j-1)\lambda_i^{j-2}\omega_i\dot{\lambda}_i(0) + \lambda_i^{j-1}\dot{\omega}_i(0) \right)^2 \right) \\
&\leq \sum_{i=1}^{n} \omega_i^2 \left(\frac{1}{4}\theta_i^2 + \sum_{j=2}^{n} \lambda_i^{2(j-2)} \left(j - 1 + \frac{1}{2}|\lambda_i|\theta_i \right)^2 \right) \\
&= \eta.
\end{aligned} \tag{2.4.14}$$

Combining (2.4.12) and (2.4.14) gives

$$\|E\|_F \leq \sqrt{\eta}\,\varepsilon + O(\varepsilon^2). \tag{2.4.15}$$

Substituting (2.4.15) into (2.4.13), we obtain

$$||W||_F \le \sqrt{2\eta}||A^{-1}||_2\varepsilon + O(\varepsilon^2). \tag{2.4.16}$$

Notice that $||\Lambda||_2 = \beta$. The inequality (2.4.4) immediately follows from (2.4.8) and (2.4.16). This completes the proof of Theorem 2.4.1. □

From (2.4.6) we know that if we want to compute the condition number $\kappa(\{\lambda_i\}_1^n, \{\mu_i\}_1^{n-1})$, we have to compute A^{-1} and its spectral norm. However, it is well known that the spectral norm of a matrix is hard to compute. One way to avoid the computation of $||A^{-1}||_2$ is to give a reasonable upper bound estimation for $||A^{-1}||_2$. The following theorem gives an upper bound on $||A^{-1}||_2$.

Theorem 2.4.3 *Let A be defined as in* (2.4.3). *Then A satisfies*

$$||A^{-1}||_2 \le \alpha, \tag{2.4.17}$$

where

$$\alpha = \sqrt{n} \max_{1\le k\le n} \prod_{\substack{i=1\\ i\ne k}}^{n} \frac{1+|\lambda_i|}{|\lambda_k - \lambda_i|} \Big/ \min_{1\le i\le n} \omega_i. \tag{2.4.18}$$

Proof For any $u = (\alpha_1, \dots, \alpha_n)^T \in \mathbf{R}^n$ with $||u||_2 = 1$, using (2.4.9) we have

$$\begin{aligned} ||Au||_2^2 &= ||(\alpha_1 I + \alpha_2\Lambda + \dots + \alpha_n\Lambda^{n-1})q||_2^2 \\ &= \sum_{i=1}^{n}\Big(\omega_i \sum_{k=1}^{n} \lambda_i^{k-1}\alpha_k\Big)^2 \\ &\ge \big(\min_{1\le i\le n}\omega_i\big)^2 \sum_{i=1}^{n}\Big(\sum_{k=1}^{n}\lambda_i^{k-1}\alpha_k\Big)^2 \\ &= \big(\min_{1\le i\le n}\omega_i\big)^2 ||u^T V(\lambda_1, \dots, \lambda_n)||_2^2. \end{aligned} \tag{2.4.19}$$

Since λ_i are distinct we know from Theorem 1.6.2 that Vandermonde matrix $V(\lambda_1, \dots, \lambda_n)$ is nonsingular and

$$\begin{aligned} ||V(\lambda_1, \dots, \lambda_n)^{-1}||_2 &\le \sqrt{n}||V(\lambda_1, \dots, \lambda_n)^{-1}||_\infty \\ &\le \sqrt{n} \max_{1\le k\le n} \prod_{\substack{i=1\\ i\ne k}}^{n} \frac{1+|\lambda_i|}{|\lambda_k - \lambda_i|}. \end{aligned} \tag{2.4.20}$$

It follows from (2.4.19) and (2.4.20) that

$$||Au||_2 \ge \big(\min_{1\le i\le n}\omega_i\big) / ||V(\lambda_1, \dots, \lambda_n)^{-1}||_2 \ge \frac{1}{\alpha}, \tag{2.4.21}$$

where α is defined by (2.4.18). Noting that u is arbitrary, we know from (2.4.21) that A is invertible and

$$\|A^{-1}\|_2 \le \alpha,$$

which is the desired result. □

By Theorem 2.4.3, we can take the scalar

$$\overline{\kappa}(\{\lambda_i\}_1^n, \{\mu_i\}_1^{n-1}) = \sqrt{2}(1 + 2\alpha\beta\sqrt{2\eta}) \tag{2.4.22}$$

as a condition estimator for Problem 2.1. In many cases this estimator can produce a good order-of-magnitude estimate of the condition number, as is the case in the following example.

Example 2.4.1 Suppose that

$$\lambda_1 = -2, \quad \lambda_2 = -1, \quad \lambda_3 = 0, \ \lambda_4 = 1, \quad \lambda_5 = 2,$$
$$\mu_1 = -1.5, \quad \mu_2 = -0.4, \quad \mu_3 = \alpha, \quad \mu_4 = 1 + \alpha,$$

where α is a positive number. For $\alpha = 10^{-k}$, $k = 2, 4, 6, 8, 10$, we obtain

α	$\kappa(\{\lambda_i\}_1^5, \{\mu_i\}_1^4)$	$\overline{\kappa}(\{\lambda_i\}_1^5, \{\mu_i\}_1^4)$
10^{-2}	2.656641×10^4	3.247061×10^5
10^{-4}	6.493975×10^5	7.946062×10^6
10^{-6}	5.954891×10^7	7.286528×10^8
10^{-8}	5.949252×10^9	7.279630×10^{10}
10^{-10}	5.949196×10^{11}	7.279561×10^{12}

Observe that the condition number is increasing along with the decrease of the separation of the given eigenvalues.

2.5 Numerical Methods

In this section, we will develop several efficient and reliable algorithms for solving Problem 2.1 in Section 2.1. The basic techniques of these methods can be used to solve some other inverse problems.

2.5.1 Lanczos Method

Theorem 2.3.1 gives us a recipe for finding the solution of Problem 2.1. For given data $\lambda_1, \ldots, \lambda_n$ and $\mu_1, \ldots, \mu_{n-1}$, satisfying $\lambda_1 < \mu_1 < \cdots < \mu_{n-1} < \lambda_n$, if we can compute an orthogonal matrix $Q \in \mathbf{R}^{n\times n}$ and a Jacobi matrix $J \in \mathbf{R}^{n\times n}$ such that

$$Q^T JQ = \Lambda \equiv \operatorname{diag}(\lambda_1, \dots, \lambda_n) \tag{2.5.1}$$

and the first row elements of Q satisfy (2.3.2), then such a Jacobi matrix J must be the unique solution of Problem 2.1 for the given data. Let $Q^T = U$. Then (2.5.1) can be rewritten as

$$\Lambda U = UJ. \tag{2.5.2}$$

Thus, in order to solve Problem 2.1, we should find a method to compute an orthogonal matrix U with its first column elements satisfying (2.3.2) and a Jacobi J so that (2.5.2) holds. This reminds us to realize this idea by using Lanczos Algorithm. Therefore, let us first recall the basic idea of the Lanczos Algorithm. Let $A \in \mathbf{R}^{n\times n}$ be symmetric, and assume that an orthogonal matrix $U = [u_1, \dots, u_n] \in \mathbf{R}^{n\times n}$ satisfies

$$U^T AU = T = \begin{bmatrix} \alpha_1 & \beta_2 & & & & \\ \beta_2 & \alpha_2 & \beta_3 & & 0 & \\ & \ddots & \ddots & \ddots & & \\ & & \ddots & \ddots & \ddots & \\ & 0 & & \beta_{n-1} & \alpha_{n-1} & \beta_n \\ & & & & \beta_n & \alpha_n \end{bmatrix}.$$

Equating the columns in $AU = UT$, we find

$$\begin{aligned} Au_1 &= \alpha_1 u_1 + \beta_2 u_2, \\ Au_j &= \beta_j u_{j-1} + \alpha_j u_j + \beta_{j+1} u_{j+1}, \quad j = 2, \dots, n-1, \\ Au_n &= \beta_n u_{n-1} + \alpha_n u_n. \end{aligned}$$

Notice that $u_i^T u_j = \delta_{ij}$, $i, j = 1, 2, \dots, n$. It follows from the above formulae that

$$\begin{aligned} \alpha_j &= u_j^T A u_j, \\ \beta_{j+1} &= \pm \| Au_j - \beta_j u_{j-1} - \alpha_j u_j \|_2 \quad (\beta_1 u_0 \equiv 0), \\ u_{j+1} &= (Au_j - \beta_j u_{j-1} - \alpha_j u_j) / \beta_{j+1}, \quad \text{if } \beta_{j+1} \neq 0. \end{aligned}$$

This shows that α_1, β_2, and u_2 can be determined by A and u_1, in turn,α_2, β_3, and u_3 can be determined by A, u_1, u_2, α_1, and β_2, and so on. Therefore, once we have known the first column u_1 of U, we can step by step fill in all of T and U starting with A and u_1. This is the basic idea of the Lanczos Algorithm. So by properly sequencing the above formulae we obtain the **Lanczos iteration**:

$r = u_1$; $\beta_1 = 1$; $u_0 = 0$; $j = 0$;
while $(\beta_{j+1} \neq 0)$

$$u_{j+1} = r/\beta_{j+1}\ ;\ j = j+1;\ \alpha_j = u_j^T A u_j;$$
$$r = (A - \alpha_j I)u_j - \beta_j u_{j-1};\ \beta_{j+1} = \|r\|_2;$$

end

We return now to the construction of a Jacobi from the given data using Lanczos Algorithm.

First, in order to ensure that the first column elements of U to be determined as in (2.5.2) satisfy (2.3.2), we compute a unit vector u by (2.3.2) from the given data. In practice, there are better ways to compute vector u than (2.3.2). The following algorithm, for example, guards against overflow and underflow.

Algorithm 2.5.1 Given $x \in \mathbf{R}^n$ and $y \in \mathbf{R}^{n-1}$ satisfying

$$x(1) < y(1) < x(2) < y(2) < \cdots < y(n-1) < x(n),$$

this function computes a unit vector $u \in \mathbf{R}^n$ so that

$$u(j) = \left(\prod_{i=1}^{n-1}\left(x(j) - y(i)\right) \Big/ \prod_{\substack{i=1\\ i\neq j}}^{n}\left(x(j) - x(i)\right)\right)^{\frac{1}{2}}.$$

```
function u = firstrow(x, y)
    n=length(x);
    for j = 1 : n
        u(j) = 1;
        for i = 1 : i - 1
            u(j) = u(j)(x(j) - y(i)) / (x(j) - x(i));
        end
        for i = i + 1 : n
            u(j) = u(j)(x(j) - y(i - 1)) / (x(j) - x(i));
        end
        u(j) = sqrt(u(j));
    end
```

This algorithm requires $4(n-1)$ flops and a single square root. Here, and here in after, a **flop** is a floating point operation.

Then we set the starting matrix $A = \Lambda = \text{diag}(\lambda_1, \ldots, \lambda_n)$, and the starting vector $u_1 = u$ determined by (2.3.2), and apply Lanczos iteration to step by step fill in all of J and U, eventually find the desired Jacobi matrix.

Noting that the starting matrix in the present situation is diagonal, the above discussion gives rise to the following algorithm.

Algorithm 2.5.2 (Lanczos Method) Given eig, $w \in \mathbf{R}^n$ satisfying $eig(1) < eig(2) < \cdots < eig(n)$, $w(i) \neq 0$, and $\|w\|_2 = 1$, this algorithm applies Lanczos algorithm to compute a Jacobi matrix $J \in \mathbf{R}^{n\times n}$ such that its eigenvalues are

$eig(1), \ldots, eig(n)$ and the first row of the orthogonal matrix of its eigenvectors is w^T, and the diagonal and the subdiagonal elements of J are respectively stored in $d \in \mathbf{R}^n$ and $sd(2:n)$, where $sd \in \mathbf{R}^n$.

```
n = length(eig);
for j = 1 : n
        d(1) = d(1) + eig(j)w(j)^2;
end
for j = 1 : n - 1
        if j = 1
                for i = 1 : n
                        y(i) = w(i)(eig(i) - d(1));
                end
        else
                for i = 1 : n
                        y(i) = (eig(i) - d(j))w(i) - sd(j)z(i);
                end
        end
        for i = 1 : n
                sd(j + 1) = sd(j + 1) + y(i)^2;
        end
        sd(j + 1) = sqrt(sd(j + 1)); z = w; w = y/sd(j + 1);
        for i = 1 : n
                d(j + 1) = d(j + 1) + eig(i)w(i)^2;
        end
end
```

This algorithm requires $10n^2$ flops and $n-1$ square roots. However, in many situation, to maintain numerical stability, the reorthogonalization must be taken, increasing the overall cost to about $O(n^3)$ flops.

Example 2.5.1 Let

$$J_n = \begin{bmatrix} 2 & 1 & & & & \\ 1 & 2 & 1 & & & \\ & \ddots & \ddots & \ddots & & \\ & & \ddots & \ddots & \ddots & \\ & & & 1 & 2 & 1 \\ & & & & 1 & 2 \end{bmatrix} \in \mathbf{R}^{n \times n}.$$

It is well known that J_n has eigenvalues

$$\lambda_k = 4 \sin^2 \frac{k\pi}{2(n+1)}, \quad k = 1, 2, \ldots, n.$$

Applying Algorithms 2.5.1 and 2.5.2 to reconstruct the Jacobi matrix J_n from the data

$$\lambda_k = 4\sin^2 \frac{k\pi}{2(n+1)}, \quad k = 1, 2, \ldots, n, \tag{2.5.3}$$

and

$$\mu_k = 4\sin^2 \frac{k\pi}{2n}, \quad k = 1, 2, \ldots, n-1, \tag{2.5.4}$$

for $n = 10 : 10 : 100$, we find

n	Diagonal Max-Errors	Off-Diagonal Max-Errors
10	1.776×10^{-15}	1.110×10^{-15}
20	7.105×10^{-15}	2.887×10^{-15}
30	7.105×10^{-15}	3.553×10^{-15}
40	1.954×10^{-14}	2.665×10^{-15}
50	1.710×10^{-14}	4.885×10^{-15}
60	1.643×10^{-14}	4.996×10^{-15}
70	2.220×10^{-14}	9.770×10^{-15}
80	4.485×10^{-14}	1.732×10^{-14}
90	4.396×10^{-14}	1.821×10^{-14}
100	2.798×10^{-14}	1.121×10^{-14}

2.5.2 Orthogonal Reduction Methods

From the proof of Theorem 2.3.2, it is natural that we should propose the following method for solving Problem 2.1 in Section 2.1:

Step 1. Compute a unit vector $q = (\omega_1, \cdots, \omega_n)^T$ by (2.3.2), and form the following bordered diagonal matrix

$$A = \begin{bmatrix} 0 & q^T \\ q & \Lambda \end{bmatrix}, \tag{2.5.5}$$

where $\Lambda = \operatorname{diag}(\lambda_1, \ldots, \lambda_n)$.

Step 2. Compute an $n \times n$ orthogonal matrix Q such that

$$\begin{bmatrix} 1 & 0 \\ 0 & Q \end{bmatrix}^T \begin{bmatrix} 0 & q^T \\ q & \Lambda \end{bmatrix} \begin{bmatrix} 1 & 0 \\ 1 & Q \end{bmatrix} = \begin{bmatrix} 0 & \beta e_1^T \\ \beta e_1 & J \end{bmatrix}, \tag{2.5.6}$$

where $\beta = ||q||_2$ and J is a Jacobi matrix, which must solve Problem 2.1.

The step 1 can be carry out by using Algorithm 2.5.1. Therefore, the key of the above method is how to realize the step 2. The basic idea of orthogonal reduction methods is to carry out this step by using elementary orthogonal similarity transformations. In the rest of this section we will devote to introducing three basic methods to reduce the above matrix A to tridiagonal form by using elementary orthogonal similarity transformations.

2.5.2a Householder Reduction Method

One of the most commonly used tools in matrix computations is Householder reflections. A **Householder reflection** is an $n \times n$ matrix H of the form

$$H = I - 2vv^T / v^T v \tag{2.5.7}$$

where $v \in \mathbf{R}^n$ is nonzero and is called a **Householder vector**. Sometimes a Householder reflection is also called a **Householder matrix** or a **Householder transformation** by others.

Householder reflections possess a number of excellent mathematical properties. For example, if H is a Househoder reflection, then $H^T = H^{-1} = H$, that is, Householder reflections are symmetric and orthogonal. Geometrically, when a Householder reflection is applied to a vector $x \in \mathbf{R}^n$, it amounts to that x is reflected in hyperplane $\{v\}^\perp$, where v is the Househoder vector.

From the numerical point of view, the main reason why Householder reflections are so important is their ability used to zero selected components of a vector and their favorable numerical properties. In practice, given a nonzero vector $x \in \mathbf{R}^n$, we frequently need to determine a Househoder reflection H so that Hx is a multiple of e_1. Indeed, for any $x \in \mathbf{R}^n$, it is easy to verify that

$$v = x \pm \|x\|_2 e_1 \quad \text{implies} \quad Hx = \left(I - 2\frac{vv^T}{v^T v}\right)x = \mp\|x\|_2 e_1. \tag{2.5.8}$$

It is this simple determination of v that makes the Householder reflections so useful.

There are a number of important practical details associated with the determination of a Householder matrix. The detailed discussion about them can be found in any text book on matrix computations(e.g., see [90]). Hence, we only present here the following basic algorithm.

Algorithm 2.5.3 (**Househoder Reflection**) Given a vector $x \in \mathbf{R}^n$, the following function computes a vector $v \in \mathbf{R}^n$ and two real numbers β and α such that $(I - \beta vv^T)x = \alpha e_1$.

function $[v, \beta, \alpha] =$ **house** (x)
  n=**length**(x); $\mu = \|x\|_\infty$; $v = x$;
  if $\mu = 0$
    $\beta = 0$; $\alpha = 0$;

```
else
    v = v / μ;
    δ = ||x||_2;
    if v(1) < 0
        δ = −δ;
    end
    v(1) = v(1) + δ;
    β = 1 /(δ * v(1));
    α = −δμ;
end
```

This Algorithm requires about $3n$ flops and a singe square root. In addition, the roundoff properties associated with this algorithm are very favorable. It can be shown that this algorithm produces a Householder vector $\hat{v}$ very near the exact v. See [221] for the details.

Now we show how to use Householder reflections to reduce the matrix A defined by (2.5.5) to tridiagonal form. Firstly, using the above algorithm, determine an $n \times n$ Householder matrix H_0 such that

$$H_0 q = \alpha e_1, \quad \text{with } |\alpha| = 1,$$

and compute

$$B = H_0 \Lambda H_0.$$

Then, using the above algorithm, we can compute $n-2$ Householder matrices $H_1, \cdots, H_{n-2}$ such that

$$T = H_{n-2} \cdots H_1 B H_1 \cdots H_{n-2}$$

is tridiagonal form. An 5×5 example illustrates the roles of the individual H_i:

$$\begin{bmatrix} \times & \times & \times & \times & \times \\ \times & \times & \times & \times & \times \\ \times & \times & \times & \times & \times \\ \times & \times & \times & \times & \times \\ \times & \times & \times & \times & \times \end{bmatrix} \xrightarrow{H_1} \begin{bmatrix} \times & \times & 0 & 0 & 0 \\ \times & \times & \times & \times & \times \\ 0 & \times & \times & \times & \times \\ 0 & \times & \times & \times & \times \\ 0 & \times & \times & \times & \times \end{bmatrix}$$

$$\xrightarrow{H_2} \begin{bmatrix} \times & \times & 0 & 0 & 0 \\ \times & \times & \times & 0 & 0 \\ 0 & \times & \times & \times & \times \\ 0 & 0 & \times & \times & \times \\ 0 & 0 & \times & \times & \times \end{bmatrix} \xrightarrow{H_3} \begin{bmatrix} \times & \times & 0 & 0 & 0 \\ \times & \times & \times & 0 & 0 \\ 0 & \times & \times & \times & 0 \\ 0 & 0 & \times & \times & \times \\ 0 & 0 & 0 & \times & \times \end{bmatrix}.$$

In general, suppose we have determined Householder reflections $H_1, \ldots, H_{k-1}$

such that

$$B_k = (H_{k-1}\cdots H_1)B(H_{k-1}\cdots H_1)^T = \begin{bmatrix} T_{k-1} & \beta_k e_k & 0 \\ \beta_k e_k^T & \alpha_k & b_k^T \\ 0 & b_k & \widetilde{B}_k \end{bmatrix},$$

where T_{k-1} is a $(k-1)\times(k-1)$ tridiagonal matrix, $b_k \in \mathbf{R}^{n-k}$. Next step is first to determine an $(n-k)\times(n-k)$ Householder reflection $\widetilde{H}_k$ so that $\widetilde{H}_k b_k = \beta_{k+1}e_1$, $\beta_{k+1} \in \mathbf{R}$, and then to make the similarity transformation $B_{k+1} = H_k B_k H_k$, where $H_k = \operatorname{diag}(I_k, \widetilde{H}_k)$. It is easy to see that

$$B_{k+1} = \begin{bmatrix} T_{k-1} & \beta_k e_k & 0 \\ \beta_k e_k & \alpha_k & \beta_{k+1}e_1^T \\ 0 & \beta_{k+1}e_1 & \widetilde{H}_k\widetilde{B}_k\widetilde{H}_k \end{bmatrix}.$$

Clearly, after $n-2$ such steps, B is reduced to a tridiagonal matrix. Now let $Q = H_0H_1\cdots H_{n-2}$ and $T = B_{n-2}$. Then it is easy to verify that such Q and T satisfy

$$\begin{bmatrix} 1 & 0 \\ 0 & Q \end{bmatrix}^T \begin{bmatrix} 0 & q^T \\ q & \Lambda \end{bmatrix} \begin{bmatrix} 1 & 0 \\ 0 & Q \end{bmatrix} = \begin{bmatrix} 0 & \alpha e_1^T \\ \alpha e_1 & T \end{bmatrix}.$$

Noting that some of the subdiagonal elements of the matrix T obtained above may be negative, we must take the absolute value of them so that we can obtain the desired Jacobi matrix J. In fact, taking the absolute value of the negative subdiagonal elements of T amounts to performing a similarity transformation to T using a transformation matrix of form $\operatorname{diag}(1, d_1, \cdots, d_{n-1})$, where $d_i = 1$ or -1. Hence, the eigenstructure of T is not changed.

In summary, we obtain the following algorithm.

Algorithm 2.5.4 (Householder Reduction Method) Given eig, $u \in \mathbf{R}^n$ with $eig(1) < \cdots < eig(n)$, $u(i) \neq 0$, and $\|u\|_2 = 1$, this algorithm applies Householder reflections to compute a Jacobi matrix J such that its eigenvalues are $eig(1), \ldots, eig(n)$ and the first row of the orthogonal matrix of its eigenvectors is u^T.

$n =$ **length**(eig)
$[v, \beta, \alpha] =$ **house** (u);
$J = \operatorname{diag}(eig(1), \ldots, eig(n))$; $J = (I_n - \beta vv^T)J(I_n - \beta vv^T)$;
for $k = 1 : n-2$
    $[v, \beta, \alpha] =$ **house** $(J(k+1:n, k:k))$;

$$J(k+1:n,k+1:n) = (I_{n-k}-\beta vv^T)J(k+1:n,k+1:n)(I_{n-k}-\beta vv^T);$$

end
for $i = 2:n$
$$J(i,i-1) = |J(i,i-1)|;\; J(i-1,i) = |J(i-1,i)|;$$
end

This algorithm requires about $4n^3/3$ flops and $n-1$ square roots when symmetry is exploited in calculating the matrix product.

2.5.2b Fast Orthogonal Reduction Method

Although Algorithm 2.5.4 is particularly easy to implement, the user need only call the appropriate EISPACK subroutines, A's rich structure is ignored throughout Algorithm 2.5.4. This wastes computer time and storage. A way round these practical problems is to make use of Givens rotations.

Givens rotations are rank two corrections to the identity of the form

$$G(i,k,\theta) = \begin{bmatrix} 1 & & & & & & & & & \\ & \ddots & & & & & & 0 & & \\ & & 1 & & & & & & & \\ & & & c & \cdots & \cdots & \cdots & s & & \\ & & & \vdots & 1 & & & \vdots & & \\ & & & \vdots & & \ddots & & \vdots & & \\ & & & \vdots & & & 1 & \vdots & & \\ & & & -s & \cdots & \cdots & \cdots & c & & \\ & & & & & & & & 1 & \\ & 0 & & & & & & & & \ddots \\ & & & & & & & & & & 1 \end{bmatrix} \begin{matrix} \\ \\ \\ i\text{th row} \\ \\ \\ \\ k\text{th row} \\ \\ \\ \\ \end{matrix}$$

where $c = \cos(\theta)$ and $s = \sin(\theta)$ for some θ.

It is clear that Givens rotations are orthogonal. If $x \in \mathbf{R}^n$ and $y = G(i,k,\theta)^T x$, then we have

$$\begin{aligned} y_i &= cx_i + sx_k, \\ y_k &= -sx_i + cx_k, \\ y_j &= x_j, \qquad j \neq i \text{ or } k. \end{aligned}$$

These formulae show that y is just the vector obtained by rotating x counter clockwise through an angle θ in the (i,k) coordinate plane. Moreover, from these formulae it is clear that we can force y_k to be zero by setting

$$c = \frac{x_i}{(x_i^2+x_k^2)^{\frac{1}{2}}}, \qquad s = \frac{x_k}{(x_i^2+x_k^2)^{\frac{1}{2}}}.$$

Thus, we can zero any nonzero component of a given vector via a Givens rotation. In practice, in order to avoid overflow and underflow, we use the following algorithm rather than the above formulae to compute c and s .

Algorithm 2.5.5 (Givens Rotation) Given two real numbers x and y, this function computes two real numbers $s = \sin(t)$ and $c = \cos(t)$ such that

$$\begin{bmatrix} c & s \\ -s & c \end{bmatrix} \begin{bmatrix} x \\ y \end{bmatrix} = \begin{bmatrix} \gamma \\ 0 \end{bmatrix}$$

function$[c, s] =$ **givens** (x, y)
  if $|y| = 0$
    $c = 1;\ s = 0;$
  elseif $|y| \geq |x|$
    $d = x\,/y\ ;\ s = 1\big/\sqrt{1+d^2}\ ;\ c = sd;$
  else
    $d = y\,/x\ ;\ c = 1\big/\sqrt{1+d^2}\ ;\ s = cd;$
  end

This algorithm requires 5 flops and a single square root. The numerical properties of Givens rotations are as favorable as those for Householder reflections. It can be shown that the computed $\hat{c}$ and $\hat{s}$ are very near to the exact c and s. See [221] for the details.

Since Givens rotations are able to zero elements more selectively and the bordered diagonal matrix A defined by (2.5.5) is a matrix of very special form, we can reduce such matrix to tridiagonal form using a scheme that is fast than the method based on Householder transformation described above. In the following, we present such an elegant method which takes only $O(n^2)$ operations and at most $3n$ storage locations. This method consists of applying a certain sequence of Givens rotations in a very particular order, and in part parallels the extensively used and numerically reliable symmetric QR algorithm for symmetric eigenproblem. The essence of the algorithm can be conveyed by a small example. Suppose $n = 4$. We have

$$\begin{bmatrix} \times & \times & \times & \times & \times \\ \times & \times & 0 & 0 & 0 \\ \times & 0 & \times & 0 & 0 \\ \times & 0 & 0 & \times & 0 \\ \times & 0 & 0 & 0 & \times \end{bmatrix} \xrightarrow{(4,5)} \begin{bmatrix} \times & \times & \times & \times & 0 \\ \times & \times & 0 & 0 & 0 \\ \times & 0 & \times & 0 & 0 \\ \times & 0 & 0 & \times & + \\ 0 & 0 & 0 & + & \times \end{bmatrix}$$

$$\xrightarrow{(3,4)} \begin{bmatrix} \times & \times & \times & 0 & 0 \\ \times & \times & 0 & 0 & 0 \\ \times & 0 & \times & + & + \\ 0 & 0 & + & \times & \times \\ 0 & 0 & + & \times & \times \end{bmatrix} \xrightarrow{(4,5)} \begin{bmatrix} \times & \times & \times & 0 & 0 \\ \times & \times & 0 & 0 & 0 \\ \times & 0 & \times & \times & 0 \\ 0 & 0 & \times & \times & \times \\ 0 & 0 & 0 & \times & \times \end{bmatrix}$$

$$\xrightarrow{(2,3)} \begin{bmatrix} \times & \times & 0 & 0 & 0 \\ \times & \times & + & + & 0 \\ 0 & + & \times & \times & 0 \\ 0 & + & \times & \times & \times \\ 0 & 0 & 0 & \times & \times \end{bmatrix} \xrightarrow{(3,4)} \begin{bmatrix} \times & \times & 0 & 0 & 0 \\ \times & \times & \times & 0 & 0 \\ 0 & \times & \times & \times & + \\ 0 & 0 & \times & \times & \times \\ 0 & 0 & + & \times & \times \end{bmatrix}$$

$$\xrightarrow{(4,5)} \begin{bmatrix} \times & \times & 0 & 0 & 0 \\ \times & \times & \times & 0 & 0 \\ 0 & \times & \times & \times & 0 \\ 0 & 0 & \times & \times & \times \\ 0 & 0 & 0 & \times & \times \end{bmatrix}.$$

Where "$\times$" denotes a nonzero element, "$+$" a new nonzero element, and "(i,j)" the Givens orthogonal similarity transformation in (i,j) coordinate plane.

In the above process, when we zero a bordered element, it will appear one unwanted nonzero element outside the tridiagonal band. Then we must chase the unwanted nonzero element out of the matrix. For example, when we zero the $(3,1)$ element, it will appear an unwanted nonzero element in position $(4,2)$. Then we use Givens orthogonal similarity transformations in $(3,4)$ and $(4,5)$ coordinate planes to chase the unwanted nonzero element out of the matrix.

In general, we start with zeroing the bordered element ω_n in position $(n+1,1)$ using Givens orthogonal similarity transformation in $(n,n+1)$ coordinate plane. That is, we first compute $c = \cos\theta$ and $s = \sin\theta$ such that

$$\begin{bmatrix} c & s \\ -s & c \end{bmatrix} \begin{bmatrix} \omega_{n-1} \\ \omega_n \end{bmatrix} = \begin{bmatrix} \widetilde{\omega}_{n-1} \\ 0 \end{bmatrix},$$

then make the transformation $A_1 = G_1 A G_1^T$, where $G_1 = G(n, n+1, \theta)$. Since this transformation is only to change the last two columns and rows of A, it is clear that A_1 has the form

$$A_1 = \begin{bmatrix} 0 & q_1^T & \widetilde{\omega}_{n-1} & 0 \\ q_1 & \Lambda_1 & 0 & 0 \\ \widetilde{\omega}_{n-1} & 0 & \multicolumn{2}{c}{\multirow{2}{*}{T_1}} \\ 0 & 0 & & \end{bmatrix}, \qquad G_1 e_1 = e_1,$$

where $q_1 = (\omega_1, \ldots, \omega_{n-2})^T$, $\widetilde{\omega}_{n-1} \in \mathbf{R}$, $\Lambda_1 = \mathrm{diag}(\lambda_1, \ldots, \lambda_{n-2})$, and T_1 is a 2×2 symmetric matrix. Suppose we have determined $k-1$ orthogonal matrices

$$G_1, G_2, \ldots, G_{k-1} \in \mathbf{R}^{(n+1)\times(n+1)}$$

with $G_j e_1 = e_1$, $j = 1, \ldots, k-1$, such that

$$A_{k-1} = (G_{k-1} \cdots G_1) A (G_{k-1} \cdots G_1)^T$$

has the following form

$$A_{k-1} = \begin{bmatrix} 0 & q_{k-1}^T & \widetilde{\omega}_{n-k+1} & 0 \\ q_{k-1} & \Lambda_{k-1} & 0 & 0 \\ \widetilde{\omega}_{n-k+1} & 0 & & \\ 0 & 0 & \multicolumn{2}{c}{T_{k-1}} \end{bmatrix}$$

where $q_{k-1} = (\omega_1, \dots, \omega_{n-k})^T$, $\widetilde{\omega}_{n-k+1} \in \mathbf{R}$, $\Lambda_{k-1} = \text{diag}(\lambda_1, \dots, \lambda_{n-k})$, and T_k is a tridiagonal matrix of order k.

Next first compute $c = \cos\theta_0$ and $s = \sin\theta_0$ such that

$$\begin{bmatrix} c & s \\ -s & c \end{bmatrix} \begin{bmatrix} \omega_{n-k} \\ \widetilde{\omega}_{n-k+1} \end{bmatrix} = \begin{bmatrix} \widetilde{\omega}_{n-k} \\ 0 \end{bmatrix},$$

and then make the transformation $\widetilde{A}_k = G_{k0}A_{k-1}G_{k0}^T$, where

$$G_{k0} = G(n-k+1, n-k+2, \theta_0) \in \mathbf{R}^{(n+1)\times(n+1)}.$$

Note the transformation $\widetilde{A}_k = G_{k0}A_{k-1}G_{k0}^T$ is only to modify the $(n-k+1)$th and $(n-k+2)$th columns and rows of A_{k-1}, and so $\widetilde{A}_k$ has the form

$$\widetilde{A}_k = \begin{bmatrix} 0 & q_k^T & \widetilde{\omega}_{n-k} & 0 \\ q_k & \Lambda_k & 0 & 0 \\ \widetilde{\omega}_{n-k} & 0 & & \\ 0 & 0 & \multicolumn{2}{c}{\widetilde{T}_k} \end{bmatrix}$$

where $q_k = (\omega_1, \dots, \omega_{n-k-1})^T$, $\widetilde{\omega}_{n-k} \in \mathbf{R}$, $\Lambda_k = \text{diag}(\lambda_1, \dots, \lambda_{n-k-1})^T$, and $\widetilde{T}_k$ is a $(k+1)\times(k+1)$ symmetric matrix having the form

$$\begin{bmatrix} \alpha_1 & \gamma_1 & \eta_1 & 0 & \cdots & 0 \\ \gamma_1 & \tau_1 & \omega_1 & 0 & & \vdots \\ \eta_1 & \omega_1 & \times & \times & \ddots & \vdots \\ 0 & 0 & \times & \times & \ddots & 0 \\ \vdots & & \ddots & \ddots & \ddots & \times \\ 0 & \cdots & \cdots & 0 & \times & \times \end{bmatrix}$$

where $\times$ denote these nonzero elements that are not changed during the transformation. We now need to reduce $\widetilde{T}_k$ to tridiagonal form. This can be accomplished by making use of the zero-chasing technique, which has been well developed in the establishment of the well-known symmetric QR algorithm.

The first step of the reduction is to zero the unwanted element η_1. For this, we first compute $c = \cos\theta_1$ and $s = \sin\theta_1$ so that

$$\begin{bmatrix} c & s \\ -s & c \end{bmatrix} \begin{bmatrix} \gamma_1 \\ \eta_1 \end{bmatrix} = \begin{bmatrix} \beta_2 \\ 0 \end{bmatrix},$$

and then make the transformation $\widetilde{G}_{k1}\widetilde{T}_k\widetilde{G}_{k1}^T$, where $\widetilde{G}_{k1} = G(2,3,\theta_1)$ is a Givens rotation of order $k+1$. It is clear that the matrix $\widetilde{G}_{k1}\widetilde{T}_k\widetilde{G}_{k1}^T$ has the form

$$\begin{bmatrix} \alpha_1 & \beta_2 & 0 & 0 & 0 & \cdots & 0 \\ \beta_2 & \alpha_2 & \gamma_2 & \eta_2 & 0 & & \vdots \\ 0 & \gamma_2 & \tau_2 & \omega_2 & 0 & & \vdots \\ 0 & \eta_2 & \omega_2 & \times & \times & \ddots & \vdots \\ 0 & 0 & 0 & \times & \times & \ddots & 0 \\ \vdots & & & \ddots & \ddots & \ddots & \times \\ 0 & \cdots & \cdots & \cdots & 0 & \times & \times \end{bmatrix}$$

where

$$\begin{bmatrix} \alpha_2 & \gamma_2 \\ \gamma_2 & \tau_2 \end{bmatrix} = \begin{bmatrix} c & s \\ -s & c \end{bmatrix} \begin{bmatrix} \tau_1 & \omega_1 \\ \omega_2 & \widetilde{T}_k(3,3) \end{bmatrix} \begin{bmatrix} c & s \\ -s & c \end{bmatrix}^T,$$

$$\begin{bmatrix} \eta_2 \\ \omega_2 \end{bmatrix} = \begin{bmatrix} c & s \\ -s & c \end{bmatrix} \begin{bmatrix} 0 \\ \widetilde{T}_k(3,4) \end{bmatrix}.$$

The second step of the reduction is to zero the unwanted element η_2. As before, we first compute $c = \cos\theta_2$ and $s = \sin\theta_2$ so that

$$\begin{bmatrix} c & s \\ -s & c \end{bmatrix} \begin{bmatrix} \gamma_2 \\ \eta_2 \end{bmatrix} = \begin{bmatrix} \beta_3 \\ 0 \end{bmatrix},$$

and then make the transformation $\widetilde{G}_{k2}\left(\widetilde{G}_{k1}\widetilde{T}_k\widetilde{G}_{k1}^T\right)\widetilde{G}_{k2}^T$, where $\widetilde{G}_{k2} = G(3,4,\theta_2)$. Obviously, the matrix $\widetilde{G}_{k2}\left(\widetilde{G}_{k1}\widetilde{T}_k\widetilde{G}_{k1}^T\right)\widetilde{G}_{k2}^T$, has the form

$$\begin{bmatrix} \alpha_1 & \beta_2 & 0 & 0 & 0 & 0 & \cdots & 0 \\ \beta_2 & \alpha_2 & \beta_3 & 0 & 0 & 0 & & \vdots \\ 0 & \beta_3 & \alpha_3 & \gamma_3 & \eta_3 & 0 & & \vdots \\ 0 & 0 & \gamma_3 & \tau_3 & \omega_3 & 0 & & \vdots \\ 0 & 0 & \eta_3 & \omega_3 & \times & \times & \ddots & \vdots \\ 0 & 0 & 0 & 0 & \times & \times & \ddots & 0 \\ \vdots & & & & \ddots & \ddots & \ddots & \times \\ 0 & \cdots & \cdots & \cdots & \cdots & 0 & \times & \times \end{bmatrix}$$

The third step is to zero the unwanted element η_3, and so on. After $k-1$ steps, the reduction is complete. The result is a tridiagonal matrix T_k that is orthogonal similar to $\widetilde{T}_k$:

$$T_k = \left(\widetilde{G}_{k,k-1} \cdots \widetilde{G}_{k1}\right) \widetilde{T}_k \left(\widetilde{G}_{k,k-1} \cdots \widetilde{G}_{k1}\right)^T,$$

where $\widetilde{G}_{ki} = G(i+1, i+2, \theta_i) \in \mathbf{R}^{(k+1)\times(k+1)}$ for $i = 1, 2, \ldots, k-1$.

Now let

$$G_k = G_{k,k-1} G_{k,k-2} \cdots G_{k1} G_{k0},$$

where $G_{ki} = \operatorname{diag}(I, \widetilde{G}_{ki})$, with I the identity matrix of order $n-k$. Then G_k is a product of k Givens rotations with $G_k e_1 = e_1$, and $A_k = G_k A_{k-1} G_k^T$ has the form

$$A_k = \begin{bmatrix} 0 & q_k^T & \widetilde{\omega}_{n-k} & 0 \\ q_k & \Lambda_k & 0 & 0 \\ \widetilde{\omega}_{n-k} & 0 & & \\ 0 & 0 & \multicolumn{2}{c}{T_k} \end{bmatrix}$$

where $q_k = (\omega_1, \ldots, \omega_{n-k-1})^T$, $\widetilde{\omega}_{n-k} \in \mathbf{R}$, $\Lambda_{k-1} = \operatorname{diag}(\lambda_1, \ldots, \lambda_{n-k-1})$, and T_k is a tridiagonal matrix of order $k+1$.

Thus, after $n-1$ such major steps, eventually we reduce A to tridiagonal form:

$$A_{n-1} = G^T A G = \begin{bmatrix} 0 & \beta e_1^T \\ \beta e_1 & T \end{bmatrix},$$

where $G = G_1 G_2 \cdots G_{n-1}$ is an $(n+1)\times(n+1)$ orthogonal matrix with $Ge_1 = e_1$, T is an $n \times n$ tridiagonal matrix. Similar to Householder method, to obtain the desired Jacobi matrix, we must take the absolute value of the negative subdiagonal elements of T. The overall process is summarized as follows:

Algorithm 2.5.6 (Fast Orthogonal Reduction Method) Given eig, $w \in \mathbf{R}^n$ with $eig(1) < eig(2) < \cdots < eig(n)$, $w(i) \neq 0$, and $\|w\|_2 = 1$, this function applies Givens rotations to compute a Jacobi matrix $J \in \mathbf{R}^{n\times n}$ such that its eigenvalues are $eig(1), \ldots, eig(n)$ and the first row of the orthogonal matrix of its eigenvectors is w^T, and the diagonal and the subdiagonal elements of J are respectively stored in $d \in \mathbf{R}^n$ and $sd(2:n)$, where $sd \in \mathbf{R}^n$.

function $[d, sd] =$ **Jacobi** (eig, w)
$n =$ **length** (eig)
$d = eig;\ sd = 0;$
for $i = n : -1 : 2$
  $[c, s] =$ **givens** $(w(i-1), w(i));$
  $w(i-1) = w(i-1)c + w(i)s;$
  $\alpha = d(i-1);\ \beta = d(i);$
  $d(i-1) = \alpha c^2 + \beta s^2;\ d(i) = \alpha s^2 + \beta c^2;$
  $sd(i) = cs(\beta - \alpha);$

```
        if i < n
            for k = i : n − 1
                α = sd(k + 1)s; sd(k + 1) = sd(k + 1)c;
                [c, s] = givens (sd(k), α);
                sd(k) = sd(k)c + αs;
                α = d(k); γ = sd(k + 1); β = d(k + 1);
                d(k) = αc² + βs² + 2γcs;
                d(k + 1) = αs² + βc² − 2γcs;
                sd(k + 1) = cs(β − α) + γ(c² − s²);
            end
        end
    end
    for k = 2 : n
        sd(k) = |sd(k)|;
    end
```

This algorithm requires about $15n^2$ flops and $\frac{1}{2}n^2$ square roots and can be implemented with at most $3n$ storage locations.

The roundoff properties of this algorithm is essentially the same as the symmetric QR iteration. Consequently this algorithm is numerically stable.

Example 2.5.2 Applying Algorithm 2.5.6 to reconstruct the Jacobi matrix J_n as in Example 2.5.1, we find

n	Diagonal Max-Errors	Off-Diagonal Max-Errors
10	6.217×10^{-15}	3.331×10^{-15}
20	1.510×10^{-14}	6.217×10^{-15}
30	2.442×10^{-14}	6.883×10^{-15}
40	1.865×10^{-14}	1.443×10^{-14}
50	2.087×10^{-14}	9.770×10^{-15}
60	4.619×10^{-14}	2.731×10^{-14}
70	3.975×10^{-14}	2.065×10^{-14}
80	5.018×10^{-14}	2.842×10^{-14}
90	6.395×10^{-14}	3.597×10^{-14}
100	4.263×10^{-14}	2.065×10^{-14}

2.5.2c Rutishauser Method

Rutishauser method is essentially the same as the fast orthogonal reduction method described above. The difference is only that this method will zero the bordered elements of A defined by (2.5.5) from the top to the bottom, whereas the previous method zeroed these elements from the bottom to the top. To see how the process proceed, let us consider a simple example. As has done before, suppose $n = 4$. We have

$$\begin{bmatrix} \times & \times & \times & \times & \times \\ \times & \times & 0 & 0 & 0 \\ \times & 0 & \times & 0 & 0 \\ \times & 0 & 0 & \times & 0 \\ \times & 0 & 0 & 0 & \times \end{bmatrix} \xrightarrow{(2,3)} \begin{bmatrix} \times & \times & 0 & \times & \times \\ \times & \times & + & 0 & 0 \\ 0 & + & \times & 0 & 0 \\ \times & 0 & 0 & \times & 0 \\ \times & 0 & 0 & 0 & \times \end{bmatrix}$$

$$\xrightarrow{(2,4)} \begin{bmatrix} \times & \times & 0 & 0 & \times \\ \times & \times & \times & + & 0 \\ 0 & \times & \times & + & 0 \\ 0 & + & + & \times & 0 \\ \times & 0 & 0 & 0 & \times \end{bmatrix} \xrightarrow{(3,4)} \begin{bmatrix} \times & \times & 0 & 0 & \times \\ \times & \times & \times & 0 & 0 \\ 0 & \times & \times & \times & 0 \\ 0 & 0 & \times & \times & 0 \\ \times & 0 & 0 & 0 & \times \end{bmatrix}$$

$$\xrightarrow{(2,5)} \begin{bmatrix} \times & \times & 0 & 0 & 0 \\ \times & \times & \times & 0 & + \\ 0 & \times & \times & \times & + \\ 0 & 0 & \times & \times & 0 \\ 0 & + & + & 0 & \times \end{bmatrix} \xrightarrow{(3,5)} \begin{bmatrix} \times & \times & 0 & 0 & 0 \\ \times & \times & \times & 0 & 0 \\ 0 & \times & \times & \times & \times \\ 0 & 0 & \times & \times & + \\ 0 & 0 & \times & + & \times \end{bmatrix}$$

$$\xrightarrow{(4,5)} \begin{bmatrix} \times & \times & 0 & 0 & 0 \\ \times & \times & \times & 0 & 0 \\ 0 & \times & \times & \times & 0 \\ 0 & 0 & \times & \times & \times \\ 0 & 0 & 0 & \times & \times \end{bmatrix}.$$

For general n, we start with zeroing the $(2,3)$ element by using Givens rotation of the form $G(2,3,\theta_2)$. Suppose we have reduce the matrix A to the following form by using a certain sequence of Givens rotations:

$$A_k = \begin{bmatrix} J_k & B_k \\ B_k^T & \Lambda_k \end{bmatrix},$$

where J_k is a $k \times k$ real symmetric tridiagonal matrix, $\Lambda_k = \operatorname{diag}(\lambda_k, \ldots, \lambda_n)$, and

$$B_k = \begin{bmatrix} \omega_k & \omega_{k+1} & \cdots & \omega_n \\ 0 & 0 & \cdots & 0 \\ \vdots & \vdots & & \vdots \\ 0 & 0 & \cdots & 0 \end{bmatrix}, \quad \omega_i = e_i^T q.$$

The kth major step is first to zero ω_k using Givens rotation of the form $G(2,k,\theta_k)$. It will appear an unwanted nonzero element in position $(k+1,2)$. Then we use the Givens rotations $G(3,k+1,\theta_3)$, $G(4,k+1,\theta_4), \cdots, G(k,k+1,\theta_k)$ to successively zero the elements in positions $(k+1,2)$, $(k+1,3), \ldots, (k+1,k-1)$, so that A_k is reduced to the following form

$$A_{k+1} = \begin{bmatrix} J_{k+1} & B_{k+1} \\ B_{k+1}^T & \Lambda_{k+1} \end{bmatrix},$$

where $\Lambda_k = \operatorname{diag}(\lambda_{k+1}, \dots, \lambda_n)$, J_{k+1} is a $(k+1) \times (k+1)$ real symmetric tridiagonal matrix, and

$$B_k = \begin{bmatrix} \omega_{k+1} & \omega_{k+2} & \cdots & \omega_n \\ 0 & 0 & \cdots & 0 \\ \vdots & \vdots & & \vdots \\ 0 & 0 & \cdots & 0 \end{bmatrix}.$$

Clearly, after $n-1$ such major steps we reduce A to a tridiagonal matrix. As before, in order to obtain the desired Jacobi matrix, we need to change the signs of the negative subdiagonal elements.

The details of this algorithm are left as an exercise for the reader.

Summarizing, we see that all the methods we have described consist of two major steps. The first step is to compute the first row w of the orthogonal matrix of eigenvectors of the desired Jacobi matrix J from the given initial eigenvalue data $\{\lambda_i\}_1^n$ and $\{\mu_i\}_1^{n-1}$. We have described an algorithm to carry out this step. The second step is to construct the Jacobi matrix J from the computed unit vector w and the given eigenvalue data $\{\lambda_i\}_1^n$. For this step we have introduced four algorithms to carry out it: Lanczos Method, Householder Method, Fast Orthogonal Reduction Method, and Rutishauser Method.

In fact, we may also start by computing the first row of the orthogonal matrix of eigenvectors of the first lower principal submatrix $J_{\hat{1}}$ of the desired Jacobi matrix J, and then apply those matrix reduction methods described before to generate the desired J. Indeed, if we are given a Jacobi matrix $J \in \mathbf{R}^{n \times n}$ with eigenvalues $\lambda_1 < \cdots < \lambda_n$, and if the first lower principal submatrix $J_{\hat{1}}$ of J has the spectral decomposition

$$J_{\hat{1}} = QMQ^T, \quad M = \operatorname{diag}(\mu_1, \dots, \mu_{n-1}),$$

where Q is an $(n-1) \times (n-1)$ orthogonal matrix, $\mu_1 < \cdots < \mu_{n-1}$, then it is clear that

$$\begin{bmatrix} 1 & 0 \\ 0 & Q^T \end{bmatrix} \begin{bmatrix} \alpha_1 & \beta_2 e_1^T \\ \beta_2 e_1 & J_{\hat{1}} \end{bmatrix} \begin{bmatrix} 1 & 0 \\ 0 & Q \end{bmatrix} = \begin{bmatrix} \alpha_1 & b^T \\ b & M \end{bmatrix} \equiv \widetilde{A}, \tag{2.5.9}$$

where $b = \beta_2 Q^T e_1$, and we have partitioned J as

$$\begin{bmatrix} \alpha_1 & \beta_2 e_1^T \\ \beta_2 e_1 & J_{\hat{1}} \end{bmatrix}.$$

By expanding the determinant, we find

$$\prod_{j=1}^{n}(\lambda-\lambda_j) = \det(\lambda I - \widetilde{A})$$

$$= (\lambda-\alpha_1)\prod_{j=1}^{n-1}(\lambda-\mu_j) - \sum_{k=1}^{n-1} b(k)^2 \prod_{\substack{j=1\\ j\neq k}}^{n-1}(\lambda-\mu_j).$$

Setting $\lambda=\mu_1,\dots,\mu_{n-1}$ we have

$$b(i)^2 = -\frac{\prod_{j=1}^{n}(\mu_i-\lambda_j)}{\prod_{\substack{j=1\\ j\neq i}}^{n-1}(\mu_i-\mu_j)}. \tag{2.5.10}$$

Conversely, if we are given $2n-1$ real numbers $\{\lambda_i\}_1^n$ and $\{\mu_i\}_1^{n-1}$ as in (2.1.2), then we may compute a vector $b \in \mathbf{R}^{n-1}$ by (2.5.10) and form the bordered diagonal matrix

$$\widetilde{A} = \begin{bmatrix} \alpha_1 & b^T \\ b & M \end{bmatrix},$$

where $\alpha_1 = \sum_{j=1}^{n}\lambda_j - \sum_{j=1}^{n-1}\mu_j$ and $M = \operatorname{diag}(\mu_1,\dots,\mu_{n-1})$. After this, we can generate the desired Jacobi matrix by applying Algorithm 2.5.2 or 2.5.4 or 2.5.6 to $\widetilde{A}$.

2.6 Rank One Update Problem

In the previous sections we have established the theory of existence and uniqueness of the solution for Problem 2.1, done a sensitivity analysis of the problem, and developed several numerical methods for solving it. In the remainder of this chapter we will show how the foregoing methods and theory can be used to solve some other inverse problems. In this section, we are concerned with only the following problem.

Problem 2.2 *Given two strictly increasing sequences $\{\lambda_i\}_1^n$ and $\{\widetilde{\lambda}_i\}_1^n$, with $\lambda_i < \widetilde{\lambda}_i$ for all i, find an $n \times n$ Jacobi matrix J such that J has $\{\lambda_i\}_1^n$ as its eigenvalues and $\widetilde{J} = J + \alpha e_1 e_1^T$ has eigenvalues $\{\widetilde{\lambda}_i\}_1^n$, where $\alpha = \sum_{j=1}^{n}(\widetilde{\lambda}_j - \lambda_j)$.*

Suppose that Jacobi matrix J is a solution to Problem 2.2. Define

$$p_n(t) = \det(tI - J), \quad p_{n-1}(t) = \det\big(tI - J(2:n,2:n)\big),$$

$$p_{n-2}(t) = \det\big(tI - J(3:n,3:n)\big), \quad \widetilde{p}_n(t) = \det(tI - \widetilde{J}).$$

Then, by expanding the determinants, we find

$$p_n(t) = (t - \alpha_1)p_{n-1}(t) - \beta_2^2 p_{n-2}(t), \tag{2.6.1}$$

$$\widetilde{p}_n(t) = (t - \widetilde{\alpha}_1)p_{n-1}(t) - \beta_2^2 p_{n-2}(t), \tag{2.6.2}$$

where $\widetilde{\alpha}_1 = \alpha_1 + \alpha$. Subtracting (2.6.2) from (2.6.1) yields

$$p_n(t) - \widetilde{p}_n(t) = \alpha p_{n-1}(t). \tag{2.6.3}$$

Notice that $p_n(t) = \prod_{i=1}^n (t-\lambda_i)$, $\widetilde{p}_n(t) = \prod_{i=1}^n (t-\widetilde{\lambda}_i)$, and $\alpha = \sum_{i=1}^n (\widetilde{\lambda}_i - \lambda_i) > 0$. It follows from (2.6.3) that

$$p_{n-1}(t) = \frac{\prod_{i=1}^n (t-\lambda_i) - \prod_{i=1}^n (t-\widetilde{\lambda}_i)}{\sum_{i=1}^n (\widetilde{\lambda}_i - \lambda_i)}. \tag{2.6.4}$$

This formula shows that p_{n-1} is completely determined by the eigenvalues of J and $\widetilde{J}$. Consequently by Theorem 2.2.3 the first row elements of the orthogonal matrix of eigenvectors of J are uniquely determined up to signs by the given data $\{\lambda_i\}_i^n$ and $\{\widetilde{\lambda}_n\}_i^n$. Thus, by Theorem 2.3.5, we have prove

Theorem 2.6.1 *Problem 2.2 has a unique solution.*

The above discussion also gives us a recipe for finding the unique solution to Problem 2.2. Summarizing, we have

Algorithm 2.6.1 Given $eig1, eig2 \in \mathbf{R}^n$ with $eig1(1) < eig2(1) < eig1(2) < eig2(2) < \cdots < eig1(n) < eig2(n)$, this algorithm computes an $n \times n$ Jacobi matrix J such that its eigenvalues are $\{eig1(i)\}_1^n$ and $J + \alpha e_1 e_1^T$ has eigenvalues $\{eig2(i)\}_1^n$, where $\alpha = \sum_{i=1}^n (eig2(i) - eig2(i))$, and the diagonal and the subdiagonal elements of J are respectively stored in $d \in \mathbf{R}^n$ and $sd(2:n)$, where $sd \in \mathbf{R}^n$.

```
n = length(eig1); α = 0;
for i = 1 : n
        α = α + eig2(i) − eig1(i);
end
for j = 1 : n
      w(j) = (eig2(j) − eig1(j)) /α ;
      for i = 1 : j − 1
            w(j) = w(j)(eig1(j) − eig2(i)) /(eig1(j) − eig1(i)) ;
      end
      for i = j + 1 : n
```

$w(j) = w(j)\big(eig1(j) - eig2(i)\big) / \big(eig1(j) - eig1(i)\big)$;
end
$w(j) = \sqrt{w(j)}$;
end
$[d, sd] =$ **jacobi**$(eig1, w)$;

The cost of this algorithm is essentially the same as Algorithm 2.5.6. Of course, we may replace Algorithm 2.5.6 in the last step and instead generate J by applying Algorithm 2.5.2 or 2.5.4 to the data w and $eig1$.

2.7 Persymmetric Problem

In this section, we will study the inverse eigenvalue problem for persymmetric Jacobi matrices. An $n \times n$ Jacobi matrix J is called a **persymmetric Jacobi matrix** if it satisfies

$$\alpha_i = \alpha_{n+1-i}, \quad \beta_i = \beta_{n+2-i},$$

that is, J is symmetric around its anti-diagonal. The problem we will discuss is as follows.

Problem 2.3 *Given a set of real number $\{\lambda_i\}_1^n$, with $\lambda_1 < \lambda_2 < \cdots < \lambda_n$, find a persymmetric Jacobi matrix J such that J has $\{\lambda_i\}_1^n$ as its eigenvalues.*

To solve this problem, it is useful to give some basic properties of a persymmetric Jacobi matrix. If we are given an $n \times n$ Jacobi matrix J, if we set $k = [\frac{n}{2}]$, where $[x]$ denotes the greatest integer less than or equal to x, and set $S = E_k S^p E_k$, where E_k denotes the $k \times k$ **reversal matrix**, that is the $k \times k$ identity matrix with its columns reversed, and

$$S^p = J(1:k, 1:k) = \begin{bmatrix} \alpha_1 & \beta_2 & & & \\ \beta_2 & \ddots & \ddots & & \\ & \ddots & \ddots & \ddots & \\ & & \ddots & \ddots & \beta_k \\ & & & \beta_k & \alpha_k \end{bmatrix},$$

then it is easy to see that J is persymmetric if and only if

$$J = \begin{bmatrix} S^p & \beta_{k+1} e_k e_1^T \\ \beta_{k+1} e_1 e_k^T & S \end{bmatrix}, \quad \text{if } n = 2k$$

or

$$J = \begin{bmatrix} S^p & \beta e_k & 0 \\ \beta e_k^T & \alpha_{k+1} & \beta e_1^T \\ 0 & \beta e_1 & S \end{bmatrix}, \quad \text{if } n = 2k+1,$$

where $\beta = \beta_{k+1} = \beta_{k+2}$. Thus, by a short calculation, we find that if $n = 2k$, then

$$J = \frac{1}{2}\begin{bmatrix} E_k & -E_k \\ I & I \end{bmatrix}\begin{bmatrix} S + \beta e_1 e_1^T & 0 \\ 0 & S - \beta e_1 e_1^T \end{bmatrix}\begin{bmatrix} E_k & I \\ -E_k & I \end{bmatrix}, \tag{2.7.1}$$

where $\beta = \beta_{k+1}$; if $n = 2k + 1$, then

$$J = \frac{1}{2}\begin{bmatrix} 0 & E_k & -E_k \\ \sqrt{2} & 0 & 0 \\ 0 & I & I \end{bmatrix}\begin{bmatrix} \alpha & \sqrt{2}\beta e_1^T & 0 \\ \sqrt{2}\beta e_1 & S & 0 \\ 0 & 0 & S \end{bmatrix}\begin{bmatrix} 0 & \sqrt{2} & 0 \\ E_k & 0 & I \\ -E_k & 0 & I \end{bmatrix}, \tag{2.7.2}$$

where $\alpha = \alpha_{k+1}$, $\beta = \beta_{k+1} = \beta_{k+2}$.

These formulae show that we can easily reduce the eigenproblem for a persymmetric Jacobi matrix J to two real symmetric tridiagonal eigenproblems of half the size.

We now turn our attention to Problem 2.3. We first consider the case of $n = 2k + 1$. Set

$$J_k = \begin{bmatrix} \alpha & \sqrt{2}\beta e_1^T \\ \sqrt{2}\beta e_1 & S \end{bmatrix} \tag{2.7.3}$$

and assume that the eigenvalues of J, J_k, and S are respectively

$$\lambda_1 < \lambda_2 < \cdots < \lambda_n, \quad \mu_1 < \mu_2 < \cdots < \mu_{k+1},$$

and

$$\nu_1 < \nu_2 < \cdots < \nu_k.$$

Then (2.7.2) implies that

$$\begin{cases} \mu_i = \lambda_{2i-1}, & i = 1, 2, \dots, k, k+1, \\ \nu_i = \lambda_{2i}, & i = 1, 2, \dots, k. \end{cases} \tag{2.7.4}$$

Thus, we have proved that in the $n = 2k + 1$ case Problem 2.3 is equivalent to finding a $(k+1)\times(k+1)$ Jacobi matrix J_k of form (2.7.3) such that its eigenvalues are $\{\lambda_{2i-1}\}_1^{k+1}$, and its first lower principal submatrix S has eigenvalues $\{\lambda_{2i}\}_1^k$, which is just Problem 2.1 with given data $\{\lambda_{2i-1}\}_1^{k+1}$ and $\{\lambda_{2i}\}_1^k$. As a consequence in the $n = 2k + 1$ case Problem 2.3 has one and only has one solution and the solution can be efficiently computed by applying those algorithms developed in Section 2.5 to Problem 2.1 with the given data $\{\lambda_{2i-1}\}_1^{k+1}$ and $\{\lambda_{2i}\}_1^k$.

For the $n = 2k$ case, we need the following result.

Lemma 2.7.1 *Let $A \in \mathbf{R}^{n\times n}$ be symmetric with eigenvalues $\lambda_1 \le \lambda_2 \le \ldots \le \lambda_n$, and let $A - \beta e_1 e_1^T$ have eigenvalues $\widetilde{\lambda}_1 \le \widetilde{\lambda}_2 \le \ldots \le \widetilde{\lambda}_n$, where β is a positive number. Then we have*

$$\widetilde{\lambda}_1 \le \lambda_1 \le \widetilde{\lambda}_2 \le \ldots \le \widetilde{\lambda}_n \le \lambda_n. \tag{2.7.5}$$

Proof For any $x \in \mathbf{R}^n$ we have

$$x^T(A - \beta e_1 e_1^T)x = x^T A x - \beta x(1)^2 \le x^T A x,$$

by Theorem 1.5.3 (i.e., Minimax Theorem), which implies that

$$\widetilde{\lambda}_i \le \lambda_i, \quad i = 1, 2, \ldots, n. \tag{2.7.6}$$

Let $A(2:n,\, 2:n)$ have eigenvalues $\mu_1 \le \cdots \le \mu_{n-1}$. Note $A(2:n,\, 2:n)$ is also the first lower principal submatrix of $A - \beta e_1 e_1^T$. By Theorem 1.5.2, we know that

$$\lambda_1 \le \mu_1 \le \lambda_2 \le \cdots \le \mu_{n-1} \le \lambda_n$$

and

$$\widetilde{\lambda}_1 \le \mu_1 \le \widetilde{\lambda}_2 \le \cdots \le \mu_{n-1} \le \widetilde{\lambda}_n.$$

This, together with (2.7.6), gives (2.7.5). □

Remark 2.7.1 From the proof we easily see that if $\beta < 0$ in Lemma 2.7.1, then we have

$$\lambda_1 \le \widetilde{\lambda}_1 \le \lambda_2 \le \cdots \le \widetilde{\lambda}_n.$$

we now return to Problem 2.3 for the ease of $n = 2k$. Let $S + \beta e_1 e_1^T$ and $S - \beta e_1 e_1^T$ have eigenvalues

$$\mu_1 < \mu_2 < \cdots < \mu_k \quad \text{and} \quad \nu_1 < \nu_2 < \cdots < \nu_k,$$

respectively. Then, by Lemma 2.7.1, it follows that

$$\nu_1 \le \mu_1 \le \nu_2 \le \mu_2 \le \cdots \le \nu_k \le \mu_k.$$

Notice that (2.7.1) implies that

$$\lambda(J) = \lambda(S + \beta e_1 e_1^T) \cup \lambda(S - \beta e_1 e_1^T).$$

Thus, we have

$$\nu_i = \lambda_{2i-1}, \quad \mu_i = \lambda_{2i}, \quad i = 1, 2, \ldots, k,$$

where $\lambda_1 < \cdots < \lambda_n$ are the eigenvalues of J. This, together with (2.7.1), shows that in the $n = 2k$ case Problem 2.3 is equivalent to finding an $k \times k$ Jacobi matrix $S - \beta e_1 e_1^T$ such that its eigenvalues are $\{\lambda_{2i-1}\}_1^k$, and $S + \beta e_1 e_1^T$ has eigenvalues $\{\lambda_{2i}\}_1^k$, where $\beta = \frac{1}{2}\sum_{i=1}^k(\lambda_{2i} - \lambda_{2i-1})$, which is just Problem 2.2 in Section 2.6 with given data $\{\lambda_{2i-1}\}_1^k$ and $\{\lambda_{2i}\}_1^k$. Consequently, for the $n = 2k$ case Problem 2.3 also has a unique solution, and the solution may be efficiently computed by using Algorithm 2.7.1 with starting data $\{\lambda_{2i-1}\}_1^k$ and $\{\lambda_{2i}\}_1^k$.

2.8 Periodic Problem

In this section we will show how to apply the techniques developed in the previous sections to the problem of reconstructing a periodic Jacobi matrix from some given eigenvalue data. A **periodic Jacobi matrix** is an $n \times n$ real symmetric matrix of the form

$$P = J + \beta_1(e_1 e_n^T + e_n^T e_1), \tag{2.8.1}$$

where J is an $n \times n$ Jacobi matrix as in (2.1.1) and $\beta_1 > 0$, that is,

$$P = \begin{bmatrix} \alpha_1 & \beta_2 & 0 & \cdots & 0 & \beta_1 \\ \beta_2 & \alpha_2 & \beta_3 & \ddots & & 0 \\ 0 & \beta_3 & \ddots & \ddots & \ddots & \vdots \\ \vdots & \ddots & \ddots & \ddots & \ddots & 0 \\ 0 & & \ddots & \ddots & \ddots & \beta_n \\ \beta_1 & 0 & \cdots & 0 & \beta_n & \alpha_n \end{bmatrix},$$

where $\beta_i > 0$ for $i = 1, 2, \ldots, n$. We are concerned only with the following periodic Jacobi matrix inverse eigenvalue problem.

Problem 2.4 *Given two sets of real numbers $\{\lambda_i\}_1^n$ and $\{\mu_i\}_1^{n-1}$ which satisfy the interlacing property*

$$\lambda_1 \le \mu_1 \le \lambda_2 \le \cdots \le \mu_{n-1} \le \lambda_n \tag{2.8.2}$$

and $\mu_i \ne \mu_j$ if $i \ne j$, and given a positive number β, find a periodic Jacobi matrix P such that P has eigenvalues $\{\lambda_i\}_1^n$, its first lower principal submatrix $J_{\hat{1}}$ has eigenvalues $\{\mu_i\}_1^{n-1}$, and $\beta_1\beta_2\cdots\beta_n = \beta$.

Note that in this problem we only require that the given eigenvalue data satisfy the interlacing property (2.8.2) rather than the strictly interlacing property as it was in Problem 2.1 . This is because the eigenvalues of the first lower principal submatrix of a periodic Jacobi matrix need not strictly separate its eigenvalues.

Example 2.8.1 Consider the following 3×3 periodic Jacobi matrix

$$P = \begin{bmatrix} 2 & 1 & 1 \\ 1 & 2 & 1 \\ 1 & 1 & 2 \end{bmatrix}.$$

With a bit of algebra we find the eigenvalues of P are 1, 1, and 4, and its first lower principal submatrix $J_{\hat{1}} = \begin{bmatrix} 2 & 1 \\ 1 & 2 \end{bmatrix}$ has eigenvalues 1, 3.

Observe that this example also shows that a periodic Jacobi matrix need not have all eigenvalues distinct, whereas a Jacobi matrix must have all eigenvalues distinct.

In order to solve Problem 2.4, we need some special properties of the periodic Jacobi matrix. Suppose we are given a periodic Jacobi matrix P defined by (2.8.1). Define

$$\beta = \prod_{j=1}^{n} \beta_i \quad \text{and} \quad \Delta(\lambda) = 2 + \frac{1}{\beta}\det(\lambda I - P). \tag{2.8.3}$$

$\Delta(x)$ is called the **discriminant** of P. Firstly, by using elementary properties of determinant, it is easy to establish the following lemma.

Lemma 2.8.1 *For any nonzero real number ρ we have*

$$\det\left(\lambda I - \left(J + \frac{\beta_1}{\rho}e_1 e_n^T + \rho\beta_1 e_n e_1^T\right)\right) = \beta\left(\Delta(\lambda) - \left(\rho + \frac{1}{\rho}\right)\right).$$

The proof of this lemma is left as an exercise for the reader.

Next, we establish a fundamental theorem, in which some of the interesting special properties of a Periodic Jacobi matrix are described.

Theorem 2.8.2 *Let the first lower principal submatrix $J_{\hat{1}}$ of P have the spectral decomposition*

$$J_{\hat{1}} = QMQ^T, \quad M = \operatorname{diag}(\mu_1, \dots, \mu_{n-1}),$$

where $Q = [q_1, q_2, \dots, q_{n-1}] = [q_{ij}] \in \mathbf{R}^{(n-1)\times(n-1)}$ is orthogonal, and $\mu_1 < \mu_2 < \cdots < \mu_{n-1}$. If we define

$$\rho_j = -\beta \Bigg/ \left(\beta_2^2 q_{1j}^2 \prod_{\substack{k=1 \\ k\neq j}}^{n-1} (\mu_j - \mu_k)\right) \tag{2.8.4}$$

for each $j = 1, 2, \dots, n-1$, then we have

$$\Delta(\mu_j) = \rho_j + \frac{1}{\rho_j}, \tag{2.8.5}$$

which implies that

$$(-1)^{n-j}\Delta(\mu_j) \geq 2, \tag{2.8.6}$$

for each $j = 1, 2, \dots, n-1$, where $\Delta(\lambda)$ is defined by (2.8.3).

Proof Using Theorem 2.2.3 we have

$$\prod_{k=3}^{n} \beta_k = q_{1j}q_{n-1,j}\prod_{\substack{k=1\\k\neq j}}^{n-1}(\mu_j - \mu_k),$$

for $j = 1, 2, \dots, n-1$. This, together with (2.8.4), yields

$$\rho_j\beta_2 q_{1j} = -\beta_1 q_{n-1,j}.$$

Combining this with $J_{\hat{1}}q_j = \mu_j g_j$ gives

$$\left(J + \frac{\beta_1}{\rho_j}e_1e_n^T + \rho_j\beta_1 e_n e_1^T\right)\begin{bmatrix} 0 \\ q_j \end{bmatrix} = \mu_j \begin{bmatrix} 0 \\ q_j \end{bmatrix},$$

and so μ_j is an eigenvalue of $J + \frac{\beta_1}{\rho_j}e_1e_n^T + \rho_j\beta_1 e_n e_1^T$ for each $j = 1, 2, \dots, n-1$. Consequently,

$$\det\left(\mu_j I - \left(J + \frac{\beta_1}{\rho_j}e_1e_n^T + \rho_j\beta_1 e_n e_1^T\right)\right) = 0$$

for $j = 1, 2, \dots, n-1$. By Lemma 2.8.1 it follows that

$$\Delta(\mu_j) = \rho_j + \frac{1}{\rho_j}$$

for $j = 1, 2, \dots, n-1$, which is the desired result.

Note that (2.8.4) implies

$$\rho_j \prod_{\substack{k=1\\k\neq j}}^{n-1}(\mu_j - \mu_k) < 0.$$

This, together with $\mu_1 < \cdots < \mu_{n-1}$, yields

$$\rho_j = (-1)^{n-j}|\rho_j|.$$

Hence we have

$$(-1)^{n-j}\Delta(\mu_j) = \frac{1}{|\rho_j|} + |\rho_j| \geq 2$$

for $j = 1, 2, \ldots, n-1$. □

With the previous preparations we may now establish the existence theorem of the solution to Problem 2.4.

Theorem 2.8.3 *Problem 2.4 has a solution if and only if*

$$\prod_{k=1}^{n} |\mu_j - \lambda_k| \geq 2\beta\big(1 + (-1)^{n-j+1}\big) \tag{2.8.7}$$

for all $j = 1, 2, \ldots, n-1$.

Proof We first prove the necessity. Assume that P is a solution to Problem 2.4. Then, by Theorem 2.8.2, we know that

$$(-1)^{n-j}\Delta(\mu_j) \geq 2 \quad \text{for } j = 1, 2, \ldots, n-1.$$

where $\Delta(\lambda)$ is the discriminant of P, i.e.,

$$\Delta(\lambda) = 2 + \frac{1}{\beta}\det(\lambda I - P) = 2 + \frac{1}{\beta}\prod_{k=1}^{n}(\lambda - \lambda_k).$$

This implies that (2.8.7) holds.

We next prove the sufficiency. Assume that the given data in Problem 2.4 satisfies conditions (2.8.7). Define $\rho_1, \rho_2, \ldots, \rho_{n-1}$ to be solutions of

$$2 + \frac{1}{\beta}\prod_{k=1}^{n}(\mu_j - \lambda_k) = \frac{1}{\rho_j} + \rho_j \tag{2.8.8}$$

for $j = 1, 2, \ldots, n-1$. Conditions (2.8.7) imply that all ρ_j are real and

$$\rho_j = (-1)^{n-j}|\rho_j|,$$

in turn, this ensures that

$$\rho_j \prod_{\substack{k=1 \\ k \neq j}}^{n-1} (\mu_j - \mu_k) < 0.$$

for each $j = 1, 2, \ldots, n-1$. Thus we can define β_2 and q_{1j} by

$$\beta_2 = \left(\sum_{j=1}^{n-1} \frac{-\beta}{\rho_j \prod_{k=1,k\neq j}^{n-1}(\mu_j - \mu_k)}\right)^{\frac{1}{2}} \tag{2.8.9}$$

and

$$q_{1j} = \frac{1}{\beta_2}\left(\frac{-\beta}{\rho_j \prod_{k=1,k\neq j}^{n-1}(\mu_j - \mu_k)}\right)^{\frac{1}{2}} \tag{2.8.10}$$

for $j = 1, 2, \ldots, n-1$. After this, by Theorem 2.3.5, there exists a unique Jacobi matrix $J_{\hat{1}}$ of order $n-1$ such that $J_{\hat{1}}$ has $\{\mu_j\}_1^{n-1}$ as its eigenvalues and the first row elements of the orthogonal matrix of eigenvectors of $J_{\hat{1}}$ are exactly the q_{1j}. Finally, we define

$$\alpha_1 = \sum_{j=1}^{n} \lambda_j - \sum_{j=1}^{n-1} \mu_j \tag{2.8.11}$$

and

$$\beta_1 = \beta \Bigg/ \prod_{j=2}^{n} \beta_j, \tag{2.8.12}$$

where $\beta_3, \ldots, \beta_n$ are the subdiagonal elements of $J_{\hat{1}}$. At this point, we have enough information to define a periodic Jacobi matrix:

$$P = J + \beta_1(e_1 e_n^T + e_n e_1^T), \tag{2.8.13}$$

where β_1 is defined by (2.8.12), and

$$J = \begin{bmatrix} \alpha_1 & \beta_2 e_1^T \\ \beta_2 e_1 & J_{\hat{1}} \end{bmatrix}$$

with $J_{\hat{1}}$, α_1, β_2 defined as above. Obviously, to show such a P is a solution, it is enough to prove that P has $\{\lambda_j\}_1^n$ as its eigenvalues.

Since the first lower principal submatrix $J_{\hat{1}}$ of P defined by (2.8.13) has the q_{1j} defined by (2.8.10) as its first row elements of the orthogonal matrix of eigenvectors of $J_{\hat{1}}$, it follows from Theorem 2.8.2 that

$$2 + \frac{1}{\beta}\det(\mu_j I - P) = \rho_j + \frac{1}{\rho_j} \tag{2.8.14}$$

for $j = 1, 2, \ldots, n-1$. Now let

$$q(\lambda) = \det(\lambda I - P) - \prod_{k=1}^{n}(\lambda - \lambda_k).$$

Then $q(\lambda)$ is a polynomial of degree at most $n-2$ because the trace of P is $\operatorname{tr} P = \alpha_1 + \operatorname{tr} J_{\hat{1}} = \alpha_1 + \sum_{j=1}^{n-1} \mu_j = \sum_{j=1}^{n} \lambda_j$ by the definitions of α_1 and $J_{\hat{1}}$. The definition of ρ_j and (2.8.14) imply that

$$q(\mu_j) = 0, \quad \text{for } j = 1, 2, \ldots, n-1,$$

and so $q \equiv 0$ because the only polynomial of degree at most $n-2$ which is zero at $n-1$ distinct points is the zero polynomial. Therefore, P has $\{\lambda_i\}_1^n$ as its eigenvalues. □

It is worthwhile to point out that Theorem 2.8.3 does not assert the uniqueness of the solution of Problem 2.4, but the existence, if the given data satisfies (2.8.7). In fact, generally speaking, the solution of Problem 2.4 is not unique if it exists. This is because for each j the equation (2.8.8) has two possible solutions, and so there are two possible choices for each ρ_j. However, once the ρ_j have been selected, then the periodic Jacobi matrix P will be uniquely determined.

Example 2.8.2 Suppose

$$\lambda_1 = 2 - \sqrt{10}, \quad \lambda_2 = \lambda_3 = 2, \quad \lambda_4 = 2 + \sqrt{10},$$
$$\mu_1 = 2 - \sqrt{2}, \quad \mu_2 = 2, \quad \mu_3 = 2 + \sqrt{2}, \quad \beta = 2.$$

It is easy to verify that these given numbers satisfy conditions (2.8.7). Moreover, a short calculation verifies that both

$$P_1 = \begin{bmatrix} 2 & 2 & 0 & 2 \\ 2 & 1 & \frac{\sqrt{2}}{2} & 0 \\ 0 & \frac{\sqrt{2}}{2} & 2 & \frac{\sqrt{2}}{2} \\ 2 & 0 & \frac{\sqrt{2}}{2} & 3 \end{bmatrix}$$

and

$$P_2 = \begin{bmatrix} 2 & \sqrt{4+2\sqrt{2}} & 0 & \sqrt{4-2\sqrt{2}} \\ \sqrt{4+2\sqrt{2}} & 2 & \frac{1}{2}\sqrt{4+2\sqrt{2}} & 0 \\ 0 & \frac{1}{2}\sqrt{4+2\sqrt{2}} & 2 & \frac{1}{2}\sqrt{4-2\sqrt{2}} \\ \sqrt{4-2\sqrt{2}} & 0 & \frac{1}{2}\sqrt{4-2\sqrt{2}} & 2 \end{bmatrix}$$

have the given values $2-\sqrt{10}, 2, 2$, and $2+\sqrt{10}$ as their eigenvalues and the eigenvalues of their first lower principal submatrices are the given values $2-\sqrt{2}, 2$, and $2+\sqrt{2}$. In addition, it is clear that their nonzero off-diagonal elements satisfy that

$$2 \times 2 \times \frac{\sqrt{2}}{2} \times \frac{\sqrt{2}}{2} = 2 = \sqrt{4+2\sqrt{2}} \times \sqrt{4-2\sqrt{2}} \times \frac{1}{2}\sqrt{4+2\sqrt{2}} \times \frac{1}{2}\sqrt{4-2\sqrt{2}}.$$

Besides, it is worth mentioning that even if the given eigenvalue data satisfies a strictly interlacing property, Problem 2.4 need not have a solution. For example, if $\lambda_1 = 1$, $\lambda_2 = 3$, $\lambda_3 = 5$, $\mu_1 = 2$, $\mu_2 = 4$, and $\beta = 1$, then they satisfy the strictly interlacing property: $\lambda_1 < \mu_1 < \lambda_2 < \mu_2 < \lambda_3$, but

$$|\mu_2 - \lambda_1||\mu_2 - \lambda_2||\mu_2 - \lambda_3| = 3 \not\geq 4 = 2\beta\left(1 + (-1)^{3-2+1}\right),$$

which shows that there is no solution for Problem 2.4 with these given values by Theorem 2.8.3.

We now turn our attention to numerically constructing a periodic Jacobi matrix from the given data. Indeed, the proof of Theorem 2.8.3 has given us a complete recipe for finding a solution of Problem 2.4. In summary, we obtain the following algorithm.

Algorithm 2.8.1 Given $eig1 \in \mathbf{R}^n$ and $eig2 \in \mathbf{R}^{n-1}$ with $eig1(1) \leq eig2(1) \leq \cdots \leq eig2(n-1) \leq eig1(n)$, and $eig2(1) < eig2(2) < \cdots < eig2(n-1)$, and given a positive number β, this algorithm computes a periodic Jacobi matrix P such that P has eigenvalues $\{eig1(i)\}_1^n$, its first lower principal submatrix has eigenvalues $\{eig2(i)\}_1^{n-1}$, and the product of all its nonzero off-diagonal elements is β, and the diagonal and the nonzero off-diagonal elements of P are stored in d and sd, respectively.

```
n = length(eig1); α = 0; d(1) = eig(n); sd(1) = 1;
for j = 1 : n − 1
    w(j) = 1; d(1) = d(1) + (eig1(j) − eig2(j));
    for k = 1 : n
        w(j) = w(j)(eig2(j) − eig1(k));
    end
    w(j) = 2 + w(j) /β ;
    if |w(j)| ≥ 2
        w(j) = ½(w(j) + √(w(j)² − 4)) or w(j) = ½(w(j) − √(w(j)² − 4));
    else
        return (Problem 2.4 has no solution);
    end
    for k = 1 : j − 1
        w(j) = w(j)(eig2(j) − eig2(k));
    end
    for k = j + 1 : n − 1
        w(j) = w(j)(eig2(j) − eig2(k));
    end
    w(j) = β/(−w(j));
    α = α + w(j); w(j) = √w(j);
end
α = √α; w = w /α;
[d(2 : n), sd(2 : n)] = Jacobi(eig2, w); sd(2) = α;
```

```
for  j = 2 : n
        sd(1) = sd(1) sd(j);
end
sd(1) = β / sd(1);
```

This algorithm requires about $19n^2$ flops and $\frac{1}{2}n^2$ square roots.

2.9 Double Dimension Problem

In this section, we will discuss a class of inverse eigenvalue problem for Jacobi matrices as follows.

Problem 2.5 *Given an $n \times n$ Jacobi matrix J_n and a set of distinct eigenvalue $\{\lambda_i\}_1^{2n}$, construct a $2n \times 2n$ Jacobi matrix J_{2n} whose eigenvalues are the given values $\{\lambda_i\}_1^{2n}$ and whose leading $n \times n$ principal submatrix is exactly J_n.*

Sometimes this problem is referred to as **double dimension problem**. The continuum version of this problem is concerned with the following practical question: given a violin string of variable density in $0 \leq x \leq L$, can the string be extended to $0 \leq x \leq 2L$ and vibrate with given tones $\lambda_1, \lambda_2, \ldots$?

We first give some preliminary results which play a fundamental role in this section.

Lemma 2.9.1 *Let J_m be an $m \times m$ Jacobi matrix defined by (2.1.1) ($m \geq 2$). Then for any integer k with $1 \leq k < m$ the vector*

$$(\xi_1^{(k)}, \ldots, \xi_m^{(k)})^T \equiv J_m^k e_1$$

satisfies that $\xi_{k+2}^{(k)} = \cdots = \xi_m^{(k)} = 0$, $\xi_{k+1}^{(k)} = \beta_{k+1} \cdots \beta_2$ and $\xi_1^{(k)}, \ldots, \xi_k^{(k)}$ are determined completely by $\alpha_1, \ldots, \alpha_k$ and $\beta_2, \ldots, \beta_k$.

Proof The proof is by induction on k. The lemma is trivially true for $k = 1$. Noting that $J_m^l e_1 = J_m J_m^{l-1} e_1$, we have

$$\begin{aligned}
\xi_1^{(l)} &= \alpha_1 \xi_1^{(l-1)} + \beta_2 \xi_2^{(l-1)}, \\
\xi_i^{(l)} &= \beta_i \xi_{i-1}^{(l-1)} + \alpha_i \xi_i^{(l-1)} + \beta_{i+1} \xi_{i+1}^{(l-1)}, \quad i = 2, 3, \ldots m-1, \\
\xi_m^{(l)} &= \beta_m \xi_{m-1}^{(l-1)} + \alpha_m \xi_m^{(l-1)},
\end{aligned}$$

where $2 \leq l < m$. Thus it follows from these relations that the lemma is also true for $k = l$, assuming that it is true for $k = l - 1$. By mathematical induction, the lemma is established. □

Lemma 2.9.2 *Let J_{2n} be a $2n \times 2n$ Jacobi matrix, and let J_n be the leading principal submatrix of J_{2n} of order n. Then*

$$e_1^T J_{2n}^k e_1 = e_1^T J_n^k e_1 \tag{2.9.1}$$

for each $k = 1, 2, \ldots, 2n-1$.

Proof Partition J_{2n} as follows:

$$J_{2n} = \begin{bmatrix} J_n & \beta_{n+1} e_n e_1^T \\ \beta_{n+1} e_1 e_n^T & J \end{bmatrix}.$$

By Lemma 2.9.1 it is easy to know that

$$J_{2n}^k e_1 = \begin{bmatrix} J_n^k e_1 \\ 0 \end{bmatrix} \tag{2.9.2}$$

for each $k = 1, 2, \ldots, n-1$, and

$$J_{2n}^n e_1 = \begin{bmatrix} J_n^n e_1 \\ \beta e_1 \end{bmatrix}, \quad \beta = \beta_{n+1}\beta_n \cdots \beta_2. \tag{2.9.3}$$

Notice that

$$e_1^T J_{2n}^{n+k} e_1 = (J_{2n}^k e_1)^T (J_{2n}^n e_1)$$

for each $k = 1, 2, \ldots, n-1$. The lemma immediately follows from (2.9.2) and (2.9.3) . □

Lemma 2.9.3 *Let J_n and $\widetilde{J}_n$ be two $n \times n$ Jacobi matrices. If*

$$e_1^T J_n^k e_1 = e_1^T \widetilde{J}_n^k e_1 \tag{2.9.4}$$

for $k = 1, 2, \ldots, 2n-1$, then $J_n = \widetilde{J}_n$.

Proof Using (2.9.4) for $k = 1, 2$, we have

$$\alpha_1 = e_1^T J_n e_1 = e_1^T \widetilde{J}_n e_1 = \widetilde{\alpha}_1$$

and

$$\alpha_1^2 + \beta_2^2 = e_1^T J_n^2 e_1 = e_1^T \widetilde{J}_n^2 e_1 = \widetilde{\alpha}_1^2 + \widetilde{\beta}_2^2,$$

and so $\beta_2 = \widetilde{\beta}_2$ since β_2, $\widetilde{\beta}_2$ are positive. Assume that we have proved that

$$\begin{aligned} \alpha_i &= \widetilde{\alpha}_i, \quad i = 1, 2, \ldots, m-1, \\ \beta_i &= \widetilde{\beta}_i, \quad i = 2, 3, \ldots, m, \end{aligned} \tag{2.9.5}$$

where $2 \le m \le n$. Then, by Lemma 2.9.1, we have that

$$J_n^{m-1}e_1 = \tilde{J}_n^{m-1}e_1 = (\xi_1, \dots, \xi_m, 0, \dots, 0)^T \tag{2.9.6}$$

with $\xi_m = \beta_m \cdots \beta_2 > 0$, $\xi_1, \dots, \xi_{m-1}$ determined completely by $\alpha_1, \dots, \alpha_{m-1}$ and $\beta_2, \dots, \beta_{m-1}$.

Now we first prove that $\alpha_m = \tilde{\alpha}_m$. Using (2.9.4) for $k = 2m - 1$ we have

$$\begin{aligned}(J_n^{m-1}e_1)^T J_n (J_n^{m-1}e_1) &= e_1^T J_n^{2m-1} e_1 = e_1^T \tilde{J}_n^{2m-1} e_1 \\ &= (\tilde{J}_n^{m-1}e_1)^T \tilde{J}_n (\tilde{J}_n^{m-1}e_1).\end{aligned} \tag{2.9.7}$$

Substituting (2.9.6) into (2.9.7) and using (2.9.5) we obtain

$$\begin{aligned}&x^T J_{m-1} x + 2\xi_m \beta_m \xi_{m-1} + \xi_m^2 \alpha_m \\ &\qquad = x^T J_{m-1} x + 2\xi_m \beta_m \xi_{m-1} + \xi_m^2 \tilde{\alpha}_m,\end{aligned}$$

where

$$x = (\xi_1, \dots, \xi_{m-1})^T,$$

$$J_{m-1} = \begin{bmatrix} \alpha_1 & \beta_2 & & & \\ \beta_2 & \alpha_2 & \beta_3 & & \\ & \ddots & \ddots & \ddots & \\ & & \beta_{m-2} & \alpha_{m-2} & \beta_{m-1} \\ & & & \beta_{m-1} & \alpha_{m-1} \end{bmatrix}.$$

Thus $\xi_m^2 \alpha_m = \xi_m^2 \tilde{\alpha}_m$, and so it must have $\alpha_m = \tilde{\alpha}_m$ since $\xi_m > 0$.

Next we prove that $\beta_{m+1} = \tilde{\beta}_{m+1}$. It follows from (2.9.6) that

$$\begin{aligned}J_n^m e_1 &= J_n J_n^{m-1} e_1 = (\eta_1, \dots, \eta_m, \eta_{m+1}, 0, \dots, 0)^T, \\ \tilde{J}_n^m e_1 &= \tilde{J}_n \tilde{J}_n^{m-1} e_1 = (\eta_1, \dots, \eta_m, \tilde{\eta}_{m+1}, 0, \dots, 0)^T,\end{aligned}$$

where $\eta_{m+1} = \xi_m \beta_{m+1}$, $\tilde{\eta}_{m+1} = \xi_m \tilde{\beta}_{m+1}$, and $\eta_1, \dots, \eta_m$ are determined completely by $\alpha_1, \dots, \alpha_m$ and $\beta_2, \dots, \beta_m$. Hence, using (2.9.4) for $k = 2m$ we get

$$\begin{aligned}\eta_1^2 + \cdots + \eta_m^2 + \eta_{m+1}^2 &= e_1^T J_n^{2m} e_1 = e_1^T \tilde{J}_n^{2m} e_1 \\ &= \eta_1^2 + \cdots + \eta_m^2 + \tilde{\eta}_{m+1}^2,\end{aligned}$$

and so

$$\xi_m^2 \beta_{m+1}^2 = \eta_{m+1}^2 = \tilde{\eta}_{m+1}^2 = \xi_m^2 \tilde{\beta}_{m+1}^2,$$

which implies that $\beta_{m+1} = \widetilde{\beta}_{m+1}$ since β_{m+1}, $\widetilde{\beta}_{m+1}$, and ξ_m are positive. By mathematical induction, the lemma is established. □

With the above preparations we now establish some necessary and sufficient conditions for the solvability of Problem 2.5. The following theorem is the main result in this section.

Theorem 2.9.4 *Suppose an $n \times n$ Jacobi matrix J_n and a set of distinct real numbers $\lambda_1 < \lambda_2 < \cdots < \lambda_{2n}$ are given. Then there exists a unique $2n \times 2n$ Jacobi matrix J_{2n} such that J_{2n} has the given eigenvalues λ_i and its leading $n \times n$ principal submatrix is exactly J_n if and only if*

$$\Delta_i \equiv \det \begin{bmatrix} 1 & \cdots & 1 & 1 & 1 & \cdots & 1 \\ \lambda_1 & \cdots & \lambda_{i-1} & c_1 & \lambda_{i+1} & \cdots & \lambda_{2n} \\ \lambda_1^2 & \cdots & \lambda_{i-1}^2 & c_2 & \lambda_{i+1}^2 & \cdots & \lambda_{2n}^2 \\ \vdots & & \vdots & \vdots & \vdots & & \vdots \\ \lambda_1^{2n-1} & \cdots & \lambda_{i-1}^{2n-1} & c_{2n-1} & \lambda_{i+1}^{2n-1} & \cdots & \lambda_{2n}^{2n-1} \end{bmatrix} > 0 \qquad (2.9.8)$$

for $i = 1, 2, \ldots, 2n$, where

$$c_k = e_1^T J_n^k e_1, \quad k = 1, 2, \ldots, 2n-1. \qquad (2.9.9)$$

Proof Let us first prove that conditions (2.9.8) are necessary. Assume that J_{2n} is the unique solution of Problem 2.5. Then by Lemma 2.9.2 we have

$$e_1^T J_{2n}^k e_1 = e_1^T J_n^k e_1 \qquad (2.9.10)$$

for $k = 1, 2, \ldots 2n-1$ and

$$J_{2n} = Q \Lambda Q^T, \qquad (2.9.11)$$

where $\Lambda = \operatorname{diag}(\lambda_1, \ldots, \lambda_{2n})$ and Q is the orthogonal matrix of eigenvectors of J_{2n}.

Substituting (2.9.11) into (2.9.10), we have

$$\omega_1^2 \lambda_1^k + \omega_2^2 \lambda_2^k + \cdots + \omega_{2n}^2 \lambda_{2n}^k = c_k, \quad k = 1, 2, \ldots, 2n-1, \qquad (2.9.12)$$

where c_k are defined by (2.9.9) and

$$(\omega_1, \ldots, \omega_{2n})^T = Q^T e_1, \qquad (2.9.13)$$

i.e., ω_i is the first component of the ith normalized eigenvector. Hence, we have

$$\omega_1^2 + \omega_2^2 + \cdots + \omega_{2n}^2 = 1. \qquad (2.9.14)$$

Writing (2.9.12) and (2.9.14) in matrix-vector form, we have

$$V(\lambda_1, \dots, \lambda_{2n})x = c, \tag{2.9.15}$$

where $V(\lambda_1, \dots, \lambda_{2n})$ denotes the Vandermonde matrix, i.e.,

$$V(\lambda_1, \dots, \lambda_{2n}) = \begin{bmatrix} 1 & 1 & \cdots & 1 \\ \lambda_1 & \lambda_2 & \cdots & \lambda_{2n} \\ \vdots & \vdots & & \vdots \\ \lambda_1^{2n-1} & \lambda_2^{2n-1} & \cdots & \lambda_{2n}^{2n-1} \end{bmatrix}$$

and

$$x = (\omega_1^2, \dots, \omega_{2n}^2)^T, \quad c = (1, c_1, \dots, c_{2n-1})^T.$$

Let Δ be the determinant of $V(\lambda_1, \dots, \lambda_{2n})$. By Gramer's rule we have

$$\omega_i^2 = \Delta_i \,/ \Delta\,, \qquad i = 1, 2, \cdots, 2n-1, \tag{2.9.16}$$

where Δ_i are defined by (2.9.8). Since

$$\Delta = \det(V(\lambda_1, \dots, \lambda_{2n})) = \prod_{1 \le i < j \le 2n} (\lambda_j - \lambda_i) > 0$$

and the first component of any eigenvector of a Jacobi matrix is nonzero, we know that

$$\Delta_i = \Delta\omega_i^2 > 0, \qquad i = 1, \dots, 2n.$$

This shows that conditions (2.9.8) are necessary for the solvability of Problem 2.5.

We now prove that conditions (2.9.8) are also sufficient. Assume that conditions (2.9.8) are true. Then the linear system (2.9.15) has a positive solution x, i.e., if $x = (x_1, \dots, x_{2n})^T$ satisfies (2.9.15), then $x_i > 0$ for $i = 1, \dots, 2n$. Thus we can define the first components of the ith normalized eigenvectors of J_{2n}, which we want to find as

$$\omega_i = \sqrt{x_i}. \tag{2.9.17}$$

By Theorem 2.3.5 there exists a unique $2n \times 2n$ Jacobi matrix J_{2n} such that its eigenvalues are the given values $\lambda_1, \lambda_2, \dots, \lambda_{2n}$ and the first components of its eigenvectors are exactly ω_i. Let $\widetilde{J}_n$ be the leading $n \times n$ principal submatrix of J_{2n}. Then, by Lemma 2.9.2 and the definition of ω_i, we have

$$e_1^T \widetilde{J}_n^k e_1 = e_1^T J_{2n}^k e_1 = \sum_{i=1}^{2n} \lambda_i^k \omega_i^2 = \sum_{i=1}^{2n} \lambda_i^k x_i = c_k = e_1^T J_n^k e_1$$

for $k = 1, \dots, 2n-1$. Thus, by Lemma 2.9.3, we know that $J_n = \widetilde{J}_n$. Hence, J_{2n} is the solution of Problem 2.5. This completes the proof of Theorem 2.9.4. □

The process of the proof of Theorem 2.9.4 provides us with a recipe for finding the solution of Problem 2.5 if it exists. In summary, the method is as follows.

Step 1. Compute $c_k = e_1^T J_n^k e_1$ for $k = 1, \ldots, 2n-1$.

Step 2. Solve the Vandermonde system (2.9.15) for $x = (\omega_1^2, \ldots, \omega_{2n}^2)^T$.

Step 3. Construct the Jacobi matrix J_{2n} from the spectrum data $\{\lambda_i\}_1^{2n}$ and $\{\omega_i\}_1^{2n}$, i.e., compute a $2n \times 2n$ Jacobi matrix J_{2n} such that J_{2n} has eigenvalues λ_i and eigenvectors with the first components ω_i.

As has been known, step 3 can be completed by using the algorithms described in Section 2.5. Therefore, the rest of this section is devoted to showing how to carry out steps 1 and 2 in practice.

Clearly, there are many methods for carrying out step 1. From the numerical standpoint we adopt the spectral method. Let the spectral decomposition of J_n be

$$J_n = QMQ^T, \quad M = \operatorname{diag}(\mu_1, \ldots, \mu_n), \tag{2.9.18}$$

where $Q = [q_{ij}]$ is an orthogonal matrix of order n, $\mu_1 < \cdots < \mu_n$. Then it follows that

$$c_k = e_1^T J_n^k e_1 = e_1^T Q M^k Q^T e_1 = \sum_{j=1}^{n} \mu_j^k q_{1j}^2 \tag{2.9.19}$$

for $k = 1, 2, \ldots, 2n-1$.

Thus, once (2.9.18) has been computed, we may use (2.9.19) to compute c_k. it is well known that the decomposition (2.9.18) can be completed by using the extensively used and numerically reliable symmetric QR algorithm.

We now show how to solve the system (2.9.15). In general this system is very ill-conditioned. Therefore, instead of solving it by a standard linear system solver, we apply Lagrange interpolation to solve it.

Substituting (2.9.19) into (2.9.15) and writing in component form we have

$$\sum_{i=1}^{2n} \lambda_i^k \omega_i^2 = \sum_{j=1}^{n} \mu_j^k q_{1j}^2, \quad k = 0, 1, \ldots, 2n-1. \tag{2.9.20}$$

Let

$$L_i(\lambda) = \prod_{\substack{j=1 \\ j \neq i}}^{2n} \frac{(\lambda - \lambda_j)}{(\lambda_i - \lambda_j)}, \quad i = 1, 2, \ldots, 2n, \tag{2.9.21}$$

which are Lagrange elementary interpolating polynomials. Then it is easy to verify that

$$\lambda^k = \sum_{i=1}^{2n} L_i(\lambda) \lambda_i^k, \quad k = 1, 2, \ldots, 2n-1.$$

Setting $\lambda = \mu_l, l = 1, 2, \ldots, n$, we have

$$\mu_l^k = \sum_{i=1}^{2n} L_i(\mu_l)\lambda_i^k, \quad k = 1, 2, \ldots, 2n-1. \tag{2.9.22}$$

Substituting (2.9.22) into (2.9.20) we obtain

$$\sum_{i=1}^{2n} \lambda_i^k \omega_i^2 = \sum_{l=1}^{n} \left(\sum_{i=1}^{2n} L_i(\mu_l)\lambda_i^k \right) q_{1l}^2 = \sum_{i=1}^{2n} \lambda_i^k \left(\sum_{l=1}^{n} L_i(\mu_l) q_{1l}^2 \right).$$

From this formula, it is clear to see that the solution to the system (2.9.15) is given by

$$\omega_i^2 = \sum_{l=1}^{n} L_i(\mu_l) q_{1l}^2, \quad i = 1, 2, \ldots, 2n.$$

Overall we obtain

Algorithm 2.9.1 Given an $n \times n$ Jacobi matrix J and a $2n$ dimensional vector eig with $eig(1) < eig(2) < \cdots < eig(2n)$, this algorithm computes a $2n \times 2n$ jacobi matrix J_{2n} such that its eigenvalues are $\{eig(i)\}_1^{2n}$ and its leading $n \times n$ principal submatrix is J, and the diagonal and the subdiagonal elements of J_{2n} are stored in d and $sd(2:2n)$, respectively.

```
m = length(eig); n = m/2 ;
Compute the spectral decomposition J = QMQ^T as in (2.9.18)
                                              using QR algorithm;
for i = 1 : n
    x(i) = Q(1,i)^2;
end
for i = 1 : m
    w(i) = 0;
    for l = 1 : n
        α = x(l);
        for j = 1 : i-1
            α = α(M(l,l) - eig(j)) / (eig(i) - eig(j));
        end
        for j = i+1 : m
            α = α(M(l,l) - eig(j)) / (eig(i) - eig(j));
        end
        w(i) = w(i) + α;
    end
    if w(i) > 0
        w(i) = sqrt(w(i));
```

```
        else
            return (Problem 2.5 has no solution);
        end
    end
    [d, sd] = jacobi(eig, w);
```

We may replace Algorithm 2.5.6 in last step and instead generate J_{2n} by applying Algorithms 2.5.2 or 2.5.4 to the data w and eig.

2.10 Miscellaneous Problems

In this section, we briefly introduce three miscellaneous problems which can be solved by applying the techniques developed in previous sections.

2.10.1 Banded Matrix Inverse Eigenvalue Problem

First of all we consider the problem of reconstructing banded matrices from the given spectral data.

Let $B = [b_{ij}] \in \mathbf{R}^{n\times n}$ be symmetric. B is called a p-**banded matrix** if $b_{ij} = 0$ whenever $|i-j| > p$, that is, B has the from

$$B = \begin{bmatrix} b_{11} & \cdots & b_{1,p+1} & & \\ \vdots & \ddots & & \ddots & \\ b_{p+1,1} & & \ddots & & b_{n-p,n} \\ & \ddots & & \ddots & \vdots \\ & & b_{n,n-p} & \cdots & b_{nn} \end{bmatrix}, \quad b_{ij} = b_{ji}, \tag{2.10.1}$$

and say such a matrix has half bandwidth p. For a given B with half bandwidth p we use B_k to denote the lower $(n-k)\times(n-k)$ principal submatrix of B, i.e., $B_k = B(k+1:n,\ k+1:n)$. Especially, $B_0 = B$.

Now, the problem can be stated as follows.

Problem 2.6 *Given the $p+1$ sets of real numbers $\{\lambda_j^{(k)}\}$, $k = 0,1,\dots,p$, satisfying the interlacing property*

$$\lambda_j^{(k)} \le \lambda_j^{(k+1)} \le \lambda_{j+1}^{(k)}, \quad j = 1,\dots,n-k-1,\ k = 0,\dots,p-1, \tag{2.10.2}$$

find a B with half bandwidth p such that each submatrix B_k has the given values $\{\lambda_j^{(k)}\}$ as its eigenvalues for $k = 0,1,\dots,p$.

This problem can be solved by several methods which are very similar to the methods described in Section 2.5. In fact, we may extend all those methods developed in Section 2.5 to this problem. Here we will only show how a variant of the orthogonal reduction method based on Householder reflections can be used to generate the solutions to Problem 2.6.

As has been done in Section 2.5, for each k, we first defined a bordered diagonal matrix $A_k \in \mathbf{R}^{(n-k)\times(n-k)}$ of the from

$$A_k = \begin{bmatrix} \alpha_{k+1} & b_{k+1}^T \\ b_{k+1} & \Lambda_{k+1} \end{bmatrix}, \quad \Lambda_{k+1} = \operatorname{diag}\left(\lambda_1^{(k+1)}, \ldots, \lambda_{n-k-1}^{(k+1)}\right), \tag{2.10.3}$$

such that A_k has eigenvalues $\lambda_1^{(k)}, \ldots, \lambda_{n-k}^{(k)}$. In fact, as in (2.5.9), the value α_{k+1} is uniquely determined by

$$\alpha_{k+1} = \sum_{j=1}^{n-k} \lambda_j^{(k)} - \sum_{j=1}^{n-k-1} \lambda_j^{(k+1)}; \tag{2.10.4}$$

and the vector b_{k+1} can be determined by the formula

$$[b_{k+1}(i)]^2 = -\frac{\prod_{j=1}^{n-k} (\lambda_i^{(k+1)} - \lambda_j^{(k)})}{\prod_{\substack{j=1 \\ j\neq i}}^{n-k-1} (\lambda_i^{(k+1)} - \lambda_j^{(k+1)})}. \tag{2.10.5}$$

Note that this formula requires the $\lambda_j^{(k+1)}$ be distinct.

Let the spectral decomposition of A_k be

$$A_k = Q_k \Lambda_k Q_k^T, \quad \Lambda_k = \operatorname{diag}(\lambda_1^{(k)}, \ldots, \lambda_{n-k}^{(k)}), \tag{2.10.6}$$

where Q_k is an orthogonal matrix of order $n-k$. Set $G_0 = A_0$, and then define G_k iteratively by

$$G_k = \begin{bmatrix} I & 0 \\ 0 & Q_k \end{bmatrix} G_{k-1} \begin{bmatrix} I & 0 \\ 0 & Q_k \end{bmatrix}^T \in \mathbf{R}^{n\times n} \tag{2.10.7}$$

for $k = 1, 2, \ldots, p-1$. Then $G \equiv G_{p-1}$ has the form

$$G = \begin{bmatrix} D & F^T \\ F & \Lambda_p \end{bmatrix}, \quad D \in \mathbf{R}^{p\times p}, \quad F \in \mathbf{R}^{(n-p)\times p}, \tag{2.10.8}$$

and satisfies that

$$\lambda\big(G(k+1:n, k+1:n)\big) = \{\lambda_1^{(k)}, \ldots, \lambda_{n-k}^{(k)}\}, \qquad k = 0, \ldots, p. \tag{2.10.9}$$

In fact, by the definition it is clear that the matrix G_1 has the form

$$G_1 = \begin{bmatrix} 1 & 0 \\ 0 & Q_1 \end{bmatrix} \begin{bmatrix} \alpha_1 & b_1^T \\ b_1 & \Lambda_1 \end{bmatrix} \begin{bmatrix} 1 & 0 \\ 0 & Q_1 \end{bmatrix}^T = \begin{bmatrix} \alpha_1 & b_1^T Q_1^T \\ Q_1 b_1 & A_1 \end{bmatrix} = \begin{bmatrix} D_1 & F_1^T \\ F_1 & \Lambda_2 \end{bmatrix},$$

where $D_2 \in \mathbf{R}^{2\times 2}, F_1 \in \mathbf{R}^{(n-2)\times 2}$, and so $G_1, G_1(2:n,2:n)$, and $G_1(3:n,3:n)$ have eigenvalues $\{\lambda_1^{(1)}\}, \{\lambda_1^{(2)}\}$, and $\{\lambda_1^{(3)}\}$, respectively. Suppose that the matrix G_l has the form

$$G_l = \begin{bmatrix} D_l & F_l^T \\ F_l & \Lambda_{l+1} \end{bmatrix}, \quad D_l \in \mathbf{R}^{(l+1)\times(l+1)}, \quad F_l \in \mathbf{R}^{(n-l-1)\times(l+1)},$$

and $G_l(k+1:n,k+1:n)$ has eigenvalues $\{\lambda_j^{(k)}\}$ for $k = 0, 1, \ldots, l+1$. Then

$$G_{l+1} = \begin{bmatrix} I & 0 \\ 0 & Q_{l+1} \end{bmatrix} \begin{bmatrix} D_l & F_l^T \\ F_l & \Lambda_{l+1} \end{bmatrix} \begin{bmatrix} I & 0 \\ 0 & Q_{l+1} \end{bmatrix}^T = \begin{bmatrix} D_l & F_l^T Q_{l+1}^T \\ Q_{l+1} F_l & A_{l+1} \end{bmatrix}.$$

This shows that

$$\lambda\big(G_{l+1}(k+1:n,k+1:n)\big) = \lambda\big(G_l(k+1:n,k+1:n)\big) = \{\lambda_1^{(k)}, \ldots, \lambda_{n-k}^{(k)}\},$$

for $k = 0, 1, \ldots, l+1$, and $G_{l+1}(l+3:n, l+3:n) = \Lambda_{l+2}$, which implies that $G_{l+1}(l+2:n, l+2:n)$ has eigenvalues $\{\lambda_j^{(l+2)}\}$. Moreover, it is clear that G_{l+1} has the form

$$G_{l+1} = \begin{bmatrix} D_{l+1} & F_{l+1}^T \\ F_{l+1} & \Lambda_{l+2} \end{bmatrix}, \quad D_{l+1} \in \mathbf{R}^{(l+2)\times(l+2)}, \quad F_{l+1} \in \mathbf{R}^{(n-l-2)\times(l+2)}.$$

By induction, we know that $G = G_{p-1}$ has the form of (2.10.8) and satisfies (2.10.9).

Thus, we have defined a matrix G that satisfies all requirements of Problem 2.6 except for the p-banded form. Therefore if we can find an orthogonal matrix $Q \in \mathbf{R}^{(n-p)\times(n-p)}$ such that

$$B = \begin{bmatrix} I & 0 \\ 0 & Q \end{bmatrix} G \begin{bmatrix} I & 0 \\ 0 & Q \end{bmatrix}^T = \begin{bmatrix} D & F^T Q^T \\ QF & Q\Lambda_p Q^T \end{bmatrix}$$

is a p-banded matrix, then B must solve Problem 2.6. This can be carried out by applying a series of Givens rotations in plane $p+1, \cdots, n$ or Householder reflections in a manner very similar to the reduction algorithm 2.5.4.

In summary, the above discussion gives rise to a method for solving Problem 2.6 as follows:

Step 1. Compute A_k by (2.10.3), (2.10.4), and (2.10.5), $k = 0, \ldots, p-1$;

Step 2. Compute Q_k as in (2.10.6), $k = 1, \ldots, p-1$;

Step 3. Compute G_k by (2.10.7), $k = 0, 1, \ldots, p-1$;

Step 4. Reduce G_{p-1} to a banded matrix B with half bandwidth p.

Clearly, the key steps of this method are step 2 and step 4. The simplest way to carry out step 2 is to use a standard symmetric eigenproblem solver on the $\{A_k\}$. But since A_k is a bordered diagonal matrix of form (2.10.3), one better method is first to use Algorithm 2.5.6 to reduce A_k to tridiagonal form, then to use the EISPACK routine TQL2. Thus the cost of step 2 is only $O(n^2)$.

As mentioned above, step 4 can be carried out by using Givens rotations or Householder reflections. But this will require $O(n^3)$ flops. How to reduce the cost from $O(n^3)$ to $O(n^2)$ is worth considering.

The above discussion can be formulated as a practical algorithm, The details are left as exercise for the reader.

2.10.2 Rank One Modification Problem

The following problem arises when updating the singular value decomposition or when modifying symmetric eigenvalue problems:

Problem 2.7 *Given the spectral decomposition of a real symmetric $n \times n$ matrix B : $B = Q\Lambda Q^T$, where Q is orthogonal, $\Lambda = \mathrm{diag}(\lambda_1, \ldots, \lambda_n)$, compute the spectral decomposition of the real symmetric matrix $\widetilde{B} = Q\Lambda Q^T + \rho bb^T$, where $\rho \in \mathbf{R}$, $b \in \mathbf{R}^n$.*

Observe that

$$\widetilde{B} = Q\Lambda Q^T + \rho bb^T = Q(\Lambda + \rho dd^T)Q^T, \quad d = Q^T b.$$

Therefore, to compute the spectral decomposition of $\widetilde{B}$, we only need to compute the spectral decomposition of $\Lambda + \rho dd^T$. We now show how to apply the techniques developed in Section 2.5 to compute the spectral decomposition of $\Lambda + \rho dd^T$ in $O(n^2)$ flops.

Consider the bordered diagonal matrix

$$A = \begin{bmatrix} \alpha_0 & d^T \\ d & \Lambda \end{bmatrix},$$

where α_0 is a dummy element, Λ and d defined as above. Recall from Section 2.5 that A may be reduced to tridiagonal form in $O(n^2)$ flops by using Algorithm 2.5.6, that is, we can compute an orthogonal matrix $U \in \mathbf{R}^{n\times n}$ such that

$$\begin{bmatrix} 1 & 0 \\ 0 & U^T \end{bmatrix} \begin{bmatrix} \alpha_0 & d^T \\ d & \Lambda \end{bmatrix} \begin{bmatrix} 1 & 0 \\ 0 & U \end{bmatrix} = \begin{bmatrix} \alpha_0 & \beta e_1^T \\ \beta_1 e_1 & T \end{bmatrix}, \tag{2.10.10}$$

where $\beta_1 \in \mathbf{R}, T \in \mathbf{R}^{n\times n}$ is tridiagonal. Comparing two sides of (2.10.11) we have

$$U^T \Lambda U = T, \quad \beta_1 U e_1 = d, \quad \beta_1 = \pm\|d\|_2.$$

Thus, it follows that

$$U^T(\Lambda + \rho dd^T)U = T + \rho\beta_1 e_1 e_1^T.$$

This shows that after we reduced A to tridiagonal form we have, in fact, reduced $\Lambda + \rho dd^T$ to tridiagonal form: $T + \rho\beta_1 e_1 e_1^T$. At this point, we can use the standard tridiagonal eigenproblem subroutine to compute the spectral decomposition of $T + \rho\beta_1 e_1 e_1^T$. It is well known that the spectral decomposition of a tridiagonal matrix can be computed in $O(n^2)$ plops. Thus, in this way we may compute the spectral decomposition of $\Lambda + \rho dd^T$ in $O(n^2)$ flops.

2.10.3 Construction of a Jacobi Matrix from Its Eigenpairs

So far our discussion has been on the reconstruction of a Jacobi matrix from some knowledge of its eigenvalues. As a final problem of this chapter we consider the problem of reconstructing a Jacobi matrix from certain given eigenpairs.

Problem 2.8 *Given two distinct real numbers λ, μ and two vectors $u, v \in \mathbf{R}^n$ with $\|u\|_2 = \|v\|_2 = 1$ and $v^T u = 0$, find an $n \times n$ Jacobi matrix J such that J has (λ, u) and (μ, v) as its two eigenpairs.*

Recall that for any eigenpairs (λ, x) of a Jacobi matrix J we have

$$\beta_i x(i-1) + \alpha_i x(i) + \beta_{i+1} x(i+1) = \lambda x(i)$$

where we define $\beta_1 = \beta_{n+1} = 0$ and $x(0) = x(n+1) = 0$. Applying this relation to both eigenpairs gives

$$\begin{aligned}\beta_i u(i-1) + \alpha_i u(i) + \beta_{i+1} u(i+1) &= \lambda u(i),\\ \beta_i v(i-1) + \alpha_i v(i) + \beta_{i+1} v(i+1) &= \mu v(i).\end{aligned}$$

Combining these two equations and eliminating α_i gives

$$\begin{aligned}&\beta_i\big(u(i-1)v(i) - v(i-1)u(i)\big)\\ &\quad + \beta_{i+1}\big(u(i+1)v(i) - v(i+1)u(i)\big) = (\lambda - \mu)u(i)v(i).\end{aligned} \tag{2.10.11}$$

Note that for $i = 1$ we have

$$\beta_2\big(u(2)v(1) - v(2)u(1)\big) = (\lambda - \mu)u(1)v(1). \tag{2.10.12}$$

Combining (2.10.11) and (2.10.12) we obtain the identity

$$\beta_{i+1}\big(u(i+1)v(i) - v(i+1)u(i)\big) = (\lambda - \mu)\sum_{j=1}^{i} u(j)v(j) \tag{2.10.13}$$

for $i = 1, 2, \ldots, n-1$.

Now define

$$\begin{aligned} d_i &= u(i+1)v(i) - v(i+1)u(i), \\ c_i &= \sum_{j=1}^{i} u(j)v(j), \quad i = 1, \ldots, n-1. \end{aligned} \tag{2.10.14}$$

Then (2.10.13) can be written as

$$\beta_{i+1} d_i = (\lambda - \mu) c_i, \quad i = 1, \ldots, n-1. \tag{2.10.15}$$

Therefore, we can use these equations to determine β_i provided that $d_i \neq 0$ and $(\lambda - \mu)c_i / d_i > 0$ for any $i = 1, \ldots, n-1$. If this is true, then

$$\beta_{i+1} = \frac{(\lambda - \mu)c_i}{d_i}, \quad i = 1, \ldots, n-1. \tag{2.10.16}$$

Once the β_i determined, we can determine α_i by

$$\alpha_i = \begin{cases} \lambda - \beta_i \dfrac{u(i-1)}{u(i)} - \beta_{i+1} \dfrac{u(i+1)}{u(i)}, & \text{if } u(i) \neq 0, \\ \\ \mu - \beta_i \dfrac{v(i-1)}{v(i)} - \beta_{i+1} \dfrac{v(i+1)}{v(i)}, & \text{of } v(i) \neq 0, \end{cases} \tag{2.10.17}$$

provided that not both $u(i)$ and $v(i)$ are equal to zero. It is worthwhile to point out that if $u(i)v(i) \neq 0$, then it is easy to verify that two of the formulae in (2.10.17) have the same value.

In summary, the above discussion shows that if the given data satisfies that (1) $d_i \neq 0$ and $(\lambda - \mu)c_i / d_i > 0$ and (2) $u(i)$ and $v(i)$ are not equal to zero simultaneously, then Problem 2.8 has a unique solution, and the solution can be computed by (2.10.16) and (2.10.17).

Notes and References

An excellent survey of Jacobi matrix inverse eigenvalue problems is given by Boley and Golub [21]. Most of this chapter is based on this paper.

Much of the motivation for the problems discussed in this chapter came about from an interest in the inverse Sturm-Liouville problem. Papers related to the inverse Sturm-Liouville problem include [6], [80], [99]–[105], [111], [113], [115], [117], [169], and [172].

Problem 2.1 originally appeared in [110]. Some examples of applications can be found in [9], [100], [82]–[85], and [173]. Much of the basic theory of existence and uniqueness of the solution to Problem 2.1 was developed in [112] and [100]. A simple proof given here is mainly based on Paige's formula, which is due to Paige [170].

Section 2.4 is mainly based on [233] with some significant improvement; see also [100] and [138].

The numerical methods described in Section 2.5 were developed in [13], [18], [42], [49], [92], [189], and [190]; see also [21].

Problem 2.2 (i.e., Rank one update problem) originally appeared in [49]; the content of Section 2.6 is based on this paper. Persymmetric problem (i.e, Problem 2.3) first appeared in [110]; the material of Section 2.7 is taken from [92]. Section 2.8 is mainly based on [64] with some significant changes; see also [19] and [20]. The material of Section 2.9 is mainly taken from [236]; Problem 2.5 (i.e., Double dimension problem) originally appeared in [114]; Theorem 2.9.4 is due to Xu [236]; Algorithm 2.9.1 is based on [21]; other related references include [56] and [40]. The three problems studied in the last section are mainly based on [13], [25], and [241]; see also [18], [22], [43], [74], [86], and [242].

Chapter 3

Pole Assignment Problems

3.1 Introduction

A pole assignment problem is a special inverse algebraic eigenvalue problem, which determines some aspects of a matrix so that the eigenvalues of the resulting matrix are specified. The problem arises in the theory of control of linear dynamic systems. Consider the time invariant, linear, multivariable system with dynamic state equation

$$\dot{x}(t) = Ax(t) + Bu(t) \tag{3.1.1}$$

for $t \geq 0$. Here $A \in \mathbf{R}^{n\times n}$ and $B \in \mathbf{R}^{n\times m}$ are time-invariant and determine the characteristics of the dynamic system, $x(t)$ is a vector-valued function of time and is referred to as the **state** of the system, and $u(t)$ represents the **input** to the system with the intent of controlling its state. It is often sufficient to assume that $u(t)$ is piecewise continuous. The system (3.1.1) is referred to as a **single-input system** if $m = 1$, and as a **multi-input system** if $m > 1$. Hereafter we will use the matrix pair (A, B) to stand for the system (3.1.1).

If $x(t_0) = x_0$, then (3.1.1) implies

$$x(t) = e^{(t-t_0)A}x_0 + \int_{t_0}^{t} e^{(t-\tau)A}Bu(\tau)\mathrm{d}\tau,$$

for $t_0 \geq 0$, $t > 0$, where

$$e^{sA} = \sum_{k=0}^{\infty} \frac{s^k}{k!}A^k.$$

Clearly, the solution $x(t)$ is dependent on the start time t_0, the initial state x_0, and the input u. Hence we will use $x(t; t_0, x_0, u)$ to denote the corresponding solution of (3.1.1) with $x(t_0) = x_0$.

One approach of controlling the system (3.1.1) is to choose an input which is a linear function of the current state of the system, i.e., a **state feedback control**,

$$u(t) = Fx(t)$$

for some matrix $F \in \mathbf{R}^{m \times n}$, which is referred to as the **feedback** or **gain matrix.** With this, the system (3.1.1) becomes

$$\dot{x}(t) = (A + BF)x(t). \tag{3.1.2}$$

The matrix $A + BF$ is referred to as the **closed-loop system matrix.** In applications the goal of introducing state feedback is to change the dynamic behavior of the system (3.1.1) in some desirable way: to achieve stability, say, or to speed up response.

Stability is an important concept in linear dynamic system theory, meaning intuitively that small perturbations from an equilibrium state do not produce large subsequent motions. More precisely, system(3.1.2) is said to be asymptotically stable if the state goes zero with increasing time for any given initial state. It can be shown that the system(3.1.2) possesses such stability if and only if all eigenvalues of $A + BF$ have negative real parts. This means that the requirement that the closed-loop system (3.1.2) be asymptotically stable is equivalent to the requirement that

$$\max\{\mathrm{Re}\,(\lambda) \ : \ \lambda \in \lambda(A + BF)\} < 0.$$

Similarly, for rapid response with limited frequency of oscillation it can be expressed as conditions on the spectrum of the closed-loop system matrix $A + BF$ as follows:

$$\max\{\mathrm{Re}\,(\lambda) \ : \ \lambda \in \lambda(A + BF)\} \leq -\alpha,$$
$$\max\{|\mathrm{Im}\,(\lambda)| \ : \ \lambda \in \lambda(A + BF)\} \leq \beta,$$

with suitable $\alpha > 0$ and $\beta \geq 0$.

One way of meeting such spectral requirements is to choose real matrix F such that $A + BF$ has a specified self-conjugate set of eigenvalues. Thus we are led to the following problem.

Problem PA *Given $A \in \mathbf{R}^{n \times n}$, $B \in \mathbf{R}^{n \times m}$, and a set of n complex numbers, $\mathcal{L} = \{\lambda_1, \ldots, \lambda_n\}$, closed under complex conjugation, find an $F \in \mathbf{R}^{m \times n}$ such that the eigenvalues of $A + BF$ are $\lambda_1, \lambda_2, \ldots, \lambda_n$.*

This problem is known as the **state feedback pole assignment problem** for the system (3.1.1), in reference to the fact that the eigenvalues of $AB + F$ are poles of the closed-loop system transfer matrix

$$(sI - A - BF)^{-1}B.$$

Here the system output is taken to be the state.

In this chapter we will focus our research on problem PA. We will first establish the sufficient and necessary conditions of the existence of solutions to Problem PA. Since the concept of controllability will play a key role in establishing such conditions, we will give the specified definition of controllability and discuss it in some detail in Section 3.2.

The second problem we are concerned with is the sensitivity of the solution to problem PA, which is much harder to deal with. We will use an analytic technique to derive the first order perturbation bounds for the solution to Problem PA, from which we will introduce condition numbers to measure the sensitivity of the solution to Problem PA. See Section 3.4.

How to compute a solution to Problem PA if it exists is another important problem we are primarily concerned with. Although many algorithms exist for the solution to Problem PA, most of them are numerically unstable. For example, the methods based on Lnenberger canonical form can be considered unstable since the transformation matrix leading to this canonical form can be very poorly conditioned. Methods using the characteristic polynomial of the open-loop system matrix A are also unsatisfactory owing to the absence of a reliable method for finding the characteristic polynomial of a matrix. Similar disadvantages affect the methods based on reduction to Jordan canonical form. The absence of a reliable method for Problem PA has stimulated the development of a number of computational algorithms recently. In Sections 3.5–3.7 we will devote to developing three numerically reliable methods for solving Problem PA. To illustrate the performance of the three algorithms we will give several interesting examples in Section 3.8.

In general, the solution to problem PA is underdetermined, with many degrees of freedom. Therefore, the question arises as to how these freedoms are to be exploited in practice. This will be discussed in Section 3.9.

3.2 Controllability

Intuitively, a controllable dynamic system should be a system whose states can be made to behave in some desirable way by suitable manipulation of its inputs. Thus we are led to the following definition.

Definition 3.2.1 Let $A \in \mathbf{R}^{n\times n}$, $B \in \mathbf{R}^{n\times m}$. The system ($A$, B) is said to be controllable provided that, given any time $t_1 > t_0 \geq 0$ and any states $x_0, x_1 \in \mathbf{R}^n$, there is a piecewise continuous input $u(t), t > 0$, such that $x(t_1; t_0, x_0, u) = x_1$, where $x(t; t_0, x_0, u)$ is the corresponding solution of (3.1.1) with $x(t_0) = x_0$.

Briefly, the definition says that a **controllable system** is a system that can be brought from an initial state to any desired state in a finite time using specified input.

As systems discussed here are linear and time-invariant, it is easy to prove that Definition 3.2.1 is equivalent to

Definition 3.2.2 Let $A \in \mathbf{R}^{n\times n}$, $B \in \mathbf{R}^{n\times m}$. The system $(A,\ B)$ is said to be controllable provided that, given any time $t_1 > 0$ and any state $x_0 \in \mathbf{R}^n$, there is a piecewise continuous input $u(t)$, $t > 0$, such that $x(t_1; 0, x_0, u) = 0$, where $x(t; 0, x_0, u)$ is the corresponding solution of (3.1.1) with $x(0) = x_0$.

The proof of the equivalence of the two definitions is left as an exercise for the reader.

We now present several mathematically equivalent statements on the controllability.

Theorem 3.2.1 *Let $A \in \mathbf{R}^{n\times n}$, $B \in \mathbf{R}^{n\times m}$. Then the following statements are equivalent:*

(1) *$(A,\ B)$ is controllable;*
(2) *For any $t > 0$,* **Gram matrix**

$$W(t) \equiv \int_0^t e^{-\tau A} B B^T e^{-\tau A^T} \mathrm{d}\tau \tag{3.2.1}$$

is positive definite;
(3) *For any $t > 0$, if $v \in \mathbf{R}^n$ satisfies $v^T e^{-\tau A} B = 0$ for every $\tau \in [0, t]$, then $v = 0$;*
(4) *The* **controllability matrix**

$$[B, AB, A^2B, \ldots, A^{n-1}B] \tag{3.2.2}$$

has full rank;
(5) $\mathrm{rank}([\lambda I - A, B]) = n$ *for any $\lambda \in \lambda(A)$;*
(6) *If $\mu \in \mathbf{C}$ and $v \in \mathbf{C}^n$ satisfy $v^T A = \mu v^T$ and $v^T B = 0$, then $v = 0$.*

Proof Proceeding stepwise, we prove:

$$(1) \Longrightarrow (6) \Longrightarrow (5) \Longrightarrow (4) \Longrightarrow (3) \Longrightarrow (2) \Longrightarrow (1).$$

The details are as follows.

$(1) \Longrightarrow (6)$. Suppose that (A, B) is controllable and $\mu \in \mathbf{C}$, $v \in \mathbf{C}^n$ satisfy

$$v^T A = \mu v^T \quad \text{and} \quad v^T B = 0. \tag{3.2.3}$$

By Definition 3.2.2, for any $x_0 \in \mathbf{R}^n$, there is a piecewise continuous input $u(t)$ such that

$$0 = x(1; 0, x_0, u) \equiv \int_0^1 e^{(1-\tau)A} Bu(\tau)\mathrm{d}\tau + e^A x_0. \tag{3.2.4}$$

Premultiplying (3.2.4) by v^T and using (3.2.3), we have

$$\begin{aligned} 0 &= \int_0^1 v^T e^{(1-\tau)A} Bu(\tau)\mathrm{d}\tau + v^T e^A x_0 \\ &= \int_0^1 e^{(1-\tau)\mu} v^T Bu(\tau)\mathrm{d}(\tau) + e^{\mu} v^T x_0 \\ &= e^{\mu} v^T x_0. \end{aligned}$$

Hence, it follows that $v^T x_0 = 0$ for any $x_0 \in \mathbf{R}^n$. Thus v must be zero. This shows that (6) is true.

(6)$\Longrightarrow$ (5). Assume that (6) is true. Given any $\lambda \in \lambda(A)$, let $v \in \mathbf{C}^n$ satisfy $v^T[\lambda I - A, B] = 0$. It is clear that

$$v^T A = \lambda v^T \quad \text{and} \quad v^T B = 0.$$

By (6) it follows that $v = 0$. Hence the row vectors of $[\lambda I - A, B]$ are linearly independent, that is, rank $([\lambda I - A, B]) = n$.

(5)$\Longrightarrow$ (4). Suppose that (5) is true, but rank $[B, AB, \ldots, A^{n-1}B] < n$. We will try to obtain a contradiction. By assumption, there exists a nonzero vector $v \in \mathbf{C}^n$ such that

$$v^T[B, AB, \ldots, A^{n-1}B] = 0. \tag{3.2.5}$$

Let

$$\mathcal{K}_n(A^T, v) = \text{span}\,\{v, A^T v, \ldots, (A^T)^{n-1} v\}.$$

The Cayley-Hamilton Theorem implies that $\mathcal{K}_n(A^T, v)$ is an invariant subspace of A^T. Hence, there is at least one eigenvector of A^T in $\mathcal{K}_n(A^T, v)$. Now, let $z \in \mathcal{K}_n(A^T, v)$ and $\lambda \in \lambda(A^T)(= \lambda(A))$ with

$$A^T z = \lambda z \quad \text{and} \quad z \neq 0.$$

Then (3.2.5) implies that $B^T z$ must be zero. Hence, we have found a nonzero vector $z \in \mathbf{C}^n$ such that

$$z^T A = \lambda z^T \quad \text{and} \quad z^T B = 0.$$

This is contrary to the assumption that (5) holds. Hence, the controllability matrix $[B, AB, \ldots, A^{n-1}B]$ must have full row rank. Thus it is proved that (5) implies (4).

(4)$\Longrightarrow$ (3). Assume that (4) is true. Given any $t > 0$, let $v \in \mathbf{R}^n$ satisfy

$$v^T e^{-A\tau} B = 0$$

for any $\tau \in [0,t]$. Repeated differentiation yields

$$v^T e^{-A\tau} A^k B = 0, \quad k = 1, 2, \ldots, n-1,$$

for any $\tau \in [0,t]$, so that

$$v^T e^{-A\tau}[B, AB, \ldots, A^{n-1}B] = 0, \quad 0 \leq \tau \leq t.$$

In particular, letting $\tau = 0$, we obtain

$$v^T[B, AB, \ldots, A^{n-1}B] = 0,$$

which implies that $v = 0$ since $[B, AB, \ldots, A^{n-1}B]$ has full row rank. This proves that (4) implies (3).

(3)$\Longrightarrow$ (2). Suppose that (3) is true. Given any $t > 0$, let $v \in \mathbf{R}^n$ satisfy

$$v^T W(t) v = 0, \tag{3.2.6}$$

that is

$$\int_0^t v^T e^{-A\tau} B B^T e^{-A^T \tau} v \mathrm{d}\tau = 0,$$

which implies that

$$v^T e^{-A\tau} B = 0, \quad 0 \leq \tau \leq t.$$

Hence $v = 0$ since statement (3) is true. This shows $W(t)$ is positive definite for any $t > 0$.

(2) $\Longrightarrow$ (1). Suppose that (2) is true. Given any $x_0 \in \mathbf{R}^n$ and $t_1 > 0$, define

$$u(t) = -B^T e^{-tA^T} [W(t_1)]^{-1} x_0$$

It is easy to verify that such an input satisfies

$$x(t_1; 0, x_0, u) = 0.$$

Hence (A, B) is Controllable. Thus we have proved that (2) implies (1). $\square$

From this theorem we can easily derive several useful results, which will be used frequently in the subsequence of this chapter.

Corollary 3.2.2 *Let $A \in \mathbf{R}^{n\times n}$ and $B \in \mathbf{R}^{n\times m}$ with (A, B) controllable, and let $T \in \mathbf{R}^{n\times n}$ and $G \in \mathbf{R}^{m\times m}$ be nonsingular. Then $(T^{-1}AT,\ T^{-1}BG)$ is controllable.*

Proof Let $\mu \in \mathbf{C}^n$ and $v \in \mathbf{C}^n$ with

$$v^T T^{-1} AT = \mu v^T \quad \text{and} \quad v^T T^{-1} BG = 0. \tag{3.2.7}$$

Denote $v^T T^{-1}$ by w^T. Then (3.2.7) implies

$$w^T A = \mu w^T \quad \text{and} \quad w^T B = 0.$$

Applying Theorem 3.2.1, it follows from (A, B) controllable that $w = 0$, and so $v = 0$; applying Theorem 3.2.1 again, this means that $(T^{-1}AT, T^{-1}BT)$ is controllable. Thus the statement is proved. □

Corollary 3.2.3 *Let $A \in \mathbf{R}^{n\times n}, B \in \mathbf{R}^{n\times m}$ with*

$$A = \begin{bmatrix} A_{11} & A_{12} \\ 0 & A_{22} \end{bmatrix}, \quad B = \begin{bmatrix} B_1 \\ B_2 \end{bmatrix},$$

where $A_{22} \in \mathbf{R}^{k\times k}, B_2 \in \mathbf{R}^{k\times m}$ $(k \le n)$. If (A, B) is controllable, then (A_{22}, B_2) is also controllable.

Proof The details are left as an exercise. □

Notice that Corollary 3.2.3 does not state that (A_{11}, B_1) is controllable. In fact, generally speaking, (A_{11}, B_1) need not be controllable. For example, let

$$A = \begin{bmatrix} 0 & 1 \\ 0 & 0 \end{bmatrix}, \quad B = \begin{bmatrix} 0 \\ 1 \end{bmatrix}.$$

Clearly, (A, B) is controllable, but $(A_{11}, B_1) = (0, 0)$ is not controllable. However, if $A_{12} = 0$, then (A_{11}, B_1) is still controllable; generally, this can be extended to the case where $A = \operatorname{diag}(A_{11}, \ldots, A_{kk})$. These are left to the reader.

Corollary 3.2.4 *Let $A \in \mathbf{R}^{n\times n}, B \in \mathbf{R}^{n\times m}$. Then (A, B) is controllable if and only if $(A + BF, B)$ is controllable for any $F \in \mathbf{R}^{m\times n}$.*

Proof Observe that, for any $F \in \mathbf{R}^{n\times m}$, $v \in \mathbf{C}^n$, and $\mu \in \mathbf{C}$,

$$v^T(A + BF) = \mu v^T \quad \text{and} \quad v^T B = 0$$

if and only if

$$v^T A = \mu v^T \quad \text{and} \quad v^T B = 0.$$

By Theorem 3.2.1, it follows that (A, B) is controllable if and only if $(A + BF, B)$ is controllable. □

Although the proof of this corollary is very simple, it reveals an important property that controllability is not affected by state feedback.

3.3 Pole Assignment Theorem

In this section we will explore the existence of solutions to Problem PA. Our main result is that if $(A,\ B)$ is controllable then there is a solution $F \in \mathbf{R}^{n\times m}$ to Problem PA for every set $\mathcal{L}$ of n self-conjugate complex numbers, and this property in turn implies controllability. We first consider the single-input case.

Theorem 3.3.1 *Let $A \in \mathbf{R}^{n\times n}$, $b \in \mathbf{R}^n$. If (A, b) is controllable, then, given any set $\mathcal{L}$ of n self-conjugate complex numbers, there exists a unique $f \in \mathbf{R}^n$ such that $\lambda(A + bf^T) = \mathcal{L}$.*

Briefly, the theorem says that in the single input case the solution to Problem PA exists and is unique if the system is controllable.

Proof of Theorem 3.3.1 The assumption that (A, b) is controllable implies the matrix

$$T = [b, Ab, \dots, A^{n-1}b]$$

is nonsingular by Theorem 3.2.1. Let

$$\widetilde{A} = T^{-1}AT \quad \text{and} \quad \widetilde{b} = T^{-1}b,$$

then it is easy to verify that $\widetilde{A}$ and $\widetilde{b}$ have the following forms

$$\widetilde{A} = \begin{bmatrix} 0 & 0 & 0 & \cdots & 0 & -\alpha_0 \\ 1 & 0 & 0 & \cdots & 0 & -\alpha_1 \\ & 1 & 0 & \cdots & 0 & -\alpha_2 \\ & & \ddots & & \vdots & \vdots \\ & 0 & & \ddots & 0 & -\alpha_{n-2} \\ & & & & 1 & -\alpha_{n-1} \end{bmatrix} \quad \text{and} \quad \widetilde{b} = \begin{bmatrix} 1 \\ 0 \\ \vdots \\ \vdots \\ 0 \end{bmatrix},$$

so that for any $g = (\gamma_1, \dots, \gamma_n)^T \in \mathbf{R}^n$,

$$\det(\lambda I - \widetilde{A} - \widetilde{b}g^T) = \lambda^n + \widetilde{\alpha}_{n-1}\lambda^{n-1} + \cdots + \widetilde{\alpha}_1\lambda + \widetilde{\alpha}_0,$$

where

$$\widetilde{\alpha}_k = \alpha_k - \sum_{j=1}^{n-k-1} \gamma_j \alpha_{k+j} - \gamma_{n-k}, \qquad k = 0, 1, \dots, n-2,$$

$$\widetilde{\alpha}_{n-1} = \alpha_{n-1} - \gamma_1.$$

Let $\mathcal{L} = \{\lambda_1, \dots, \lambda_n\}$ and write

$$(\lambda - \lambda_1)(\lambda - \lambda_2)\cdots(\lambda - \lambda_n) = \lambda^n + \hat{\alpha}_{n-1}\lambda^{n-1} + \cdots + \hat{\alpha}_1\lambda + \hat{\alpha}_0.$$

Then, clearly, $\lambda(\widetilde{A} + \widetilde{b}g^T) = \mathcal{L}$ if and only if

$$\hat{\alpha}_k = \widetilde{\alpha}_k, \qquad k = 0, 1, 2, \dots, n-1,$$

that is,

$$\alpha_{n-1} - \gamma_1 = \hat{\alpha}_{n-1},$$

$$\alpha_k - \sum_{j=1}^{n-k-1} \gamma_j \alpha_{k+j} - \gamma_{n-k} = \hat{\alpha}_k, \qquad k = 0, 1, \dots, n-2.$$

Obviously, the solution to the above linear system is unique, and can be determined by the following recursion formula:

$$\gamma_1 = \alpha_{n-1} - \hat{\alpha}_{n-1},$$

$$\gamma_k = \alpha_{n-k} - \hat{\alpha}_{n-k} - \sum_{j=1}^{k-1} \alpha_{n-k+j}\gamma_j, \qquad k = 2, 3, \dots, n.$$

This shows that there is a unique $g \in \mathbf{R}^n$ such that $\lambda(\widetilde{A} + \widetilde{b}g^T) = \mathcal{L}$, hence it follows that there is a unique $f \in \mathbf{R}^n (f = T^{-T}g)$ such that $\lambda(A + bf) = \mathcal{L}$, which is the desired result. □

Next, we prove a lemma, which will be used to prove the main result in this section.

Lemma 3.3.2 *Let $A \in \mathbf{R}^{n\times n}$, $B \in \mathbf{R}^{n\times m}$, and let $0 \neq b \in \mathcal{R}(B)$. If (A, B) is controllable, then there exists $F \in \mathbf{R}^{m\times n}$ such that $(A + BF, b)$ is controllable.*

Remark 3.3.1 This result says that state feedback can be used to replace a multi-input controllable system by a single-input controllable system.

Proof of Lemma 3.3.2 Let $b_1 = b$, and let n_1 be the largest integer such that the vectors

$$b_1, Ab_1, \dots, A^{n_1-1}b_1$$

are linearly independent. Define

$$x_1 = b_1 \quad \text{and} \quad x_{j+1} = Ax_j + b_1, \qquad j = 1, 2, \dots, n_1 - 1.$$

Then $x_1, \dots, x_{n_1}$ are linearly independent, and

$$\text{span}\,\{x_1, \dots, x_{n_1}\} = \text{span}\,\{b_1, Ab_1, \dots, A^{n_1-1}b_1\}.$$

If $n_1 < n$, by controllability, there exists a $b_2 \in \mathcal{R}(B)$ such that

$$b_2 \notin \text{span}\,\{b_1, Ab_1, \dots, A^{n_1-1}b_1\}.$$

Define

$$x_{n_1+j} = Ax_{n_1+j-1} + b_2, \qquad j = 1, 2, \dots, n_2,$$

where n_2 is the integer such that the vectors

$$x_1, x_2, \dots, x_{n_1}, x_{n_1+1}, \dots, x_{n_1+n_2}$$

are linearly independent, and $Ax_{n_1+n_2} + b_2$ can be expressed as a linear combination of $x_1, \dots, x_{n_1+n_2}$. It is easy to verify that

$$\text{span}\,\{x_1, \dots, x_{n_1+n_2}\} = \text{span}\,\{b_1, Ab_1, \dots, A^{n_1-1}b_1, b_2, Ab_2, \dots, A^{n_2-1}b_2\}.$$

Continuing thus, we obtain eventually $x_1, \dots, x_n$ linearly independent, and x_{j+1} has the form

$$x_{j+1} = Ax_j + \widetilde{b}_j, \qquad j = 1, 2, \dots, n-1,$$

where $\widetilde{b}_j \in \mathcal{R}(B)$. Let $c_j \in \mathbf{R}^m$ with

$$Bc_j = \widetilde{b}_j, \qquad j = 1, 2, \dots, n,$$

where $\widetilde{b}_n \in \mathcal{R}(B)$ is arbitrary. Define

$$F = [c_1, c_2, \dots, c_n]X^{-1}$$

where $X = [x_1, \dots, x_n]$. Then

$$(A + BF)x_j = Ax_j + BFx_j = Ax_j + Bc_j = Ax_j + \widetilde{b}_j = x_{j+1}$$

for $j = 1, 2, \dots, n-1$, so that

$$(A + BF)^j b = x_{j+1}, \quad j = 0, 1, \dots, n-1,$$

and, therefore, $\text{rank}[b, (A + BF)b, \dots, (A + BF)^{n-1}b] = n$. By Theorem 3.2.1, this implies that $(A + BF, b)$ is Controllable. □

We come now to Pole Assignment Theorem.

Theorem 3.3.3 *Let $A \in \mathbf{R}^{n\times n}$, $B \in \mathbf{R}^{n\times m}$. Then, for every set $\mathcal{L}$ of n self-conjugate complex numbers, there exists a matrix $F \in \mathbf{R}^{m\times n}$ such that $\lambda(A + BF) = \mathcal{L}$ if and only if (A, B) is controllable.*

Proof Sufficiency. Let $0 \neq b \in \mathcal{R}(B)$. By Lemma 3.3.2, there exists a matrix $F_1 \in \mathbf{R}^{m\times n}$ such that $(A + BF_1, b)$ is controllable. Hence, by Theorem 3.3.1, it follows that there is a vector $f \in \mathbf{R}^n$ such that $\lambda(A + BF_1 + bf^T) = \mathcal{L}$ for every set $\mathcal{L}$ of n self-conjugate complex numbers. Then

$$F = F_1 + cf^T,$$

where $Bc = b$, is a matrix with the property required.

Necessity. Let $\lambda_1, \ldots, \lambda_n$ be n distinct real numbers, with $\lambda_j \notin \lambda(A)$, $j = 1, 2, \ldots, n$, and let $F \in \mathbf{R}^{m\times n}$ with $\lambda(A + BF) = \{\lambda_1, \ldots, \lambda_n\}$. Choose nonzero vector $x_j \in \mathbf{R}^n$ so that

$$(A + BF)x_j = \lambda_j x_j \tag{3.3.1}$$

for $j = 1, 2, \ldots, n$. Then $x_1, \ldots, x_n$ are linearly independent. Now, let $v \in \mathbf{C}^n$ and $\mu \in \mathbf{C}$ with

$$v^T A = \mu v^T \quad \text{and} \quad v^T B = 0. \tag{3.3.2}$$

Then v must be zero if $\mu = \lambda_j$ for some j, since $\lambda_j \notin \lambda(A)$. Now, assume that $\mu \notin \{\lambda_1, \ldots, \lambda_n\}$. From (3.3.1) and (3.3.2), we have

$$\lambda_j v^T x_j = \mu v^T x_j, \qquad j = 1, 2, \ldots, n.$$

Hence

$$v^T x_j = 0, \quad j = 1, 2, \ldots, n,$$

and, therefore, $v = 0$. Thus, by Theorem 3.2.1, (A, B) must be controllable, which is the desired result. □

Theorem 3.3.3 shows that if (A, B) is controllable, then a solution to the corresponding Problem PA certainly exists for every set $\mathcal{L}$ of n self-conjugate complex numbers. Indeed, if (A, B) is not controllable, then, by Theorem 3.2.1, there exists a nonzero $v \in \mathbf{C}^n$ such that $v^T A = \mu v^T$ and $v^T B = 0$ for some $\mu \in \mathbf{C}$, so that $v^T(A + BF) = \mu v^T$ for all $F \in \mathbf{R}^{m\times n}$. Thus μ is an eigenvalue of $A + BF$ for all F, that is to say that μ is a pole of (A, B) cannot be modified by any state feedback. Such a pole μ is said to be uncontrollable, and we denote the set of all uncontrollable poles of a system (A, B) by $\mathcal{L}_u(A, B)$. Obviously, if a solution to Problem PA exists, then the set $\mathcal{L}$ of poles to be assigned must contain $\mathcal{L}_u(A, B)$. More precisely, we can prove the following theorem.

Theorem 3.3.4 *Let $A \in \mathbf{R}^{n\times n}$, and $B \in \mathbf{R}^{n\times m}$, and $\mathcal{L} = \{\lambda, \dots, \lambda_n\} \subset \mathbf{C}$, closed under complex conjugation. Then a solution to the corresponding Problem PA exists if and only if $\mathcal{L}_u(A, B) \subset \mathcal{L}$, i.e., $\mathcal{L}$ contains all of the uncontrollable poles of (A, B).*

Proof By Real Schur Decomposition Theorem, we can choose an orthogonal matrix $Q \in \mathbf{R}^{n\times n}$ such that

$$Q^T AQ = \begin{bmatrix} A_{11} & A_{12} \\ 0 & A_{22} \end{bmatrix} \begin{matrix} n_1 \\ n - n_1 \end{matrix}$$
$$\qquad\qquad n_1 \quad n - n_1$$

with $\lambda(A_{11}) = \mathcal{L}_u(A, B)$. Let

$$Q^T B = \begin{bmatrix} B_1 \\ B_2 \end{bmatrix} \begin{matrix} n_1 \\ n - n_1 \end{matrix}$$

Then (A_{22}, B_2) is controllable. Thus the conclusion of this theorem immediately follows from Theorem **3.3.3**. □

From the above proof we see that as long as a solution to Problem PA exists, it can be essentially turn into a pole assignment problem for a controllable system. Hence, In the remainder of this chapter we will focus on Problem PA with (A, B) controllable.

It is worthwhile to point out that for the multi-input case controllability does not guarantee the uniqueness of the solution to Problem PA, but merely ensures the existence. Generally speaking, the solution is essentially underdetermined, with many degrees of freedom. Therefore, an interesting problem that merits further exploration is how to utilize the remaining freedom of choice of F to achieve additional desirable properties. This will be the main problem to be explored in Section 3.9.

3.4 Sensitivity Issues

In this section we will consider the sensitivity problem of a solution to Problem PA, that is, for any given solution F to Problem PA with the given data A, B, and $\mathcal{L}$, we will investigate the variation in F with respect to changes in A, B, and $\mathcal{L}$.

Notice that in the previous section we have pointed out that in the single-input case the solution to Problem PA is unique, but in the multi-input case Problem PA is essentially underdetermined. Hence, we will here deal with the two different cases separately.

3.4.1 The Single-input Case

Given a single input controllable system $(A,\ b)$ with $A \in \mathbf{R}^{n\times n},\ b \in \mathbf{R}^n$, and given a set of n complex numbers, $\mathcal{L} = \{\lambda_1, \dots, \lambda_n\}$, closed under complex conjugation. Let $E \in \mathbf{R}^{n\times n},\ e_b \in \mathbf{R}^n$, and $\varepsilon_j \in \mathbf{C},\ j = 1, 2, \dots, n$. Define

$$\begin{aligned} A(t) = A + tE, &\quad b(t) = b + te_b, \\ \lambda_j(t) = \lambda_j + t\varepsilon_j, &\qquad j = 1, 2, \dots, n, \quad t \in [-1, 1]. \end{aligned}$$

For simplicity, we assume that $\lambda_i \neq \lambda_j,\ i \neq j$, and that E and e_b are sufficiently small in magnitude such that the system $(A(t), b(t))$ is controllable for every $t \in [-1, 1]$ and assume that the ε_j are sufficiently small such that $\lambda_j(t) \neq \lambda_i(t), i \neq j$, and the set $\{\lambda_1(t), \dots, \lambda_n(t)\}$ is closed under complex conjugation for any $t \in [-1, 1]$. Thus by Theorem 3.3.1, for every $t \in [-1, 1]$, there is a unique vector $f(t) \in \mathbf{R}^n$ such that

$$\lambda\big(A(t) + b(t)f(t)^T\big) = \{\lambda_1(t), \dots, \lambda_n(t)\}.$$

In particular, we write $f = f(0)$. We now analyze how the vector $f(t) - f$ is dependent on $A(t) - A,\ b(t) - b$, and $\lambda_j(t) - \lambda_j,\ j = 1, 2, \dots, n$, when $t \longrightarrow 0$.

Since $\lambda_i \neq \lambda_j,\ i \neq j$, there exists a nonsingular matrix $X = [x_1, \dots, x_n] \in \mathbf{C}^{n\times n}$ such that

$$(A + bf^T)X = X\Lambda, \tag{3.4.1}$$

where $\Lambda = \operatorname{diag}(\lambda_1, \dots, \lambda_n)$. Let

$$Y = X^{-T} = [y_1, y_2, \dots, y_n].$$

Then

$$y_i^T x_j = \delta_{ij}, \qquad i, j = 1, 2, \dots, n, \tag{3.4.2}$$

and (3.4.1) implies that

$$(A + bf^T)x_j = \lambda_j x_j \quad \text{and} \quad y_j^T(A + bf^T) = \lambda_j y_j^T, \tag{3.4.3}$$

i.e., $x_j,\ y_j^T$ are the right and left eigenvectors of $A + bf^T$ corresponding to λ_j, $j = 1, 2, \dots, n$. In addition, by Theorem 3.2.1, the controllability, together with (3.4.3), implies that

$$y_j^T b \neq 0, \quad j = 1, 2, \dots, n. \tag{3.4.4}$$

Now consider the matrix-valued function

$$A(t; x) = A(t) + b(t)x^T, \qquad t \in [-1,\ 1],\ x = (\xi_1, \dots, \xi_n)^T \in \mathbf{R}^{n\times n}.$$

It is clear that $A(t;x)$ is analytic and $A(0;f) = A + bf^T$, which has distinct eigenvalues $\lambda_1, \ldots, \lambda_n$. Without loss of generality, we may assume that $\lambda_1, \ldots, \lambda_r$ are real numbers, and $\lambda_{r+2k-1} = \overline{\lambda}_{r+2k} = \nu_k + i\mu_k$, with $\nu_k, \mu_k \in \mathbf{R}$ and $\mu_k \neq 0$, $k = 1, 2, \ldots, s$, where $r + 2s = n$. Applying Theorems 1.9.2 and 1.9.5, it follows that there exist real analytic functions $\lambda_j(t;x)$, $j = 1, 2, \ldots, r$, $\mu_k(t;x)$, $\nu_k(t;x)$, $k = 1, 2, \ldots, s$, and analytic vector-valued functions $u_j(t;x)$, $j = 1, 2, \ldots, n$ such that

$$\begin{gathered} A(t;x)u_j(t;x) = \lambda_j(t;x)u_j(t;x), \qquad x \in \mathcal{B}(f),\ |t| < \varepsilon_1, \\ \lambda_j(0;f) = \lambda_j, \quad u_j(0;f) = x_j, \end{gathered}$$

for each $j = 1, 2, \ldots, n$, where $\mathcal{B}(f)$ is some neighborhood of f, ε_1 some positive number, and

$$\begin{gathered} \lambda_{r+2k-1}(t;x) = \nu_k(t;x) + i\mu_k(t;x), \\ \lambda_{r+2k}(t;x) = \nu_k(t;x) - i\mu_k(t;x), \qquad k = 1, 2, \ldots, s. \end{gathered}$$

Notice that $\lambda_j(t)$ is real provided λ_j is real since $\lambda_j \neq \lambda_i$, $i \neq j$, and $\{\lambda_1(t), \lambda_2(t), \ldots, \lambda_n(t)\}$ is closed under complex conjugation for any $t \in [-1, 1]$. We can define $g(t;x) = \big(g_1(t;x), \ldots, g_n(t;x)\big)^T \in \mathbf{R}^n$ by

$$g_j(t;x) = \begin{cases} \lambda_j(t;x) - \lambda_j(t), & 1 \leq j \leq r, \\ \nu_k(t;x) - \operatorname{Re}\big(\lambda_j(t)\big), & j = r + 2k - 1,\ 1 \leq k \leq s, \\ \mu_k(t;x) + \operatorname{Im}\big(\lambda_j(t)\big), & j = r + 2k,\ 1 \leq k \leq s. \end{cases}$$

Clearly, $g(t;x)$ is analytic in $(-\varepsilon_1, \varepsilon_1) \times \mathcal{B}(f)$ and $g(0;f) = 0$. By Theorems 1.9.2 and 1.9.5, we have

$$\begin{aligned} \frac{\partial \lambda_k(0;\, f)}{\partial \xi_j} &= y_k^T \frac{\partial A(0;\, f)}{\partial \xi_j} x_k = y_k^T b e_j^T x_k, \quad k = 1, 2, \ldots, r, \\ \begin{bmatrix} \frac{\partial \nu_k(0;f)}{\partial \xi_j} \\ \frac{\partial \mu_k(0;f)}{\partial \xi_j} \end{bmatrix} &= \frac{1}{2} \begin{bmatrix} 1 & 1 \\ -i & i \end{bmatrix} \begin{bmatrix} \frac{\partial \lambda_{r+2k-1}(0;f)}{\partial \xi_j} \\ \frac{\partial \lambda_{r+2k}(0;f)}{\partial \xi_j} \end{bmatrix} \\ &= \frac{1}{2} \begin{bmatrix} 1 & 1 \\ -i & i \end{bmatrix} \begin{bmatrix} y_{r+2k-1}^T b e_j^T x_{r+2k-1} \\ y_{r+2k}^T b e_j^T x_{r+2k} \end{bmatrix}, \quad k = 1, 2, \ldots, s. \end{aligned}$$

From this it immediately follows that

$$g'_x(0;\, f) = DMX^T, \tag{3.4.5}$$

where

$$\begin{gathered} D = \operatorname{diag}(I_r, D_1, \ldots, D_s), \qquad M = \operatorname{diag}(y_1^T b, y_2^T b, \ldots, y_n^T b), \\ D_k = \frac{1}{2} \begin{bmatrix} 1 & 1 \\ -i & i \end{bmatrix}, \qquad k = 1, \ldots, s. \end{gathered}$$

Thus, it follows from (3.4.4) and X nonsingular that

$$\frac{\partial(g_1,\dots,g_n)}{\partial(\xi_1,\dots,\xi_n)}(0;f) = \det\big(g'_x(0;f)\big) = (-2i)^{-s}\prod_{j=1}^{n}(y_j^T b)\det X \neq 0.$$

Consequently, by Theorem 1.8.2, there exists a positive number $\varepsilon_0 < \varepsilon_1$ and a unique vector-valued function $\widetilde{f}(t) \in \mathbf{R}^n$ such that

(1) $\widetilde{f}(t)$ is analytic in $(-\varepsilon_0, \varepsilon_0)$;

(2) $\widetilde{f}(0) = f$;

(3) $g\big(t;\widetilde{f}(t)\big) = 0$ for each $t \in (-\varepsilon_0,\ \varepsilon_0)$.

By the definition of g, (3) means that for each $t \in (-\varepsilon_0,\ \varepsilon_0)$, $\widetilde{f}(t)$ is the solution to Problem PA with data $A(t)$, $B(t)$, and $\{\lambda_1(t),\dots,\lambda_n(t)\}$. Therefore, it must have $\widetilde{f}(t) = f(t)$ for each $t \in (\varepsilon_0,\ \varepsilon_0)$, and so $f(t)$ is certainly analytic in $(\varepsilon_0,\ \varepsilon_0)$. Now let

$$x_j(t) = u_j\big(t; f(t)\big).$$

Then $x_j(t)$ is analytic in $(-\varepsilon_0,\ \varepsilon_0)$ and satisfies

$$\big(A(t) + b(t)f(t)^T\big)x_j(t) = \lambda_j(t)x_j(t), \qquad t \in (\varepsilon_0,\ \varepsilon_0), \tag{3.4.6}$$

for each $j = 1, 2, \dots, n$. Differentiating (3.4.6) yields

$$\begin{aligned}\big(\dot{A}(t) + \dot{b}(t)f(t)^T + b(t)\dot{f}(t)^T\big)x_j(t) + \big(A(t) + b(t)f(t)^T\big)\dot{x}_j(t)\\ = \dot{\lambda}_j(t)x_j(t) + \lambda_j(t)\dot{x}_j(t).\end{aligned}$$

Especially, at $t = 0$ we have

$$\big(\dot{A}(0) + \dot{b}(0)f^T + b\dot{f}(0)^T\big)x_j + (A + bf^T)\dot{x}_j(0) = \dot{\lambda}_j(0)x_j + \lambda_j\dot{x}_j(0). \tag{3.4.7}$$

Substituting $\dot{A}(0) = E$, $\dot{b}(0) = e_b$, and $\dot{\lambda}_j(0) = \varepsilon_j$ into (3.4.7), premultiplying by y_j^T, and using (3.4.2) and (3.4.3), we obtain

$$y_j^T\big(E + e_b f^T + b\dot{f}(0)^T\big)x_j = \varepsilon_j,$$

and hence

$$x_j^T\dot{f}(0) = \frac{1}{y_j^T b}(\varepsilon_j - y_j^T E x_j - y_j^T e_b f^T x_j) \tag{3.4.8}$$

for $j = 1, 2, \dots, n$. Writing (3.4.8) in vector form, we get

$$X^T\dot{f}(0) = D^{-1}\begin{bmatrix} \varepsilon_1 - y_1^T E x_1 - (f^T x_1)y_1^T e_b \\ \varepsilon_2 - y_2^T E x_2 - (f^T x_2)y_2^T e_b \\ \cdots\cdots\cdots\cdots\cdots\cdots \\ \varepsilon_n - y_n^T E x_n - (f^T x_n)y_n^T e_b \end{bmatrix},$$

and so

$$\dot{f}(0) = YD^{-1} \begin{bmatrix} \varepsilon_1 - y_1^T E x_1 - (f^T x_1) y_1^T e_b \\ \varepsilon_2 - y_2^T E x_2 - (f^T x_2) y_2^T e_b \\ \cdots\cdots\cdots\cdots\cdots\cdots \\ \varepsilon_n - y_n^T E x_n - (f^T x_n) y_n^T e_b \end{bmatrix}, \tag{3.4.9}$$

where

$$D = \operatorname{diag}(y_1^T b, y_2^T b, \ldots, y_n^T b). \tag{3.4.10}$$

Applying Lemma 1.7.2, in terms of the vec operator and the Kronecker product we may rewrite (3.4.9) as

$$\dot{f}(0) = W_A \operatorname{vec}(E) + W_b e_b + W_\lambda e_\lambda, \tag{3.4.11}$$

where

$$\begin{aligned} W_\lambda &= YD^{-1}, \qquad e_\lambda = (\varepsilon_1, \varepsilon_2, \ldots, \varepsilon_n)^T, \\ W_A &= -YD^{-1} \begin{bmatrix} x_1^T \otimes y_1^T \\ x_2^T \otimes y_2^T \\ \cdots\cdots \\ x_n^T \otimes y_n^T \end{bmatrix} \\ &= -YD^{-1}[D_1(X)Y^T, D_2(X)Y^T, \ldots, D_n(X)Y^T], \\ W_b &= -YD^{-1} \operatorname{diag}(x_1^T f, x_2^T f, \ldots, x_n^T f) Y^T, \end{aligned} \tag{3.4.12}$$

with $D_i(X) = \operatorname{diag}(x_{i1}, x_{i2}, \ldots, x_{in})$, $i = 1, 2, \ldots, n$. Consequently, we have

$$\begin{aligned} \|f(t) - f\|_2 &= \|\dot{f}(0)t + O(t^2)\|_2 \\ &\le \|(W_A \operatorname{vec}(E) + W_b e_b + W_\lambda e_\lambda)t\|_2 + O(t^2) \\ &\le \|W_A\|_2 \|\operatorname{vec}(E)t\|_2 + \|W_b\|_2 \|e_b t\|_2 + \|W_\lambda\|_2 \|e_\lambda t\|_2 + O(t^2) \\ &= \|W_A\|_2 \|Et\|_F + \|W_b\|_2 \|e_b t\|_2 + \|W_\lambda\|_2 \|e_\lambda t\|_2 + O(t^2). \end{aligned}$$

Thus, we have proved the following theorem.

Theorem 3.4.1 *Let a controllable system (A, b) and a set $\mathcal{L}$ of self-conjugate complex numbers $\lambda_1, \ldots, \lambda_n$, where $\lambda_i \neq \lambda_j, i \neq j$ be given. Suppose that (A, b) is slightly perturbed to a controllable system $(\widetilde{A}, \widetilde{b})$, and $\mathcal{L}$ is slightly perturbed to a set $\widetilde{\mathcal{L}}$ of self-conjugate complex numbers $\widetilde{\lambda}_1, \widetilde{\lambda}_2, \ldots, \widetilde{\lambda}_n$. Let $f, \widetilde{f} \in \mathbf{R}^n$ be the* *solutions to Problem PA with the data $A, b, \mathcal{L}$ and $\widetilde{A}, \widetilde{b}, \widetilde{\mathcal{L}}$, respectively. Then*

$$\begin{aligned}
\frac{\|\widetilde{f}-f\|_2}{\|f\|_2} &\le \frac{1}{\|f\|_2}\left(\|W_A \operatorname{vec}(\widetilde{A}-A) + W_b(\widetilde{b}-b) + W_\lambda(\widetilde{\lambda}-\lambda)\|_2\right) \\
&\quad + O\left(\|\operatorname{vec}(\widetilde{A}-A)\|_2^2 + \|\widetilde{b}-b\|_2^2 + \|\widetilde{\lambda}-\lambda\|_2^2\right) \\
&\le \kappa_A(f)\frac{\|\widetilde{A}-A\|_F}{\|A\|_F} + \kappa_b(f)\frac{\|\widetilde{b}-b\|_2}{\|b\|_2} + \kappa_\lambda(f)\frac{\|\widetilde{\lambda}-\lambda\|_2}{\|\lambda\|_2} \\
&\quad + O\left(\frac{\|\widetilde{A}-A\|_F^2}{\|A\|_F^2} + \frac{\|\widetilde{b}-b\|_2^2}{\|b\|_2^2} + \frac{\|\widetilde{\lambda}-\lambda\|_2^2}{\|\lambda\|_2^2}\right),
\end{aligned} \tag{3.4.13}$$

where $\lambda = (\lambda_1, \ldots, \lambda_n)^T$, $\widetilde{\lambda} = (\widetilde{\lambda}_1, \ldots, \widetilde{\lambda}_n)^T$, W_A, W_b, W_λ are defined in (3.4.12), *and $\kappa_a(f), \kappa_b(f)$, and $\kappa_\lambda(f)$ are defined by*

$$\kappa_A(f) = \frac{\|W_A\|_2\,\|A\|_F}{\|f\|_2}, \quad \kappa_b(f) = \frac{\|W_b\|_2\,\|b\|_2}{\|f\|_2}, \quad \kappa_\lambda(f) = \frac{\|W_\lambda\|_2\,\|\lambda\|_2}{\|f\|_2}. \tag{3.4.14}$$

Based on Theorem 3.4.1, we may refer to the scalars $\kappa_A(f)$, $\kappa_b(f)$, $\kappa_\lambda(f)$ defined in (3.4.14) as the **condition numbers** of the solution f with respect to A, b, and λ, respectively. That is, $\kappa_A(f)$, $\kappa_b(f)$, and $\kappa_\lambda(f)$ can be used to measure the sensitivity of the solution f to small changes in A, b, and $\mathcal{L}$, respectively. Moreover, the scalar $\kappa(f)$ defined by

$$\kappa(f) = \sqrt{\kappa_A(f)^2 + \kappa_b(f)^2 + \kappa_\lambda(f)^2}$$

can be called the condition number of f. If $\kappa(f)$ is large, then we will call the corresponding Problem PA **ill-conditioned.**

It is worthwhile to point out that the condition numbers $\kappa_A(f), \kappa_b(f), \kappa_\lambda(f)$ are independent of the lengths of the right eigenvectors $x_1, \ldots, x_n$ of $A + bf^T$. Therefore, if we take x_j with $\|x_j\|_2 = 1$ for all j, then from (3.4.12) and (3.4.14) we can easily derive the following upper bounds on the condition numbers:

$$\begin{aligned}
\kappa_A(f) &\le \|YD^{-1}\|_2\,\|Y\|_2\,\|A\|_F\,/\|f\|_2\,, \\
\kappa_b(f) &\le \|YD^{-1}\|_2\,\|Y\|_2\,\|b\|_2, \\
\kappa_\lambda(f) &= \|YD^{-1}\|_2\,\|\lambda\|_2\,/\|f\|_2\,.
\end{aligned} \tag{3.4.15}$$

3.4.2 The Multi-input Case

We now extend the foregoing idea to the multi-input case. Given $A \in \mathbf{R}^{n\times n}$ and $B \in \mathbf{R}^{n\times m} (m > 1)$ with (A, B) controllable, and given a set of n distinct complex numbers, $\mathcal{L} = \{\lambda_1, \ldots, \lambda_n\}$, closed under complex conjugation. Then by Theorem 3.3.4, a solution to the corresponding Problem PA certainly exists, and in general the number of the solutions is infinite. Now, assume that $\widetilde{F} \in \mathbf{R}^{m\times n}$ is any given solution, written $\widetilde{F}^T = [\widetilde{f}_1, \dots, \widetilde{f}_m]$, and assume that $x_j \in \mathbf{C}^n$ is a right eigenvector of $A + B\widetilde{F}$ associated with λ_j, i.e.,

$$(A + B\widetilde{F})x_j = \lambda_j x_j, \qquad j = 1, 2, \dots, n.$$

Let $X = [x_1, \dots, x_n]$. Then X is nonsingular. As before, let

$$Y = X^{-T} = [y_1, \dots, y_n].$$

Then it follows that

$$y_j^T(A + B\widetilde{F}) = \lambda_j y_j^T \quad \text{and} \quad y_j^T x_j = 1$$

for $j = 1, 2, \dots, n$.

Let $E_a \in \mathbf{R}^{n\times n}$, $E_b \in \mathbf{R}^{n\times m}$, $e_\lambda = (\varepsilon_1, \dots, \varepsilon_n)^T \in \mathbf{C}^n$. Again define

$$A(t) = A + tE_a, \quad B(t) = B + tE_b,$$
$$\lambda_j(t) = \lambda_j + t\varepsilon_j, \quad j = 1, 2, \dots, n, \quad t \in [-1, 1].$$

As before, we assume that E_a and E_b are sufficiently small in magnitude such that $\big(A(t), B(t)\big)$ are still controllable for every $t \in [-1, 1]$, and assume that the ε_j are small enough so that $\lambda_i(t) \neq \lambda_j(t), i \neq j$, and the set $\{\lambda_1(t), \dots, \lambda_n(t)\}$ is closed under complex conjugation for any $t \in [-1, 1]$. Moreover, we suppose once again that $\lambda_j \in \mathbf{R}$, $j = 1, \dots, r$, and $\lambda_{r+2k+1} = \bar{\lambda}_{r+2k} = \nu_k + i\mu_k$ with ν_k, $\mu_k \in \mathbf{R}$ and $\mu_k \neq 0$, $k = 1, 2, \dots, s$, where $r + 2s = n$.

In a manner similar to the single-input case, we define $\varphi(t;\, z) = (\varphi_1(t;\, z), \dots, \varphi_n(t;\, z))^T \in \mathbf{R}^n$ by

$$\varphi_j(t;\, z) = \begin{cases} \lambda_j(t;\, z) - \lambda_j(t), & 1 \leq j \leq r, \\ \nu_k(t;\, z) - \mathrm{Re}\,(\lambda_j(t)), & j = r + 2k - 1,\ 1 \leq k \leq s, \\ \mu_k(t;\, z) + \mathrm{Im}\,(\lambda_j(t)), & j = r + 2k,\ 1 \leq k \leq s, \end{cases}$$

where $\lambda_j(t;\, z)(j = 1, \dots, r)$ and $\nu_k(t;\, z) \pm i\mu_k(t;\, z)$ $(k = 1, \dots, s)$ are the eigenvalues of $A(t) + B(t)Z$, $Z \in \mathbf{R}^{m\times n}$, and $z = \mathrm{vec}\,(Z)$. Moreover, by Theorems 1.9.2 and 1.9.5, we may assume that the $\lambda_j(t;\, z)$, $\nu_k(t;\, z)$, and $\mu_k(t;\, z)$ are real analytic in some neighborhood of $(0;\, \widetilde{z}_0)$, and satisfy

$$\lambda_j(0, \widetilde{z}_0) = \lambda_j, \quad \nu_k(0, \widetilde{z}_0) = \nu_k, \quad \mu_k(0, \widetilde{z}_0) = \mu_k,$$

where $\widetilde{z}_0 = \mathrm{vec}\,(\widetilde{F})$. Thus, $\varphi(t,\, z)$ is real analytic and $\varphi(0;\, z_0) = 0$. Similar to the argument of (3.4.5), we can get

$$\varphi_z'(0;\, z_0) = DV_f$$

and

$$\varphi_t'(0;\, z_0) = -D\big(V_a \mathrm{vec}\,(E_a) + V_b \mathrm{vec}\,(E_b) + e_\lambda\big),$$

where

$$D = \mathrm{diag}\,(I_r, D_1, \dots, D_s), \quad D_k = \frac{1}{2}\begin{bmatrix} 1 & 1 \\ -i & i \end{bmatrix}, \quad k = 1, 2, \dots, s,$$

$$V_f = \begin{bmatrix} x_1^T \otimes (y_1^T B) \\ x_2^T \otimes (y_2^T B) \\ \cdots\cdots \\ x_n^T \otimes (y_n^T B) \end{bmatrix}, \quad V_a = -\begin{bmatrix} x_1^T \otimes y_1^T \\ x_2^T \otimes y_2^T \\ \cdots\cdots \\ x_n^T \otimes y_n^T \end{bmatrix}, \quad V_b = -\begin{bmatrix} (x_1^T \widetilde{F}^T) \otimes y_1^T \\ (x_2^T \widetilde{F}^T) \otimes y_2^T \\ \cdots\cdots \\ (x_n^T \widetilde{F}^T) \otimes y_n^T \end{bmatrix}. \tag{3.4.16}$$

It is easy to verify that

$$\begin{aligned} V_f &= [D_1(X)Y^T B, D_2(X)Y^T B, \dots, D_n(X)Y^T B], \\ V_a &= -[D_1(X)Y^T, D_2(X)Y^T, \dots, D_n(X)Y^T], \\ V_b &= -[T_1 Y^T, T_2 Y^T, \dots, T_m Y^T], \end{aligned} \tag{3.4.17}$$

where

$$D_i(X) = \mathrm{diag}\,(x_{i1}, x_{i2}, \dots, x_{in}), \qquad i = 1, 2, \dots, n,$$

$$T_j = \mathrm{diag}\,(x_1^T \widetilde{f}_j, x_2^T \widetilde{f}_j, \dots, x_n^T \widetilde{f}_j), \qquad j = 1, 2, \dots, m.$$

Let $\widetilde{B} = [\beta_{ij}] = [b_1, \dots, b_m] = Y^T B$. By rearranging the columns of V_f we can find an $nm \times nm$ permutation matrix P such that

$$\begin{aligned} V_f P &= \big[D_1(x)b_1, \dots, D_n(x)b_1, \dots, D_1(x)b_m, \dots, D_n(x)b_m\big] \\ &= [S_1, S_2, \dots, S_m]\mathrm{diag}\,(X^T, \dots, X^T) \end{aligned}$$

where

$$S_j = \mathrm{diag}\,(\beta_{1j}, \dots, \beta_{nj}), \quad j = 1, 2, \dots, m.$$

By Corollaries 3.2.2 and 3.2.4, we know that

$$(\Lambda,\, \widetilde{B}) = \big(Y^T(A + B\widetilde{F})X,\; Y^T B\big)$$

is controllable, where $\Lambda = \mathrm{diag}\,(\lambda_1, \dots, \lambda_n)$. Hence, by Theorem 3.2.1, it follows that $e_i^T \widetilde{B} \neq 0$ for all i. This implies that

$$\mathrm{rank}\,[S_1, S_2, \dots, S_m] = n,$$

and so

$$\operatorname{rank}\varphi'_z(0;\, z_0) = \operatorname{rank} V_f = \operatorname{rank}\,[S_1, S_2, \dots, S_m] = n,$$

since D, P, and diag$(X^T, \dots, X^T)$ are nonsingular.

Consequently, by Theorem 1.8.3, it follows that for any $W \in \mathbf{R}^{m\times n}$ satisfying

$$V_f \operatorname{vec}(W) = V_a \operatorname{vec}, (E_a) + V_b \operatorname{vec}(E_b) + e_\lambda, \tag{3.4.18}$$

there exists a sufficient small $\varepsilon < 0$ and an analytic matrix-valued function $\widetilde{F}(t) \in \mathbf{R}^{m\times n}$ with $\dot{\widetilde{F}}(0) = W$ and $\widetilde{F}(0) = \widetilde{F}$ such that

$$\lambda\big(A(t) + B(t)\widetilde{F}(t)\big) = \{\lambda_1(t), \dots, \lambda_n(t)\}, \qquad t \in (-\varepsilon,\, \varepsilon), \tag{3.4.19}$$

i.e., $\widetilde{F}(t)$ is a solution to Problem RA with data $A(t)$, $B(t)$, and $\{\lambda_1(t), \dots, \lambda_n(t)\}$ for every $t \in (-\varepsilon, \varepsilon)$.

Notice that (3.4.18) has an infinity of real solutions. Therefore, for a given t sufficient small we must choose one $\widetilde{F}_0(t)$ from those $\widetilde{F}(t)$ as the solution to Problem PA with the data $A(t), B(t)$, and $\{\lambda_1(t), \dots, \lambda_n(t)\}$ corresponding to $\widetilde{F}$. Naturally, we should require such $\widetilde{F}_0(t)$ to be closest to $\widetilde{F}$ in some sense. Recall from linear algebra that the underdetermined system

$$V_f w = V_a \operatorname{vec}(E_a) + V_b \operatorname{vec}(E_b) + e_\lambda$$

has a unique minimal 2-norm solution:

$$w = V_f^\dagger\big(V_a \operatorname{vec}(E_a) + V_b \operatorname{vec}(E_b) + e_\lambda\big).$$

Note the w is certainly an mn-dimensional real vector, though it is entirely possible that V_f, V_a, V_b, and e_λ are complex. Consequently, it is natural to choose $\widetilde{F}_0(t)$ satisfying that vec $\big(\dot{\widetilde{F}}_0(0)\big) = w$. In this case, we have

$$\begin{aligned}
\|\widetilde{F}_0(t) - \widetilde{F}\|_F &\le \|\dot{\widetilde{F}}_0(0)t\|_F + O(t^2) \\
&= \|\operatorname{vec}(\dot{\widetilde{F}}_0(0))\|_2\, |t| + O(t^2) \\
&\le \|V_f^\dagger V_a\|_2\, \|\operatorname{vec}(E_a)t\|_2 + \|V_f^\dagger V_b\|_2\, \|\operatorname{vec}(E_b)t\|_2 \\
&\quad + \|V_f^\dagger\|_2\, \|e_\lambda t\|_2 + O(t^2) \\
&= \|V_f^\dagger V_a\|_2\, \|tE_a\|_F + \|V_f^\dagger V_b\|_2\, \|tE_b\|_F + \|V_f^\dagger\|_2\, \|te_\lambda\|_2 + O(t^2).
\end{aligned}$$

Thus, we have proved the following theorem.

Theorem 3.4.2 *Let $A \in \mathbf{R}^{n\times n}$ and $B \in \mathbf{R}^{n\times m}$ with (A, B) controllable be given, and let a set $\mathcal{L}$ of n self-conjugate complex numbers $\lambda_1, \dots, \lambda_n$ be given, where $\lambda_i \neq \lambda_j, i \neq j$. Let $\widehat{A} \in \mathbf{R}^{n\times n}$ and $\widehat{B} \in \mathbf{R}^{n\times m}$ with $(\widehat{A}, \widehat{B})$ controllable, and let $\widehat{\mathcal{L}} = \{\widehat{\lambda}_1, \dots, \widehat{\lambda}_n\}$ be a set of self-conjugate complex numbers. If*

$$\delta \equiv \left(\|\widehat{A} - A\|_F^2 + \|\widehat{B} - B\|_F^2 + \sum_{j=1}^{n} |\widehat{\lambda}_j - \lambda_j|^2\right)^{\frac{1}{2}}$$

is sufficiently small, then for every given solution $\widetilde{F}$ to Problem PA with data A, B, and $\mathcal{L}$ there is a solution $\widehat{F}$ to Problem PA with data $\widetilde{A}$, $\widetilde{B}$, and $\widetilde{\mathcal{L}}$ such that

$$\begin{aligned}\frac{\|\widetilde{F} - \widehat{F}\|_F}{\|\widetilde{F}\|_F} &\leq \frac{1}{\|\widetilde{F}\|_F}\|V_f^\dagger\left(V_a \mathrm{vec}\,(\widehat{A} - A) + V_b \mathrm{vec}\,(\widehat{B} - B) + e_\lambda\right)\|_2 + O(\delta^2) \\ &\leq \kappa_a(\widetilde{F})\frac{\|\widehat{A} - A\|_F}{\|A\|_F} + \kappa_b(\widetilde{F})\frac{\|\widehat{B} - B\|_F}{\|B\|_F} + \kappa_\lambda(\widetilde{F})\frac{\|e_\lambda\|_2}{\|\lambda\|_2} + O(\delta^2),\end{aligned} \tag{3.4.20}$$

where V_f, V_a, V_b are defined as in (3.4.16), $\lambda = (\lambda_1, \dots, \lambda_n)$, $e_\lambda = (\widehat{\lambda}_1 - \lambda_1, \dots, \widehat{\lambda}_n - \lambda_n)^T$, and

$$\kappa_a(\widetilde{F}) = \|V_f^\dagger V_a\|_2 \frac{\|A\|_F}{\|\widetilde{F}\|_F}, \quad \kappa_b(\widetilde{F}) = \|V_f^\dagger V_b\|_2 \frac{\|B\|_F}{\|\widetilde{F}\|_F}, \quad \kappa_\lambda(\widetilde{F}) = \|V_f^\dagger\|_2 \frac{\|\lambda\|_2}{\|\widetilde{F}\|_F}. \tag{3.4.21}$$

In view of Theorem 3.4.2, the scalars $\kappa_a(\widetilde{F}), \kappa_b(\widetilde{F}), \kappa_\lambda(\widetilde{F})$ defined by (3.4.21) are referred to as the **condition numbers** of the solution $\widetilde{F}$ to Problem PA with data A, B and $\mathcal{L}$. Moreover, the scalar $\kappa(\widetilde{F})$ defined by

$$\kappa(F) = \sqrt{\kappa_a(\widetilde{F})^2 + \kappa_b(\widetilde{F})^2 + \kappa_\lambda(\widetilde{F})^2}$$

can be regarded as the condition number of $\widetilde{F}$.

From (3.4.21) and (3.4.16) it is easy to see that the condition numbers $\kappa_a(\widetilde{F})$, $\kappa_b(\widetilde{F})$, $\kappa_\lambda(\widetilde{F})$ are independent of the lengths of the right eigenvectors $x_1, \dots, x_n$ of $A + B\widetilde{F}^T$. As have done in preceding subsection, if we take x_j with $\|x_j\|_2 = 1$ for all j, then it is easy to derive the following upper bounds on the condition numbers:

$$\begin{aligned} \kappa_a(\widetilde{F}) &\le \frac{\|A\|_F}{\|\widetilde{F}\|_F}\|V_f^\dagger\|_2\,\|Y\|_2, \\ \kappa_b(\widetilde{F}) &\le \frac{\|B\|_F}{\|\widetilde{F}\|_F}\max_{1\le j\le m}\|\widetilde{f}_j\|_2\,\|V_f^\dagger\|_2\,\|Y\|_2, \\ \kappa_\lambda(\widetilde{F}) &= \frac{\|\lambda\|_2}{\|\widetilde{F}\|_F}\|V_f^\dagger\|_2. \end{aligned} \tag{3.4.22}$$

Finally, it is worth pointing out that the conditioning of the solution to Problem PA and the conditioning of the resulting closed-loop eigenvalues are two entirely different things.

Let $F \in \mathbf{R}^{m\times n}$ be a solution to Problem PA with data A, B, and $\mathcal{L}$, and let the resulting closed loop matrix $A + BF$ have the spectral decomposition

$$X^{-1}(A + BF)X = \operatorname{diag}(\lambda_1, \dots, \lambda_n),$$

where $X = [x_1, \dots, x_n]$ and $X^{-T} = [y_1, \dots, y_n]$. It is well-known that the scalars $c(\lambda_j)$ defined by

$$c(\lambda_j) = \|x_j\|_2\,\|y_j\|_2$$

are the absolute condition number of the eigenvalues λ_j. For the details see [221]. Moreover, if $\lambda_j \ne 0$ the scalar $c^{(r)}(\lambda_j)$ defined by

$$c^{(r)}(\lambda_j) = c(\lambda_j)\frac{\|A + BF\|_2}{|\lambda_j|},$$

is referred to as the relative condition number of λ_j. If $\lambda_j \ne 0$ for all j, then the scalar $c^{(r)}(\lambda)$ defined by

$$c^{(r)}(\lambda) = \sqrt{c^{(r)}(\lambda_1)^2 + \cdots + c^{(r)}(\lambda_n)^2}$$

can be regarded as the relative condition number of the resulting closed-loop eigenvalues $\lambda_1, \dots, \lambda_n$. See [81] for further details.

The following example illustrates that there is no relation between the conditioning of the solution to Problem PA and the conditioning of the resulting closed-loop eigenvalues.

Example 3.4.1 Let

$$A = \operatorname{diag}(0.1, 0.2, 0.3, 0.4, 0.5, 0.6), \quad b = (1, 2, 3, 4, 5, 6)^T.$$

Suppose that we wish to assign the poles $\lambda_1 = -6$, $\lambda_2 = -5$, $\lambda_3 = -4$, $\lambda_4 = -3$, $\lambda_5 = -1.1$, $\lambda_6 = -1$. For this given single-input pole assignment problem, Sun

[209] has found that

$$\kappa_a(f) = 2.0439 \times 10, \qquad \kappa_b(f) = 1.9456,$$
$$\kappa_\lambda(f) = 1.0535 \times 10, \qquad \kappa(f) = 2.3077 \times 10,$$

and

$$c^{(r)}(\lambda_1) = 2.3956 \times 10^{15}, \qquad c^{(r)}(\lambda_2) = 6.6419 \times 10^{15},$$
$$c^{(r)}(\lambda_3) = 6.4221 \times 10^{15}, \qquad c^{(r)}(\lambda_4) = 2.3411 \times 10^{15},$$
$$c^{(r)}(\lambda_5) = 4.5093 \times 10^{14}, \qquad c^{(r)}(\lambda_6) = 3.2512 \times 10^{14},$$
$$c^{(r)}(\lambda_7) = 9.8431 \times 10^{15}.$$

This shows that the pole assignment problem is well-conditioned but the resulting closed-loop eigenvalues are extremely ill-conditioned.

3.5 Schur Method

Having discussed the problem of the existence of the solution to Problem PA, and having done an analysis of the sensitivity of the solution, we now turn our attention to the problem of how a solution to Problem PA can be computed numerically. Given $A \in \mathbf{R}^{n\times n}$, $B \in \mathbf{R}^{n\times m}$ with (A, B) controllable, and given a set of n complex numbers, $\mathcal{L} = \{\lambda_1, \lambda_2, \dots, \lambda_n\}$, closed under complex conjugation and ordered so that if $\mathrm{Im}\lambda_{2k-1} \neq 0$, then $\lambda_{2k} = \bar{\lambda}_{2k-1}$ for $k = 1, 2, \dots, [\frac{n}{2}]$, where $[\frac{n}{2}]$ is the greatest integer less than or equal to $\frac{n}{2}$, we wish to compute a matrix $F \in \mathbf{C}^{m\times n}$ such that $\lambda(A + BF) = \mathcal{L}$, i.e., the eigenvalues of $A + BF$ are $\lambda_1, \dots, \lambda_n$.

Recall from the previous sections that for any nonsingular matrices $T \in \mathbf{R}^{n\times n}$, $G \in \mathbf{R}^{m\times m}$ the system $(T^{-1}AT, T^{-1}BG)$ is still controllable; moreover, $F \in \mathbf{R}^{m\times n}$ is a solution to Problem PA with data A, B, and $\mathcal{L}$ if and only if there exists a solution F_1 to Problem PA with data $T^{-1}AT$, $T^{-1}BG$, and $\mathcal{L}$ such that $F = GF_1T^{-1}$. These facts suggest that in order to compute a matrix $F \in \mathbf{R}^{m\times n}$ such that $\lambda(A + BF) = \mathcal{L}$ we should first seek two suitable matrices T and G which are nonsingular so that to find a matrix F_1 satisfying $\lambda(A_1 + B_1F_1) = \mathcal{L}$ is easier, where $A_1 = T^{-1}AT$, $B_1 = T^{-1}BG$. This is particularly important for us to develop numerical methods for solving Problem PA. Therefore, we introduce the following concept.

Two systems $(A,\ B)$ and (A_1, B_1) are said to be **equivalent** if there are two nonsingular matrices $T \in \mathbf{R}^{n\times n}$, $G \in \mathbf{R}^{m\times m}$ such that

$$A_1 = T^{-1}AT \quad \text{and} \quad B_1 = T^{-1}BG. \tag{3.5.1}$$

The pair of equations in (3.5.1) is called an **equivalence transformation**, and T and G are called the **transforming matrices**. In particular, if the transforming matrices T and G are orthogonal, we call that (A, B) and (A_1, B_1) are **orthogonally equivalent**.

From the numerical point of view it is preferable to reduce (A, B) to (A_1, B_1) by orthogonal equivalence transformations since it may be carried out by elementary orthogonal transformations, proved to be numerically stable.

In the present section we will develop an algorithm based on the real Schur canonical form of A. This method is first to reduce A to real Schur canonical form by orthogonal similarity transformations, that is, to compute an orthogonal matrix $Q_0 \in \mathbf{R}^{n\times n}$ such that

$$A_0 = Q_0 A Q_0^T = \begin{bmatrix} A_{11}^{(0)} & A_{12}^{(0)} & \cdots & A_{1k}^{(0)} \\ & A_{22}^{(0)} & \cdots & A_{2k}^{(0)} \\ & & \ddots & \vdots \\ & & & A_{kk}^{(0)} \end{bmatrix} \tag{3.5.2}$$

is quasi-upper triangular, having only 1×1 or 2×2 blocks on the diagonal, corresponding to real or to complex conjugate eigenvalues, respectively. It is well known that (3.5.2) can be accomplished efficiently using the QR algorithm.

To complete the equivalence transformation, we must multiply B on the left by Q_0, i.e., $B_0 = Q_0 B$.

Thus, once an $m \times n$ matrix F_0 has been found such that $\lambda(A_0 + B_0 F_0) = \mathcal{L}$, we can easily obtain a desired matrix F by calculating $F = F_0 Q_0$. Therefore, in the what follows we will devote to computing F_0 such that $\lambda(A_0 + B_0 F_0) = \mathcal{L}$. To do this we need two preliminary algorithms.

3.5.1 Preliminary Algorithms

First, consider Problem PA with $n = 2$, Let $M \in \mathbf{R}^{2\times 2}$, $N \in \mathbf{R}^{2\times m}$ with (M, N) controllable, and let $\mu_1 = \zeta_1 + i\eta$, $\mu_2 = \zeta_2 - i\eta$ be two complex numbers, with $\eta \neq 0$ implies $\zeta_1 = \zeta_2$. Our goal here is to develop an algorithm that produces an $m \times 2$ matrix K such that $\lambda(M + NK) = \{\mu_1, \mu_2\}$.

From elementary linear algebra we know that the eigenvalues of $M + NK$ are μ_1, μ_2 if and only if

$$\begin{cases} \operatorname{tr}(M + NK) = \mu_1 + \mu_2, \\ \det(M + NK) = \mu_1 \mu_2. \end{cases} \tag{3.5.3}$$

This means that we can find the desired K by solving the system (3.5.3). Note the system (3.5.3) is a system of two equations in $2m$ unknowns and the assumption that (M, N) is controllable ensures that (3.5.3) certainly has solution. It is clear that the system (3.5.3) has infinite many solutions if $m \geq 2$. This fact allows us to make various choice in determining K when $m \geq 2$.

Recall that

$$\mathrm{tr}\,(UMU^T + UNVV^T KU^T) = \mathrm{tr}\,(M + NK)$$

and

$$\det(UNVV^T KU^T) = \det(M + NK)$$

for any orthogonal matrices $U \in \mathbf{R}^{2\times 2}$ and $V \in \mathbf{R}^{m\times m}$. This fact allows us to simplify (3.5.3) by suitable choice of orthogonal matrices U and V. One way of simplifying (3.5.3) is to compute orthogonal matrices $U \in \mathbf{R}^{2\times 2}$, and $V \in \mathbf{R}^{2\times m}$ such that

$$UNV = [N_1, 0], \tag{3.5.4}$$

where

$$N_1 = \begin{cases} \begin{bmatrix} \beta_{11} \\ 0 \end{bmatrix}, & \text{if } r = 1, \\[2ex] \begin{bmatrix} \beta_{11} & \beta_{12} \\ 0 & \beta_{22} \end{bmatrix}, & \text{if } r = 2, \end{cases}$$

with $r = \mathrm{rank}\, N$. Note that if $r = m$, the null matrix in (3.5.4) does not appear. The decomposition (3.5.4) is easily obtained by applying Housebolder and Givens transformations. The details are left as exercise for the reader. Let

$$M_1 = UMU^T = \begin{bmatrix} \alpha_{11} & \alpha_{12} \\ \alpha_{21} & \alpha_{22} \end{bmatrix}, \qquad V^T KU^T = \begin{bmatrix} K_{11} \\ K_{21} \end{bmatrix},$$

where $K_{11} \in \mathbf{R}^{r\times 2}$, $K_{21} \in \mathbf{R}^{(m-r)\times 2}$. Then we have

$$\begin{aligned} U(M + NK)U^T &= UMU^T + UNVV^T KU^T \\ &= M_1 + [N_1,\, 0] \begin{bmatrix} K_{11} \\ K_{21} \end{bmatrix} \\ &= M_1 + N_1 K_{11}. \end{aligned}$$

Note that K_{21} does not affect the eigenvalues of $M + NK$, thus it can be chosen arbitrarily if $m > 1$. Here we set $K_{21} = 0$, as it decreases computation and the resulting rounding errors. It also decreases the norm of K, which can be an indicator of the amount of work the controller must do to provide the input $Kx(t)$.

To compute K_{11}, we transform (3.5.3) into

$$\begin{cases} \operatorname{tr}(M_1 + N_1 K_{11}) = \mu_1 + \mu_2, \\ \det(M_1 + N_1 K_{11}) = \mu_1 \mu_2. \end{cases} \tag{3.5.5}$$

If $r = 1$, then it follows from (3.5.5) that

$$\begin{cases} k_{11} = \beta_{11}^{-1}(\mu_1 + \mu_2 - \alpha_{11} - \alpha_{22}), \\ k_{12} = \dfrac{\alpha_{22}}{\alpha_{21}} k_{11} + \dfrac{1}{\alpha_{21}\beta_{11}}(\alpha_{11}\alpha_{22} - \alpha_{12}\alpha_{21} - \mu_1\mu_2), \end{cases}$$

where $K_{11} = [k_{11}, k_{12}]$, as the condition that (M, N) is controllable implies that $\alpha_{21}\beta_{11} \neq 0$. If $r = 2$, the system (3.5.5) is essentially underdetermined because it has two equations in 4 unknowns. To avoid taking square root, in this case we simply take K_{11} as

$$K_{11} = N_1^{-1}\left(\begin{bmatrix} \zeta_1 & \eta \\ -\eta & \zeta_2 \end{bmatrix} - M_1\right),$$

which is a solution to the system (3.5.5).

Thus we may give the following algorithm.

Algorithm 3.5.1 Given a 2×2 real matrix M, a $2 \times m$ real matrix N, and two complex numbers μ_1, μ_2, with $\overline{\mu}_1 = \mu_2$ if $\operatorname{Im}(\mu_1) \neq 0$, this function returns an $m \times 2$ matrix K such that the eigenvalues of $M + NK$ are μ_1 and μ_2 if (M, N) is controllable, and returns an empty matrix otherwise.

```
function K = pa2(M, N, μ1, μ2)
    m = size(N, 2);
    if m = 1
        [c, s] = givens(N(1), N(2));
        α = N(1)c + N(2)s; U = [ c  s ; -s  c ]; V = 1; r = 1;
    else
        [V, L, P] = qr(N^T);    (QR factorization with column pivoting)
        R = PL(1 : 2, :)^T; [c, s] = givens(R(1, 1), R(2, 1));
        U = [ c  s ; -s  c ]; R = UR;
        if R(2, 2) = 0
            r = 1; α = R(1, 1);
        else
            r = 2;
        end
    end
```

$M = UMU^T$;
if $r = 1$ and $\alpha M(2,1) = 0$
  $K = [\ ]$; (System uncontrollable)
else
  $K = \mathbf{zeros}(m, 2)$;
  if $r = 1$
    $K(1,1) = \left(\mu_1 + \mu_2 - M(1,1) - M(2,2)\right)/\alpha$;
    $K(1,2) = M(2,2)K(1,1)/M(2,1) + \left(\mathbf{det}(M) - \mu_1\mu_2\right)/(\alpha M(2,1))$;
  else
    $K(1:2,:) = R^{-1}\left(\begin{bmatrix} \mathbf{Re}\mu_1 & \mathbf{Im}\mu_1 \\ -\mathbf{Im}\mu_1 & \mathbf{Re}\mu_2 \end{bmatrix} - M\right)$;
  end
  $K = VKU$;
end

The Schur algorithm to be introduced needs to move one diagonal block of a real Schur canonical form matrix to another desired position by using orthogonal similarity transformations. This can be carried out by interchanging consecutive two diagonal blocks of the real Schur canonical form matrix successively, using a sequence of orthogonal similarity transformations. Therefore, our next task is to develop an algorithm that interchange two diagonal blocks of a matrix of the form

$$R = \begin{bmatrix} R_{11} & R_{12} \\ 0 & R_{22} \end{bmatrix},$$

where $R_{11} \in \mathbf{R}^{p\times p}$, $R_{22} \in \mathbf{R}^{q\times q}$ with $\lambda(R_{11}) \cap \lambda(R_{22}) = \phi$, p, $q = 1$ or 2. More precisely, we wish to develop an algorithm that computes an orthogonal matrix $U \in \mathbf{R}^{(p+q)\times(p+q)}$ such that

$$URU^T = \begin{bmatrix} \widetilde{R}_{11} & \widetilde{R}_{12} \\ 0 & \widetilde{R}_{22} \end{bmatrix}, \tag{3.5.6}$$

where $\widetilde{R}_{11} \in \mathbf{R}^{q\times q}$, $\widetilde{R}_{22} \in \mathbf{R}^{p\times p}$, with $\lambda(\widetilde{R}_{11}) = \lambda(R_{22})$ and $\lambda(\widetilde{R}_{22}) = \lambda(R_{11})$.

Suppose $X \in \mathbf{R}^{(p+q)\times q}$, with $\operatorname{rank}(X) = q$, satisfies

$$RX = XR_{22}$$

and assume that the QR decomposition of X is

$$X = U^T \begin{bmatrix} X_1 \\ 0 \end{bmatrix},$$

where $X_1 \in \mathbf{R}^{q\times q}$ is upper triangular, $U \in \mathbf{R}^{(p+q)\times(p+q)}$ orthogonal. Then it is not hard to verify that such an orthogonal matrix U satisfies the requirements of (3.5.6). This fact gives us a recipe for calculating U. The matrix X can be computed as follows. First, solve

$$R_{11}Y - YR_{22} = -R_{12} \tag{3.5.7}$$

for $Y \in \mathbf{R}^{p\times q}$. Note that the assumption that $\lambda(R_{11}) \cap \lambda(R_{22}) = \emptyset$ guarantees that (3.5.7) has a unique solution, and (3.5.7) is easy to be solved because it is a linear system with pq unknowns (at most 4 unknowns). Then, define

$$X = \begin{bmatrix} Y \\ I_q \end{bmatrix} \in \mathbf{R}^{(p+q)\times q},$$

where I_q denotes the identity matrix of order q. It is clear that such a matrix X satisfies that $\text{rank}(X) = q$ and $RX = XR_{22}$. Thus we have the following algorithm to compute the desired transforming matrix U.

Algorithm 3.5.2 Given matrices R, S, T, whose sizes are $p\times p$, $q\times q$ and $p\times q$, respectively, with $p, q = 1$ or 2, this function computes a $(p+q)\times(p+q)$ orthogonal matrix U so that

$$U\begin{bmatrix} R & T \\ 0 & S \end{bmatrix}U^T = \begin{bmatrix} \widetilde{S} & \widetilde{T} \\ 0 & \widetilde{R} \end{bmatrix},$$

where $\widetilde{S} \in \mathbf{R}^{q\times q}$, $\widetilde{R} \in \mathbf{R}^{p\times p}$, with $\lambda(S) = \lambda(\widetilde{S})$ and $\lambda(R) = \lambda(\widetilde{R})$, if the eigenvalues of S and R are different each other, and returns an empty matrix otherwise.

function $U = \textbf{trans}(R, S, T)$
    $p = \textbf{size}(R, 1);\ q = \textbf{size}(S, 1);$
    $K = I \otimes R - S^T \otimes I;$ (K is a $pq \times pq$ matrix)
    $[Q, L, P] = \textbf{qr}(K);$ (QR factorization with column pivoting)
    if $L(pq, pq) = 0$
        $U = [\ ];$ (S and R have a common eigenvalue)
    else
        $Y = -PL^{-1}Q^T\textbf{vec}(T);$
        if $q = 2$
            $X = \textbf{zeros}(p, 2);$
            $X(:, 1) = Y(1:p);\ X(:, 2) = Y(p+1:2p);\ Y = X;$
        end
        $X = \begin{bmatrix} Y \\ I \end{bmatrix};$ (I is the identity matrix of order q)
        $[U, L] = \textbf{qr}(X);\ U = U^T;$
    end

Applying Algorithm 3.5.2, we can obtain the following algorithm, interchanging consecutive two diagonal blocks of a real Schur canonical form matrix.

Algorithm 3.5.3 Given an $n \times n$ real Schur canonical form matrix T and the associated transforming matrix Q and input matrix B, and given a 3-dimensional integer vector $S = [s, p, q]$ with $p, q = 1$ or 2, this function interchanges consecutive two diagonal blocks of T, beginning at position (s, s), having sizes $p \times p$ and $q \times q$, respectively, and updates B and Q.

```
function [T, B, Q] = interch(T, B, Q, S)
    s = S(1);
    p = S(2);
    q = S(3);
    t = s + q + p - 1;
    n = size(T, 1);
    r1 = s + p - 1; r2 = s + p;
    U = trans(T(s : r1, s : r1), T(r2 : t, r2 : t), T(s : r1, r2 : t));
    T(s : t, s : n) = UT(s : t, s : n);
    T(1 : t, s : t) = T(1 : t, s : t)U^T;
    T(s + q : t, s : s + q - 1) = zeros(p, q);
    B(s : t, :) = UB(s : t, :);
    Q(s : t, :) = UQ(s : t, :);
```

3.5.2 Schur Algorithm

We now describe Schur algorithm in detail. Recall from section 3.2 that if (A, B) is controllable with

$$A = \begin{bmatrix} A_{11} & A_{12} \\ 0 & A_{22} \end{bmatrix}, \qquad B = \begin{bmatrix} B_{11} \\ B_{21} \end{bmatrix},$$

then (A_{22}, B_{21}) is also controllable; if we take a feedback matrix F of the form $F = [0, F_{12}]$, then it is easy to see that in the resulting matrix

$$A + BF = \begin{bmatrix} A_{11} & A_{12} + B_{11}F_{12} \\ 0 & A_{22} + B_{21}F_{12} \end{bmatrix}$$

only the eigenvalues corresponding to A_{22} are modified. We will use these properties systematically to assign two poles at a time.

Consider the pair (A_0, B_0). Let n_j denote the size of $A_{jj}^{(0)}$, $j = 1, 2, \ldots, k$. The first step of schur method is to assign a pair of poles λ_1, λ_2 to (A_0, B_0). More precisely, the first step is to compute two orthogonal matrices $U_{11}, U_{12} \in \mathbf{R}^{n \times n}$

and an $m \times n$ matrix $F_1 = [0,\ F_{12}]$, where $F_{12} \in \mathbf{R}^{m\times 2}$, such that

$$A_1 = U_{12}(U_{11}A_0U_{11}^T + U_{11}B_0F_1)U_{12}^T = \begin{bmatrix} A_{11}^{(1)} & A_{12}^{(1)} & \cdots & A_{1,k_1+1}^{(1)} \\ 0 & A_{22}^{(1)} & \cdots & A_{2,k_1+1}^{(1)} \\ & & \ddots & \vdots \\ & & & A_{k_1+1,k_1+1}^{(1)} \end{bmatrix} \tag{3.5.8}$$

with $\lambda(A_{11}^{(1)}) = \{\lambda_1, \lambda_2\}$, $\lambda(A_{jj}^{(1)}) = \lambda(A_{j-1,j-1}^{(0)})$, $j = 2, \ldots, k_1$, where $k_1 = k$ or $k-1$. To do this, we begin by checking the sizes of the last two blocks of A_0 to determine whether we need to interchange the two blocks. If $n_k = 2$ or $n_k = n_{k-1} = 1$, we need not interchange $A_{kk}^{(0)}$ and $A_{k-1,k-1}^{(0)}$, written $A_{\frac{1}{3}} = A_0$, $B_{\frac{1}{3}} = B_0$, and $U_{11} = I$. Otherwise we interchange $A_{kk}^{(0)}$ and $A_{k-1,k-1}^{(0)}$ by applying Algorithm 3.5.2, that is, to determine an 3×3 orthogonal matrix $\widetilde{U}_{11}$, written $U_{11} = \operatorname{diag}(I_{n-3}, \widetilde{U}_{11})$, such that

$$A_{\frac{1}{3}} = U_{11}A_0U_{11}^T = \begin{bmatrix} A_{11}^{(0)} & \cdots & A_{1,k-2}^{(0)} & A_{1,k-1}^{(\frac{1}{3})} & A_{1k}^{(\frac{1}{3})} \\ & \ddots & \vdots & \vdots & \vdots \\ & & A_{k-1,k-2}^{(0)} & A_{k-2,k-1}^{(\frac{1}{3})} & \vdots \\ & & & A_{k-1,k-1}^{(\frac{1}{3})} & \vdots \\ & & & & A_{kk}^{(\frac{1}{3})} \end{bmatrix}$$

where $\lambda\big(A_{k-1,k-1}^{(\frac{1}{3})}\big) = \lambda\big(A_{kk}^{(0)}\big)$, $\lambda\big(A_{k,k}^{(\frac{1}{3})}\big) = \lambda\big(A_{k-1,k-1}^{(0)}\big)$. Note that this transformation does not touch the $(n-3)\times(3-3)$ leading principal submatrix of A_0. Then we interchange n_k and n_{k-1} and compute $B_{\frac{1}{3}} = U_{11}B_0$. Thus we now have either $n_k = 2$ or $n_k = n_{k-1} = 1$. If $n_k = 2$, let $k_1 = k$, otherwise let $k_1 = k-1$.

Let $r = n-1$, $s = n-2$. By Corollaries 3.2.2 and 3.2.3 we know that $(A_{\frac{1}{3}}(r:n, r:n), B(r:n,:))$ is Controllable. Now applying Algorithm 3.5.1 to $M = A_{\frac{1}{3}}(r:n, r:n)$, $N = B_{\frac{1}{3}}(r:n,:)$, $\mu_1 = \lambda_1$, $\mu_2 = \lambda_2$, we obtain $F_{12} \in \mathbf{R}^{m\times 2}$ such that $A_{\frac{1}{3}}(r:n, r:n) + B_{\frac{1}{3}}(r:n,:)F_{12}$ has eigenvalues λ_1, λ_2. Let $F_1 = [0, F_{12}] \in \mathbf{R}^{m\times n}$. Then

$$\begin{aligned} A_{\frac{2}{3}} &= A_{\frac{1}{3}} + B_{\frac{1}{3}}F_1 \\ &= \begin{bmatrix} A_{\frac{1}{3}}(1:s, 1:s) & A_{\frac{1}{3}}(1:s, r:n) + B_{\frac{1}{3}}(1:s,:)F_{12} \\ 0 & A_{\frac{1}{3}}(r:n, r:n) + B_{\frac{1}{3}}(r:n,:)F_{12} \end{bmatrix}. \end{aligned}$$

This shows that the last 2×2 block of $A_{\frac{2}{3}}$ has a pair of desired eigenvalues λ_1, λ_2, and $\lambda\big(A_{\frac{2}{3}}(1:s, 1:s)\big) \subset \lambda(A_0)$.

In order to continue to assign the remaining poles, we must move the last 2×2 block of $A_{\frac{2}{3}}$ in position (1,1). This can be carried out by applying Algorithm 3.5.3 successively. This procedure eventually produces an orthogonal matrix U_{12} such that $A_1 = U_{12} A_{\frac{2}{3}} U_{12}^T$ has the desired form. Let $B_1 = U_{12} U_{11} B_0$. The second step is to assign a pair of poles λ_3, λ_4 to (A_1, B_1). That is, similarly the first step we can determine two orthogonal matrices U_{21}, $U_{22} \in \mathbf{R}^{n \times n}$ and an $m \times n$ matrix $F_2 = [0, F_{22}]$, where $F_{22} \in \mathbf{R}^{m \times 2}$, such that

$$A_2 = U_{22}(U_{21} A_1 U_{21}^T + U_{21} B_1 F_2) U_{22}^T$$

possesses the following properties:

(1) A_2 is still a real schur canonical form;
(2) $\lambda\big(A_2(1:4, 1:4)\big) = \{\lambda_1, \lambda_2, \lambda_3, \lambda_4\}$;
(3) $\lambda\big(A_2(5:n, 5:n)\big) \subset \lambda(A_0)$.

The third step is to assign a pair of poles λ_5, λ_6 to (A_2, B_2), where $B_2 = U_{22} U_{21} B_1$, and so on. After $l = [\frac{n}{2}]$ steps we obtain $2l$ orthogonal matrices $U_{11}, U_{12}, \ldots, U_{l1}, U_{l2}$, and l matrices $F_j = [0, F_{j2}] \in \mathbf{R}^{m \times n}$, where $F_{j2} \in \mathbf{R}^{m \times 2}$, $j = 1, 2, \ldots, l$, such that

$$A_j = U_{j2}(U_{j1} A_{j-1} U_{j1}^T + U_{j1} B_{j-1} F_j) U_{j2}^T$$

possesses the following properties:

(1) A_j is still a real Schur canonical form;
(2) $\lambda\big(A_j(1:2j, 1:2j)\big) = \{\lambda_1, \lambda_2, \ldots, \lambda_{2j-1}, \lambda_{2j}\}$;
(3) $\lambda\big(A_j(2j+1:n, 2j+1:n)\big) \subset \lambda(A_0)$.

Here $B_j = U_{j2} U_{j1} B_{j-1}, \quad j = 1, 2, \ldots, l.$

If $n = 2l$, we have obtained a matrix A_l whose eigenvalues are the given data $\lambda_1, \ldots, \lambda_n$. Otherwise one more step is needed. We need to determine an $m \times n$ matrix $F_{l+1} = [0, F_{l+1,2}]$, where $F_{l+1,2} \in \mathbf{R}^m$, such that

$$A_l(n, n) + B_l(n, :) F_{l+1,2} = \lambda_n.$$

To solve this equation, we first determine a Housebolder transformation H such that

$$B_l(n; :) H = (\beta, 0, \ldots, 0).$$

By Corollaries 3.2.2–3.2.3 and Lemma 3.3.2, it follows that $\beta \neq 0$. Hence we may take

$$F_{l+1,2} = \alpha H e_1,$$

where

$$\alpha = \frac{1}{\beta}\big(\lambda_n - A_l(n, n)\big).$$

Let $A_{l+1} = A_l + B_l F_{l+1}$, $U_{l+1,2} = U_{l+1,2} = I$. Then the eigenvalues of A_{l+1} are the given data $\lambda_1, \ldots, \lambda_n$.

Now write $s = [\frac{n+1}{2}]$ and $Q_j = U_{j2} U_{j1}, j = 1, 2, \ldots, s$. It follows that the eigenvalues of A_s are the given data $\lambda_1, \ldots, \lambda_n$, and

$$\begin{aligned}
A_s &= Q_s(A_{s-1} + B_{s-1} F_s U_{s1}) Q_s^T \\
&= Q_s\big(Q_{s-1}(A_{s-2} + B_{s-2} F_{s-1} U_{s-1,1}) Q_{s-1}^T + B_{s-1} F_s U_{s1}\big) Q_s^T \\
&= Q_s Q_{s-1}\big(A_{s-2} + B_{s-1}(F_{s-1} U_{s-1,1} + F_s U_{s1} Q_{s-1})\big) Q_{s-1}^T Q_s^T \\
&\cdots\cdots\cdots\cdots\cdots\cdots \\
&= Q(A + BF) Q^T,
\end{aligned}$$

where

$$\begin{aligned}
Q &= Q_s Q_{s-1} \cdots Q_0. \\
F &= F_1 U_{11} Q_0 + F_2 U_{21} Q_1 Q_0 + \cdots + F_s U_{s1}(Q_{s-1} \cdots Q_0).
\end{aligned}$$

This means that the eigenvalues of $A + BF$ are the given data $\lambda_1, \ldots, \lambda_n$, that is, we have obtained a desired matrix F.

It should be pointed out that different orderings of the desired poles will lead to different gain matrices.

In summarizing, we obtain the following algorithm.

Algorithm 3.5.4 (Schur Algorithm) Given an $n \times n$ matrix A, an $n \times m$ matrix B, and an n-dimensional vector E, satisfying $E(2l-1)) = \overline{E(2l)}$ if $\mathrm{Im}(E(2l-1))$ is not zero, this algorithm computes an $m \times n$ matrix F such that the eigenvalues of $A + BF$ are the values specified in E if $(A,\ B)$ is controllable, and returns an empty matrix otherwise.

```
[n, m] = size(B);
[Q, L] = schur(A);    (Real Schur decomposition)
Q = Q^T; k = 0; k_k = 0;
while k_k ≤ n − 2
    k = k + 1;
    if L(k_k + 2, k_k + 1) = 0
        N(k) = 1;
        E_g(k) = L(k_k + 1, k_k + 1);
    else
        N(k) = 2;
        X = eig(L(k_k + 1 : k_k + 2, k_k + 1 : k_k + 2));
                                        (Eigenvalues of the kth block)
        E_g(k) = X(1);
    end
    k_k = k_k + N(k);
```

```
end
if k_k = n − 1
        k = k + 1; N(k) = 1; E_g(k) = L(n, n);
end
B = QB; F = zeros(m, n);
for l = 1 : [n/2]
        if k ≥ 2
                if N(k) = 1 and N(k − 1) = 2
                        [L, B, Q] = interch(L, B, Q, [n − 2, 2, 1]);
                        N(k) = 2;
                        N(k − 1) = 1;
                end
        end
        F_l = pa2(L(n − 1 : n, n − 1 : n), B(n − 1 : n, :), E(2l − 1), E(2l));
        if isempty(F_l)
                F =[ ];    ((A, B) uncontrollable)
                return
        else
                L(:, n − 1 : n) = L(:, n − 1 : n) + BF_l;
                F = F + F_l Q(n − 1 : n, :);
        end
        if N(k) = 1
                k = k − 2;
        else
                k = k − 1;
        end
        if Im(E(2l − 1)) = 0
                if |L(n, n − 1)| > 0
                        d = −(L(n, n) − E(2l))/L(n, n − 1);
                        [c, s] = givens(d, 1); U = [ c  −s ; s  c ];
                        L(:, n − 1 : n) = L(:, n − 1 : n)U;
                        L(n − 1 : n, n − 1 : n) = U^T L(n − 1 : n, n − 1 : n);
                        B(n − 1 : n, :) = U^T B(n − 1 : n, :);
                        Q(n − 1 : n, :) = U^T Q(n − 1 : n, :);
                        L(n, n − 1) = 0;
                end
                for i_i = 2 : −1 : 1
                        r = n − i_i + 1;
                        for j = k : −1 : 1
                                if N(j) = 1 and E_g(j) − E(2l − i_i + 1) = 0
                                        r = r − N(j);
                                else
                                        r = r − N(j);
```

```
                        [L, B, Q] = interch(L, B, Q, [r, N(j), 1]);
                    end
                end
            end
        else
            r = n - 1;
            for j = k : -1 : 1
                d = min{|E_g(j) - E(2l - 1)|, |E_g(j) - E(2l)|};
                if N(j) = 2 and d = 0
                    r = r - N(j);
                else
                    r = r - N(j);
                    [L, B, Q] = interch(L, B, Q, [r, N(j), 2]);
                end
            end
        end
    end
    if n > 2 [n/2]
        [U, R] = qr(B(n, :)^T);
        if R(1, 1) = 0
            F = [ ];    ((A, B) uncontrollable)
        else
            d = (E(n) - L(n, n)) / R(1, 1);
            U = U^T; F = F + dU(:, 1)Q(n, :);
        end
    end
```

In this algorithm the reliable numerical techniques based on the Schur form of the state matrix A and orthogonal transformations are used. Hence, this algorithm is numerically reliable.

3.6 Invariant Subspace Method

In this section we will develop another numerically reliable method for solving Problem PA. Let A, B, and $\mathcal{L}$ be as in Section 3.5, we wish to compute an $m \times n$ matrix F such that $\lambda(A + BF) = \mathcal{L}$.

3.6.1 Reduction to Orthogonal Canonical Form

As has been pointed out in preceding section, Problem PA may be easily solved if the system (A, B) is preliminarily reduced to some simpler form by equivalence transformations. Hence, for this purpose we initially compute orthogonal matrices

$U \in \mathbf{R}^{n\times n}$, $V \in \mathbf{R}^{m\times m}$ such that (A, B) is transformed into an **orthogonal canonical form** as follows:

$$B_1 = UBV = \begin{bmatrix} B_{11} \\ 0 \end{bmatrix},$$

$$A_1 = UAU^T = \begin{bmatrix} A_{11} & A_{12} & \cdots & A_{1,k-1} & A_{1k} \\ A_{21} & A_{22} & \cdots & A_{2,k-1} & A_{2k} \\ & A_{32} & \cdots & A_{3,k-1} & A_{3k} \\ & & \ddots & \vdots & \vdots \\ & & & A_{k,k-1} & A_{kk} \end{bmatrix}, \tag{3.6.1}$$

with $A_{ij} \in \mathbf{R}^{n_i\times n_j}$, where $n_1 = \operatorname{rank} B$, $n_j = \operatorname{rank} A_{j,j-1}$ for $j = 2, \ldots, k$, and $n_j \geq n_{j+1}$ for $j = 1, 2, \ldots, k-1$. The matrix B_{11} is of the form $(0, R)$ with R nonsingular and upper triangular. Note that the assumption that the system (A, B) is controllable implies that $A_{k,k-1}$ is of full row rank ($A_{k,k-1} = 0$ if the system is uncontrollable). Customarily, $n_1, \ldots, n_k$ are referred as the **conjugate Kronecker indexes**, and k is referred as the **controllability index** of the system (A, B).

There are several ways to transform the system (A, B) into the orthogonal canonical form. Here we introduce a method based on the QR decomposition, which requires k major steps. The first step is to compute orthogonal matrices $U_1 \in \mathbf{R}^{n\times n}, V \in \mathbf{R}^{m\times m}$ so that

$$U_1BV = \begin{bmatrix} B_{11} \\ 0 \end{bmatrix},$$

where B_{11} is of the form $(0, R)$ with R nonsingular and upper triangular, using QR decomposition with column pivoting twice. Let

$$U_1AU_1^T = \begin{bmatrix} A_{11}^{(2)} & A_{12}^{(2)} \\ A_{21}^{(2)} & A_{22}^{(2)} \end{bmatrix},$$

with $A_{11}^{(2)} \in \mathbf{R}^{n_1\times n_1}$, where $n_1 = \operatorname{rank} B_{11}$. In the ith step, first compute an orthogonal matrix U_i such that

$$U_iA_{i,i-1}^{(i)} = \begin{bmatrix} A_{i,i-1} \\ 0 \end{bmatrix} \tag{3.6.2}$$

with $A_{i,i-1}$ full row rank by applying Householder transformations. Then check (3.6.2). If $A_{i,i-1}^{(i)}$ has full row rank or is already zero (if the system (A, B) is uncontrollable), then the algorithm stops with $k = i$. Otherwise form

$$U_iA_{ii}^{(i)}U_i^T = \begin{bmatrix} A_{ii}^{(i+1)} & A_{i,i+1}^{(i+1)} \\ A_{i+1,i}^{(i+1)} & A_{i+1,i+1}^{(i+1)} \end{bmatrix},$$

with $A_{ii}^{(i+1)} \in \mathbf{R}^{n_i \times n_i}$, where $n_i = \text{rank}\, A_{i,i-1}$, and proceed to the next step. After k steps we complete the reduction and the transforming matrix U can be obtained by accumulating these orthogonal transformations. This gives the following algorithm.

Algorithm 3.6.1 Given an $n \times n$ matrix A and an $n \times m$ matrix B, this function reduces (A, B) to an orthogonal canonical form and returns the transforming matrices U, V, the controllability index k, and the conjugate Kronecker indexes $N(1), N(2), \dots, N(k)$.

function $[A, B, U, V, N, k] = \mathbf{orthf}(A, B)$
$[n, m] = \mathbf{size}(B);\ k = 1;$
Compute orthogonal matrices $U \in \mathbf{R}^{n\times n}$, $V \in \mathbf{R}^{m\times m}$ so that

$$UBV = \begin{bmatrix} B_{11} \\ 0 \end{bmatrix},$$

where B_{11} is of the form $[0,\ R]$ with R nonsingular and upper triangular
$N(1) = \mathbf{rank}\,(R);\ A = UAU^T;$
$j = N(1);$
while $n - j > 0$
$[Q, R, P] = \mathbf{qr}(A(j+1:n, j-N(k)+1:j));$
$k = k+1;\ N(k) = \mathbf{rank}\,(R);$
$r = j + N(k);\ s = j - N(k-1) + 1;$
$A(j+1:r, s:j) = R(1:N(k), :)P^T;$
$A(r+1:n, s:j) = \mathbf{zeros}(n-r, N(k-1));$
if $N(k) = 0$
Return (System uncontrollable)
else
$U(j+1:n, :) = Q^T U(j+1:n, :);$
$A(j+1:n, j+1:n) = Q^T A(j+1:n, j+1:n)Q;$
$A(1:j, j+1:n) = A(1:j, j+1:n)Q;$
$j = j + N(k);$
end
end

The algorithm requires approximately $\frac{5}{3}n^3 + n^2m$ flops and additional n^3 flops are necessary to accumulate the orthogonal transformations. The algorithm is numerically stable, the computed orthogonal canonical form $(\widetilde{A}_1, \widetilde{B}_1)$ being orthogonally equivalent to $(A + \delta A,\ B + \delta B)$ with

$$\|\delta A\|_2 \le f_A \|A\|_2 \varepsilon,\ \|\delta B\|_2 \le f_B \|B\|_2 \varepsilon,$$

where f_A, f_B are small multiples of n^2 and nm, ε is the machine unit. See [171] for the further details.

3.6.2 The Basic Idea and Techniques

From now on we will describe in detail how to compute $F_1 = V^T F U^T$ so that $A_1 + B_1 F_1 = U(A + BF)U^T$ has $\mathcal{L}$ as its spectral set. If $m > n_1$, that is $B_{11} = [0, R]$, only the last n_1 rows of F_1 affect $A_1 + B_1 F_1$. We will set the first $m - n_1$ rows of F_1 to zero and in what follows assume $m = n_1$. Obviously there are other choices for the first $m - n_1$ rows of F_1, the one adopted here though is the most innocuous for the numerically stability of the algorithm and it also decreases the magnitude of F.

We will now cite two elementary results, the proofs of which are not difficult and hence are left as an exercise for the reader.

Lemma 3.6.1 *Let $E \in \mathbf{R}^{r\times m}$ be of full row rank, and let $G \in \mathbf{R}^{n\times s}$ be of full column rank. Then for any given $C \in \mathbf{R}^{r\times s}$ the matrix equation*

$$EXG = C$$

is consistent, i.e., there is at least one matrix $X \in \mathbf{R}^{m\times n}$ that satisfies the above equation.

Lemma 3.6.2 *Let $C \in \mathbf{R}^{n\times n}, S \in \mathbf{R}^{2\times 2}$, and $X \in \mathbf{R}^{n\times 2}$ with* $\operatorname{rank} X = 2$, *and assume that $CX = XS$. If the QR decomposition of X is $X = Q^T \begin{bmatrix} X_1 \\ 0 \end{bmatrix}$, where $Q \in \mathbf{R}^{n\times n}$ is orthogonal, and $X_1 \in \mathbf{R}^{2\times 2}$ upper triangular, then*

$$QCQ^T = \begin{bmatrix} C_{11} & C_{12} \\ 0 & C_{22} \end{bmatrix},$$

where $C_{11} \in \mathbf{R}^{2\times 2}$ with $\lambda(C_{11}) = \lambda(S)$.

Geometrically, Lemma 3.6.2 says that if we know a basis for an invariant subspace of a matrix corresponding to a pair of real or complex conjugate eigenvalues, then with the help of this basis we may construct an orthogonal basis for $\mathbf{R}^n$ so that the first two vectors of this orthogonal basis span the invariant subspace.

The invariant subspace method is based on this fact. Its basic idea is as follows. Since the closed-loop system matrix $A_1 + B_1 F_1$ is in block upper Hessenberg form, it is possible to determine a basis for the invariant subspace of this matrix corresponding to a pair of given eigenvalues knowing only the matrix A_1, and so it is possible to construct an orthogonal matrix Q such that the first columns of Q span the invariant subspace without knowing F_1. Thus using this orthogonal matrix to make transformation $Q(A_1 + B_1 F_1)Q^T$ may lead to an equation for the first two columns of $F_1 Q^T$. Moreover, this transformation may preserve the block structure of A_1 and B_1 by suitable choice of Q, which gives the possibility to proceed further by the same way.

We now consider how $X \in \mathbf{R}^{n\times 2}$ can be determined so that

$$(A_1 + B_1F_1)X = XS, \quad \operatorname{rank} X = 2, \tag{3.6.3}$$

knowing only A_1 and S, where $S = \begin{bmatrix} \zeta_1 & 1 \\ -\eta^2 & \zeta_2 \end{bmatrix}$, $\zeta_1 + i\eta$ and $\zeta_2 + i\eta$ are two given members of $\mathcal{L}$, satisfying $\zeta_1 = \zeta_2$ if $\eta \neq 0$. Note that the matrix S is chosen as this form instead of the usual form $\begin{bmatrix} \zeta_1 & \eta \\ -\eta & \zeta_2 \end{bmatrix}$ to avoid linear dependence between the two columns of X in case of small η. Partitioning X as

$$X = \begin{bmatrix} X_1 \\ X_2 \\ \vdots \\ X_k \end{bmatrix}, \quad X_i \in \mathbf{R}^{n_i\times 2}, \quad i = 1, \ldots, k,$$

it follows from (3.6.1) and (3.6.3) that

$$A_{j,j-1}X_{j-1} + \sum_{l=j}^{k} A_{jl}X_l = X_jS, \quad j = k, \ldots, 2. \tag{3.6.4}$$

Of course, here we assume that $k > 1$. As we shall soon see, it is easy to determine an $m \times n$ matrix F_1 such that $\lambda(A_1 + B_1F_1) = \mathcal{L}$ in the case of $k = 1$.

X may be determined from (3.6.4) as follows. First, we choose X_k so that $\operatorname{rank}\begin{bmatrix} X_{k-1} \\ X_k \end{bmatrix} = 2$, where X_{k-1} is a solution to the following matrix equation

$$A_{k,k-1}X_{k-1} = X_kS - A_{kk}X_k. \tag{3.6.5}$$

For example, taking $X_k = [e_1, 0]$, where e_1 denotes the first column of the identity matrix of order n_k. Writing $X_{k-1} = [x_{k-1}, y_{k-1}]$, (3.6.5) gives that

$$A_{k,k-1}y_{k-1} = e_1.$$

From this it follows that y_{k-1} must be nonzero, and so

$$\begin{bmatrix} X_{k-1} \\ X_k \end{bmatrix} = \begin{bmatrix} x_{k-1} & y_{k-1} \\ e_1 & 0 \end{bmatrix}$$

must have full column rank. Then we compute recursively the matrices X_{k-1}, $X_{k-2}, \ldots, X_1$ from (3.6.4).

After X has been determined, in order that such a matrix X satisfies (3.6.3), we must seek an $m \times n$ matrix F_1 so that

$$\sum_{l=1}^{n} A_{1l}X_l + B_{11}F_1X = X_1S. \tag{3.6.6}$$

Since B_{11} is of full row rank and X is of full column rank, by Lemma 3.6.1, it follows that (3.6.6) must be consistent. This means that for any X determined from (3.6.4) with full column rank we may find an $m \times n$ matrix F_1 such that X satisfies (3.6.3).

Let $X = Q^T \begin{bmatrix} X_{11} \\ 0 \end{bmatrix}$ be the QR decomposition of X, where $Q \in \mathbf{R}^{n\times n}$ is orthogonal, and $X_{11} \in \mathbf{R}^{2\times 2}$ upper triangular, which is nonsingular since rank $X =$ 2. Substituting $X = Q^T \begin{bmatrix} X_{11} \\ 0 \end{bmatrix}$ into (3.6.3), we have

$$(QA_1Q^T + QB_1F_1Q^T)\begin{bmatrix} X_{11} \\ 0 \end{bmatrix} = \begin{bmatrix} X_{11} \\ 0 \end{bmatrix} S. \tag{3.6.7}$$

Writing

$$QA_1Q^T = \underset{\quad 2 \qquad n-2}{\begin{bmatrix} A_{11}^{(1)} & A_{12}^{(1)} \\ A_{21}^{(1)} & A_2 \end{bmatrix}}\begin{matrix} 2 \\ n-2 \end{matrix}, \qquad QB_1 = \begin{bmatrix} B_{11}^{(1)} \\ B_2 \end{bmatrix}\begin{matrix} 2 \\ n-2 \end{matrix},$$

and

$$F_1Q^T = [\underset{2}{F_{11}}, \quad \underset{m-2}{F_2}],$$

(3.6.7) can be rewritten as

$$\begin{cases} (A_{11}^{(1)} + B_{11}^{(1)}F_{11})X_{11} = X_{11}S, \\ (A_{21}^{(1)} + B_2F_{11})X_{11} = 0, \end{cases}$$

and so

$$\begin{cases} B_{11}^{(1)}F_{11} = X_{11}SX_{11}^{-1} - A_{11}^{(1)}, \\ B_2F_{11} = -A_{21}^{(1)}, \end{cases} \tag{3.6.8}$$

which is consistent since it is equivalent to (3.6.3). Hence, it is easy to obtain F_{11} from (3.6.8) using QR decomposition. After the solution of (3.6.8), $Q(A_1 + B_1F_1)Q^T$ has the form

$$Q(A_1 + B_1F_1)Q^T = \begin{bmatrix} X_{11}SX_{11}^{-1} & * \\ 0 & A_2 + B_2F_2 \end{bmatrix} \tag{3.6.9}$$

for any $F_2 \in \mathbf{R}^{m\times(n-2)}$. This means that we have determined the first two columns F_{11} of F_1Q^T such that $A_1 + B_1F_1$ has one pair of desired eigenvalues, and that we may continue to assign the remaining poles by considering the system (A_2, B_2) without destroying the poles assigned previously. In order to continue the process, we must ensure that the system (A_2, B_2) is still an orthogonal canonical form by suitable choice of Q, which is used to transform X to upper triangular form.

For this purpose the transforming matrix Q may be computed as a product of k orthogonal matrices $Q = Q_1Q_2\dots Q_k$. First, if $n_k = 1$, let $\widetilde{Q}_k = 1$, $X_{k1} = X_k$, and compute an orthogonal matrix $\widetilde{Q}_{k-1} \in \mathbf{R}^{(n_{k-1}+1)\times(n_{k-1}+1)}$ such that

$$\widetilde{Q}_{k-1}\begin{bmatrix} X_{k-1} \\ X_k \end{bmatrix} = \begin{bmatrix} X_{k-1,1} \\ 0 \end{bmatrix},$$

where $X_{k-1,1} \in \mathbf{R}^{2\times2}$ is nonsingular and upper triangular. Otherwise compute two orthogonal matrices $\widetilde{Q}_k \in \mathbf{R}^{n_k\times n_k}$, $\widetilde{Q}_{k-1} \in \mathbf{R}^{(n_{k-1}+2)\times(n_{k-1}+2)}$ such that

$$\widetilde{Q}_kX_k = \begin{bmatrix} X_{k1} \\ 0 \end{bmatrix}, \quad \widetilde{Q}_{k-1}\begin{bmatrix} X_{k-1} \\ X_{k1} \end{bmatrix} = \begin{bmatrix} X_{k-1,1} \\ 0 \end{bmatrix},$$

where X_{k1}, $X_{k-1,1} \in \mathbf{R}^{2\times2}$ are nonsingular and upper triangular. Then for $j = k-2, k-3, \dots, 2$, compute orthogonal matrix $\widetilde{Q}_j \in \mathbf{R}^{(n_j+2)\times(n_j+2)}$ such that

$$\widetilde{Q}_j\begin{bmatrix} X_j \\ X_{j+1,1} \end{bmatrix} = \begin{bmatrix} X_{j1} \\ 0 \end{bmatrix},$$

where $X_{j1} \in \mathbf{R}^{2\times2}$ is nonsingular and upper triangular. Finally, to preserve the structure of B_1, we compute $\widetilde{Q}_1 \in \mathbf{R}^{(n_1+2)\times(n_1+2)}$ as a product of $2n_1$ Givens rotations

$$\widetilde{Q}_1 = G_{23}^{(2)}G_{34}^{(2)}\dots G_{n_1+1,n_1+2}^{(2)}G_{12}^{(1)}\dots G_{n_1,n_1+1}^{(1)}$$

so that

$$\widetilde{Q}_1\begin{bmatrix} X_1 \\ X_{21} \end{bmatrix} = \begin{bmatrix} X_{11} \\ 0 \end{bmatrix},$$

where $X_{11} \in \mathbf{R}^{2\times2}$ is nonsingular and upper triangular. The Givens rotations $G_{n_1,n_1+1}^{(1)}, \dots, G_{12}^{(1)}, G_{n_1+1,n_1+2}^{(2)}, \dots, G_{23}^{(2)}$ in corresponding plans are chosen so as to create successively zeros in the positions $(n_1+1,1), \dots, (2,1)$, $(n_1+2,2)$, $(n_1+1,2)\dots, (3,2)$ of $\begin{bmatrix} X_1 \\ X_{21} \end{bmatrix}$. Note that X_{21} is a nonsingular, upper triangular matrix. Hence all of these Givens rotations are nontrivial. Now the required orthogonal matrix Q_j may be defined as

$$Q_j = \operatorname{diag}(I_{p_j}, \widetilde{Q}_j, I_{q_j}), \tag{3.6.10}$$

where

$$p_j = n_1 + n_2 + \cdots + n_{j-1},$$
$$q_j = \begin{cases} 0, & j = k, \\ n - p_j - n_j - 2, & \text{otherwise}. \end{cases}$$

Let

$$\begin{aligned} A_1^{(k)} &= Q_k A_1 Q_k^T, & B_1^{(k)} &= Q_k B_1, \\ A_1^{(j)} &= Q_j A_1^{(j+1)} Q_j^T, & B_1^{(j)} &= Q_k B_1^{(j+1)}, \quad j = k-1, \ldots, 1, \end{aligned}$$

and write

$$A_1^{(1)} = \begin{bmatrix} A_{11}^{(1)} & A_{12}^{(1)} \\ A_{21}^{(1)} & A_2 \end{bmatrix} \begin{matrix} 2 \\ n-2 \end{matrix}, \quad B_1^{(1)} = \begin{bmatrix} B_{11}^{(1)} \\ B_2 \end{bmatrix} \begin{matrix} 2 \\ n-2 \end{matrix}.$$
$$\qquad\qquad 2 \quad n-2$$

Then $A_1^{(1)} = QA_1Q^T, B_1^{(1)} = QB_1$, and $A_{21}^{(1)}, A_2, B_2$ have the following forms:

$$A_{21}^{(1)} = \begin{bmatrix} A_{21}^{(2)} \\ 0 \end{bmatrix} \begin{matrix} m_1 \\ n-m_1-2 \end{matrix}, \quad B_2 = \begin{bmatrix} B_{11}^{(2)} \\ 0 \end{bmatrix} \begin{matrix} m_1 \\ n-m_1-2 \end{matrix},$$
$$\qquad 2$$

$$A_2 = \begin{bmatrix} A_{11}^{(2)} & A_{12}^{(2)} & \cdots & A_{1,l-1}^{(2)} & A_{1l}^{(2)} \\ A_{21}^{(2)} & A_{22}^{(2)} & \cdots & A_{2,l-1}^{(2)} & A_{2l}^{(2)} \\ & A_{32}^{(2)} & \cdots & A_{3,l-1}^{(2)} & A_{3l}^{(2)} \\ & & \ddots & \vdots & \vdots \\ & & & A_{l,l-1}^{(2)} & A_{ll}^{(2)} \end{bmatrix} \begin{matrix} m_1 \\ m_2 \\ m_3 \\ \\ m_l \end{matrix}$$
$$\qquad m_1 \quad m_2 \quad\quad m_{l-1} \quad m_l$$

with $B_{11}^{(2)}$ nonsingular and upper triangular, $\text{rank}(A_{j,j-1}^{(2)}) = m_j$ for $j = 2, 3, \ldots, l$, where

$$l = \begin{cases} k, & n_k > 2, \\ k-1, & n_k = 2, \text{ or } n_k = 1 \text{ and } n_{k-1} \geq 2, \\ k-2, & n_k = n_{k-1} = 1, \end{cases}$$
$$m_j = n_j, \quad j = 1, 2, \ldots, l-1,$$
$$m_l = \begin{cases} n_l - 1, & n_k = 1 \text{ and } n_{k-1} \geq 2, \\ n_l - 2, & n_k > 2, \\ n_l, & \text{otherwise}. \end{cases}$$

To see this consider the following example, where $n = 8, n_1 = 3, n_2 = n_3 = 2$, $n_4 = 1, m = n_1 = 3$, with $\times$ representing a possibly nonzero element:

$$A_1 = \left[\begin{array}{ccc|cc|cc|c} \times & \times & \times & \times & \times & \times & \times & \times \\ \times & \times & \times & \times & \times & \times & \times & \times \\ \times & \times & \times & \times & \times & \times & \times & \times \\ \hline \times & \times & \times & \times & \times & \times & \times & \times \\ \times & \times & \times & \times & \times & \times & \times & \times \\ \hline & & & \times & \times & \times & \times & \times \\ & & & \times & \times & \times & \times & \times \\ \hline & & & & & \times & \times & \times \end{array}\right], \quad B_1 = \left[\begin{array}{ccc} \times & \times & \times \\ & \times & \times \\ & & \times \\ \hline & & \\ & & \\ & & \\ & & \\ & & \end{array}\right].$$

In this case we have

$$\widetilde{Q}_4 = 1, \quad \widetilde{Q}_3 \in \mathbf{R}^{3\times 3}, \quad \widetilde{Q}_2 \in \mathbf{R}^{4\times 4},$$

$$\widetilde{Q}_1 = G_{32}^{(2)} G_{43}^{(2)} G_{54}^{(2)} G_{21}^{(1)} G_{32}^{(1)} G_{43}^{(1)} \in \mathbf{R}^{5\times 5}, \quad Q_4 = I_8,$$

$$Q_3 = \begin{bmatrix} I_5 & 0 \\ 0 & \widetilde{Q}_3 \end{bmatrix}, \quad Q_2 = \begin{bmatrix} I_3 & 0 & 0 \\ 0 & \widetilde{Q}_3 & 0 \\ 0 & 0 & 1 \end{bmatrix}, \quad Q_1 = \begin{bmatrix} \widetilde{Q}_4 & 0 \\ 0 & I_3 \end{bmatrix}.$$

Thus it is easy to see that $A_1^{(4)} = A_1$, $B_1^{(4)} = B_1^{(3)} = B_1^{(2)} = B_1$. The effect of transforming $A_1^{(4)}$ to $A_1^{(3)} = Q_3 A_1^{(4)} Q_3^T$ may be represented diagrammatically as follows:

$$\left[\begin{array}{ccc|cc|cc|c} \times & \times & \times & \times & \times & \times & \times & \times \\ \times & \times & \times & \times & \times & \times & \times & \times \\ \times & \times & \times & \times & \times & \times & \times & \times \\ \hline \times & \times & \times & \times & \times & \times & \times & \times \\ \times & \times & \times & \times & \times & \times & \times & \times \\ \hline & & & \times & \times & \times & \times & \times \\ & & & \times & \times & \times & \times & \times \\ \hline & & & & & \times & \times & \times \end{array}\right] \longrightarrow \left[\begin{array}{ccc|cc|cc|c} \times & \times & \times & \times & \times & \times & \times & \times \\ \times & \times & \times & \times & \times & \times & \times & \times \\ \times & \times & \times & \times & \times & \times & \times & \times \\ \hline \times & \times & \times & \times & \times & \times & \times & \times \\ \times & \times & \times & \times & \times & \times & \times & \times \\ \hline & & & \times & \times & \times & \times & \times \\ & & & \times & \times & \times & \times & \times \\ \hline & & & + & + & \times & \times & \times \end{array}\right],$$

where the elements denoted by $+$ are the new, possibly nonzero elements introduced during the transformation. Observe that this transformation simply combines the last three columns and rows of $A_1^{(4)} = A_1$. $A_2^{(2)} = Q_2 A_1^{(3)} Q_2^T$ is of the form

$$A_1^{(2)} = \left[\begin{array}{ccc|cc|cc|c} \times & \times & \times & \times & \times & \times & \times & \times \\ \times & \times & \times & \times & \times & \times & \times & \times \\ \times & \times & \times & \times & \times & \times & \times & \times \\ \hline \times & \times & \times & \times & \times & \times & \times & \times \\ \times & \times & \times & \times & \times & \times & \times & \times \\ \hline + & + & + & \times & \times & \times & \times & \times \\ + & + & + & \times & \times & \times & \times & \times \\ \hline & & & * & * & \times & \times & \times \end{array}\right],$$

where $+$ denotes a new, possibly nonzero element. It may be observed also that the elements denoted by $*$ in the last row must be zeros as a consequence of the fact that X_{21} is nonsingular. The transformation $A_1^{(1)} = Q_1 A_1^{(2)} Q_1^T$ is only to modify the first five columns and rows of $A_1^{(2)}$, and so $A_1^{(1)}$ has the form

$$A_1^{(1)} = \left[\begin{array}{ccc|cc|cc|c} \times & \times & \times & \times & \times & \times & \times & \times \\ \times & \times & \times & \times & \times & \times & \times & \times \\ \times & \times & \times & \times & \times & \times & \times & \times \\ \hline \times & \times & \times & \times & \times & \times & \times & \times \\ \times & \times & \times & \times & \times & \times & \times & \times \\ \hline * & * & \times & \times & \times & \times & \times & \times \\ * & * & \times & \times & \times & \times & \times & \times \\ \hline & & & & & \times & \times & \times \end{array}\right],$$

where $*$ denotes an element that must be zero as a result of the fact X_{11} is nonsingular. Finally, since $\widetilde{Q}_1$ is a product of six Givens rotations $G_{32}^{(2)}, G_{43}^{(2)}, G_{54}^{(2)}$, $G_{21}^{(1)}, G_{32}^{(1)}, G_{43}^{(1)}$ corresponding to (3,2), (4,3),(5,4),(2,1),(3,2),(4,3) planes, respectively, $B_1^{(1)} = Q_1 B_1^{(2)} = Q_1 B_1$ must have the following form

$$B_1^{(1)} = \left[\begin{array}{ccc} \times & \times & \times \\ \times & \times & \times \\ + & \times & \times \\ \hline & + & \times \\ & & + \\ \hline & & \\ & & \\ & & \end{array}\right],$$

where $+$ denotes an new, nonzero element introduced during the transformation. Thus we have

$$A_2 = \left[\begin{array}{ccc|cc|c} \times & \times & \times & \times & \times & \times \\ \times & \times & \times & \times & \times & \times \\ \times & \times & \times & \times & \times & \times \\ \hline \times & \times & \times & \times & \times & \times \\ \times & \times & \times & \times & \times & \times \\ \hline & & & \times & \times & \times \end{array}\right], \quad B_1 = \left[\begin{array}{ccc} \times & \times & \times \\ & \times & \times \\ & & \times \\ \hline & & \\ & & \\ & & \end{array}\right], \quad A_{21}^{(1)} = \left[\begin{array}{cc} \times & \times \\ \times & \times \\ \times & \times \\ \hline & \\ & \\ & \end{array}\right],$$

which are of the desired forms.

We now show that $B_{11}^{(2)}$ is nonsingular and rank $(A_{j,j-1}^{(2)}) = m_j$ for $j = 2, \ldots, l$. It is easy to see that

$$\left[\begin{array}{c} B_{11}^{(1)} \\ B_{11}^{(2)} \end{array}\right] = \widetilde{Q}_1 \left[\begin{array}{c} B_{11} \\ 0 \end{array}\right] = G_{23}^{(2)} \cdots G_{n_1+1,n_1+2}^{(2)} G_{12}^{(1)} \cdots G_{n_1,n_1+1}^{(1)} \left[\begin{array}{c} B_{11} \\ 0 \end{array}\right].$$

Since B_{11} is nonsingular and upper triangular and the $2n_1$ Givens rotations are nontrivial, $B_{11}^{(2)}$ must be nonsingular and upper triangular as well. Directing now our attention to (3.6.9), we may observe that

$$A_1^{(1)} + B_1^{(1)}[F_{11}\ 0] = \begin{bmatrix} X_{11}SX_{11}^{-1} & A_{12}^{(1)} \\ 0 & A_2 \end{bmatrix}.$$

It follows from Corollaries 3.2.2, 3.2.3, and 3.3.4 that (A_2, B_2) must be controllable. This requires $\text{rank}(A_{j,j+1}^{(2)}) = m_j$ for $j = 2, 3, \dots, l$. Thus (A_2, B_2) is an orthogonal canonical form and so we may continue to assign the remaining poles in the same way.

Summarizing the above discussion we may conclude that the invariant subspace method requires $[\frac{n-n_1+1}{2}] + 1$ major steps, the first being on $A_1, B_1, \lambda_1, \lambda_2$. In the jth step, orthogonal matrix U_j and the first two Columns F_{1j} of $F_jU_j^T = [F_{1j}, F_{j+1}]$ are chosen to give

$$U_j(A_j + B_jF_j)U_j^T = \begin{bmatrix} S_j & * \\ 0 & A_{j+1} + B_{j+1}F_{j+1} \end{bmatrix}$$

with (A_{j+1}, B_{j+1}) being an orthogonal canonical form and $\lambda(S_j) = \{\lambda_{2j-1}, \lambda_{2j}\}$. After $r = [\frac{n-n_1+1}{2}]$ steps, we obtain a closed-loop subsystem matrix

$$A_{r+1} + B_{r+1}F_{r+1}, \qquad A_{r+1} \in \mathbf{R}^{s\times s},\ B_{r+1} \in \mathbf{R}^{s\times m},$$

where $s = n - 2r \quad (s = n_1 \text{ or } n_1 - 1)$. The matrix F_{r+1} may be determined by setting

$$A_{r+1} + B_{r+1}F_{r+1} = S_{r+1},$$

where S_{r+1} is any matrix which has the remaining s desired poles $\lambda_{n-2r+1}, \dots, \lambda_n$ as its eigenvalues. The freedom in S_{r+1} may be used to obtain, e.g., a desired Jordan structure of $A_{r+1} + B_{r+1}F_{r+1}$. Here we choose S_{r+1} as a block diagonal matrix, having only 1×1 and 2×2 blocks on the diagonal, corresponding to real or to complex conjugate poles, respectively, so as to achieve computational efficiency and better condition for the eigenproblem of $A_{r+1} + B_{r+1}F_{r+1}$ than if some other choice were made. Thus it turns out that the desired matrix F_1 is as follows:

$$\begin{aligned} F_1 &= [F_{11}, F_2]U_1 = \big[F_{11}, [F_{12}, F_3]U_2\big]U_1 \\ &= [F_{11}, F_{12}, F_3]\begin{bmatrix} I_2 & 0 \\ 0 & U_2 \end{bmatrix} U_1 \\ &\qquad \cdots\cdots\cdots\cdots\cdots \\ &= [F_{11}, F_{12}, \dots, F_{1r}, F_{1,r+1}]Q, \end{aligned}$$

where

$$Q = \begin{bmatrix} I_{2r} & 0 \\ 0 & U_r \end{bmatrix}\begin{bmatrix} I_{2r-2} & 0 \\ 0 & U_{r-1} \end{bmatrix}\cdots\begin{bmatrix} I_2 & 0 \\ 0 & U_2 \end{bmatrix}U_1.$$

3.6.3 Invariant Subspace Algorithm

We now return to some practical details of the computations of X and Q. To reduce the number of necessary operations the computation of X_{j-1} may be done simultaneously with the transformation of X_j. Once X_k has been selected and X_{k-1} has been determined by solving the matrix equation (3.6.5), we may determine the orthogonal matrix $\widetilde{Q}_k$, and use it to make the transformation $A_1^{(k)} = Q_k A_1 Q_k^T$, where $Q_k = \text{diag}\,(I_{n-n_k}, \widetilde{Q}_k)$. Then for $j = k-1, \ldots, 2$, first solve

$$A_{j,j-1}X_{j-1} + A_{jj}X_j + A_1^{j+1}(s_j - n_j + 1 : s_j,\ s_j + 1 : s_j + r_j)X_{j+1,1} = X_j S \tag{3.6.11}$$

for X_{j-1} instead of solving (3.6.6), where $s_j = n_1 + \cdots + n_j$, r_j is the number of rows of $X_{j+1,1}$, then compute the orthogonal matrix $\widetilde{Q}_j$ and use it to make the transformation $A_1^{(j)} = Q_j A_1^{(j+1)} Q_j^T$, where Q_j is as in (3.6.10). Note that Equation (3.6.11) is underdetermined when $n_{j-1} > n_j$ and may be solved by using QR decomposition with column pivoting. Finally, compute $\widetilde{Q}_1$ and use it to make the transformations

$$A_1^{(1)} = Q_1 A_1^{(2)} Q_1, \quad B_1^{(1)} = Q_1 B_1.$$

In addition, to avoid underflows and overflows, we may do "on line" scaling of the element of X. For example, after the solution of (3.6.11), we may multiply $\begin{bmatrix} X_{j-1} \\ X_{j1} \end{bmatrix}$ by $\left\|\begin{bmatrix} X_{j-1} \\ X_{j1} \end{bmatrix}\right\|_1^{-1}$, which only leads to the resulting X_{11} is a multiple of the original one.

Thus we may summarize the preceding discussion as the following algorithm.

Algorithm 3.6.2 (Invariant subspace algorithm) Given an $n \times n$ matrix A, an $n \times m$ matrix B, and an n-dimensional vector E, satisfying $E(2l-1) = \overline{E(2l)}$ if $\text{Im}(E(2l-1))$ is not zero, this algorithm computes an $m \times n$ matrix F such that the eigenvalues of $A + BF$ are the values specified in E if (A, B) is controllable, and returns an empty matrix otherwise.

```
[n, m] = size(B);
[A, B, U, V, N, k] = orthf(A, B);
if N(k) = 0
    F =[ ]; return
end
n1 = N(1); B = B(:, m − n1 + 1 : m);
F = zeros(n1, n); nr = n;
while nr > 2 and k > 1
    sr = n − nr; rb = n − N(k) − N(k − 1) + 1;
```

$$S = \begin{bmatrix} \mathbf{Re}(E(s_r+1)) & 1 \\ -[\mathbf{Im}(E(s_r+1))]^2 & \mathbf{Re}(E(s_r+2)) \end{bmatrix};$$

if $N(k) = 1$

  $X = [1, 0]; \; i_b = r_b + N(k-1) - 1;$

  Solve $A(n : n, r_b : i_b)Z = (XS - A(n,n)X)$ for Z;

  $Y = \begin{bmatrix} Z \\ X \end{bmatrix};$

else

  $X = \begin{bmatrix} 1 & 0 \\ 0 & 1 \end{bmatrix};$

  $C = -A\big(n - N(k) + 1 : n, r_b + N(k-1) : r_b + N(k-1) + 1\big)X;$

  $C(1:2,:) = C(1:2,:) + XS;$

  Solve $A(n - N(k) + 1 : n, r_b : i_b)Z = C$ for Z;

  $Y = \begin{bmatrix} Z \\ X \end{bmatrix};$

end

for $j = k : -1 : 2$

  if $j > 2$

    $r = \mathbf{size}(Y, 1); \; r_e = r_b + r - 1; \; i_b = r_b + N(j-1) - 1;$

    $C = ZS - A(r_b : i_b, r_b : r_e)Y;$

    Solve $A(r_b : i_b, r_b - N(j-2) : r_b - 1)Z = C$ for Z;

    $[Q, R] = \mathbf{qr}(Y); \; X = R(1:2,:);$

    $A(r_b : r_e, :) = Q^T A(r_b : r_e, :);$

    $A(:, r_b : r_e) = A(:, r_b : r_e)Q;$

    $U(r_b : r_e, :) = Q^T U(r_b : r_e, :);$

    $Y = \begin{bmatrix} Z \\ X \end{bmatrix}; \; d = \|Y\|_1;$

    $Y = \frac{1}{d}Y; \; Z = \frac{1}{d}Z; \; r_b = r_b - N(j-2);$

  else

    $r = \mathbf{size}(Y, 1); \; r_e = r_b + r - 1;$

    for $c_l = 1 : 2$

      for $r_r = r_e : -1 : r_b + c_l$

        $[cs] = \mathbf{givens}(Y(r_r - r_b, c_l), Y(r_r - r_b + 1, c_l));$

        $G = \begin{bmatrix} c & s \\ -s & c \end{bmatrix};$

        $Y(r_r - r_b : r_r - r_b + 1, :) = GY(r_r - r_b : r_r - r_b + 1, :);$

        $A(r_r - 1 : r_r, :) = GA(r_r - 1 : r_r, :);$

        $A(:, r_r - 1 : r_r) = A(:, r_r - 1 : r_r)G^T;$

        $B(r_r - 1 : r_r, :) = GB(r_r - 1 : r_r, :);$

        $U(r_r - 1 : r_r, :) = GU(r_r - 1 : r_r, :);$

      end

    end

  end

end

```
        n_e = min([r_b + n_1 + 1, n]);
        Solve B(r_b + 2 : n_e, :)F_j = -A(r_b + 2 : n_e, r_b : r_b + 1) for F_j;
        F(:, s_r + 1 : s_r + 2) = F_j;
        n_r = n_r - 2;
        if N(k) > 2
              N(k) = N(k) - 2;
        elseif N(k) = 2
              k = k - 1;
        elseif N(k) = 1 and N(k - 1) ≥ 2
              k = k - 1; N(k) = N(k) - 1;
        else
              k = k - 2;
        end
  end
  if n_r = 1
        Solve B(n, :)F(:, n) = (E(n) - A(n, n)) for F(:, n);
  elseif n_r = 2
        F(:, n-1 : n) = pa2(A(n-1 : n, n-1 : n), B(n-1 : n, :), E(n-1), E(n));
  else
        S = diag(Re(E(n - n_r - 1)), ... , Re(E(n)));
        for j = 1 : [n_r/2];
              S = (2j - 1, 2j) = Im(E(n - n_r + 2j - 1));
              S = (2j, 2j - 1) = -S(2j - 1, 2j);
        end
        W = S - A(n - n_r + 1 : n, n - n_r + 1 : n);
        Solve B(n - n_r + 1 : n, :)F_j = W for F_j;
        F(:, n - n_r + 1 : n) = F_j;
  end
  F_1 = zeros(m, n); F_1(n_1 + 1 : m, :) = F;
  F = V F_1 U;
```

The numerical properties of the algorithm are very favorable. It may be shown that the computed gain matrix $\widetilde{F}$ is exact for a system with matrices $A+\delta A$, $B+\delta B$, where δA, δB are small perturbations, i.e., the algorithm is backwardly stable in respect to the determination of the gain matrix.

3.7 QR-like Method

In this section we introduce a QR-like algorithm for Problem PA. This algorithm in part parallels the extensively used and numerically reliable QR algorithm for the eigenvalue problem. The difference here is that the desired eigenvalues are known, and the shifts in the algorithm can be chosen to be these eigenvalues, so that the QR algorithm becomes a direct, rather than an iterative method.

3.7.1 Reduction to Staircase Canonical Form

Given a system (A, B) with $A \in \mathbf{R}^{n\times n}$ and $B \in \mathbf{R}^{n\times m}$, we initially compute orthogonal matrices $U \in \mathbf{R}^{n\times n}$, $V \in \mathbf{R}^{m\times m}$ such that (A, B) is transformed into a **staircase canonical form** as follows:

$$B_1 = UBV = \begin{bmatrix} B_{11} \\ 0 \end{bmatrix}, \tag{3.7.1}$$

$$A_1 = UAU^T = \begin{bmatrix} A_{11} & A_{12} & \cdots & A_{1,k-1} & A_{1k} \\ A_{21} & A_{22} & \cdots & A_{2,k-1} & A_{2k} \\ & A_{32} & \cdots & A_{3,k-1} & A_{3k} \\ & & \ddots & \vdots & \vdots \\ & & & A_{k,k-1} & A_{kk} \end{bmatrix} \tag{3.7.2}$$

with $A_{ij} \in \mathbf{R}^{n_i \times n_j}$, where $n_1 = \operatorname{rank} B_{11}$, $n_j = \operatorname{rank} A_{j,j-1}$ for $j = 2, \dots, k-1$, and $n_j \geq n_{j+1}$ for $j = 1, 2, \dots, k-1$. The matrices B_{11} and $A_{j,j-1}$, $j = 2, \dots, k$, are of the form $[0, R]$ with R nonsingular and upper triangular, with the possible exception of $A_{k,k-1}$ which is zero if the system is uncontrollable. As before, if (A, B) is controllable we also refer $n_1, \dots, n_k$ as the conjugate Kronecker indexes, and k as the controllability index of the system, where $n_k = \operatorname{rank} A_{k,k-1}$.

As has been done in Section 3.6, the reduction into staircase canonical form may be accomplished efficiently using the QR decomposition. This requires two major steps: the first step is to use Algorithm 3.6.1 to compute orthogonal matrices $U_1 \in \mathbf{R}^{n\times n}$, $V_1 \in \mathbf{R}^{m\times m}$ such that $(\widetilde{A}, \widetilde{B}) = (U_1 A U_1^T, U_1 B V_1)$ is of the orthogonal canonical form as in (3.6.1); the second step is to use orthogonal equivalence transformations to reduce $(\widetilde{A}, \widetilde{B})$ to a staircase canonical form (A_1, B_1) as in (3.7.1) and (3.7.2). The second step can be done as follows.

Let $A_k = \widetilde{A}$. For each $j = k, k-1, \dots, 2$, we first compute an orthogonal matrix $\widetilde{Q}_{j-1} \in \mathbf{R}^{n_{j-1}\times n_{j-1}}$ such that

$$A^{(j)}_{j,j-1} \widetilde{Q}^T_{j-1} = [0, R]$$

with R upper triangular, then use it make the transformation

$$A_{j-1} = Q_{j-1} A_j Q^T_{j-1},$$

where

$$Q_{j-1} = \operatorname{diag}(I_{s_j}, \widetilde{Q}_{j-1}, I_{r_j}), \; s_j = n_1 + \cdots + n_{j-2}, \; r_j = n - s_j - n_{j-1}.$$

Because of the from of Q_{j-1}, it is easy to see that A_{j-1} has the form

$$A_{j-1} = \begin{bmatrix} A_{11}^{(j-1)} & A_{12}^{(j-1)} & \cdots & A_{1,k-1}^{(j-1)} & A_{1k}^{(j-1)} \\ A_{21}^{(j-1)} & A_{22}^{(j-1)} & \cdots & A_{2,k-1}^{(j-1)} & A_{2k}^{(j-1)} \\ & A_{32}^{(j-1)} & \cdots & A_{3,k-1}^{(j-1)} & A_{3k}^{(j-1)} \\ & & \ddots & \vdots & \vdots \\ & & & A_{k,k-1}^{(j-1)} & A_{kk}^{(j-1)} \end{bmatrix}$$

with $A_{ij}^{(j-1)} \in \mathbf{R}^{n_i \times n_j}$, where the n_i are the conjugate Kronecker indexes, and

$$A_{j,j-1}^{(j-1)} = A_{j,j-1}^{(j)} \widetilde{Q}_{j-1}^T, \quad A_{j-1,j-2}^{(j-1)} = \widetilde{Q}_{j-1} A_{j-1,j-2}^{(j)}, \quad A_{i,i-1}^{(j-1)} = A_{i,i-1}^{(j)}, \quad i \neq j, j-1.$$

As a result the resulting matrix A_1 has the form of (3.7.2).

To complete the equivalence transformation, we must multiply $\widetilde{B}$ on the left by those Q_j:

$$\widetilde{B}_1 = Q_1 Q_2 \ldots Q_{k-1} \widetilde{B} = Q_1 \widetilde{B} = \begin{bmatrix} \widetilde{Q}_1 B_{11}^{(0)} \\ 0 \end{bmatrix}.$$

Thus, in order to satisfy (3.7.1), we finally compute an orthogonal matrix $V_2 \in \mathbf{R}^{m \times m}$ such that

$$\widetilde{Q}_1 B_{11}^{(0)} V_2 = [0, R]$$

with R upper triangular. Now let

$$U = Q_1 Q_2 \ldots Q_{k-1} U_1, \quad V = V_1 V_2.$$

Then U, V are the desired orthogonal matrices.

We summarize the algorithm as follows.

Algorithm 3.7.1 Given an $n \times n$ matrix A and an $n \times m$ matrix B, this function reduces (A, B) to a staircase canonical form, and returns the transforming matrices U, V, the controllability index k, and the conjugate Kronecker indexes, stored in vector N.

```
function [A, B, U, V, N, k] = stairf(A, B)
    [n, m] = size(B);
    [A, B, U, V, N, k] = orthf(A, B); i_b = n;
    for j = k : -1 : 2
        for i = i_b : -1 : i_b - N(j) + 1
            j_b = i_b - N(j) - N(j - 1) + 1;
            j_e = i - N(j);
```

```
            x = fliplr(A(i, j_b : j_e));
            [u, β] = house(x^T);
            v = flipud(u);
            C = A(1 : i_b, j_b : j_e);
            A(1 : i_b, j_b : j_e) = C − βCvv^T;
            C = A(j_b : j_e, :);
            A(j_b : j_e, :) = C − βv(C^T v)^T;
            U(j_b : j_e, :) = U(j_b : j_e, :) − βv(U(j_b : j_e, :)^T v)^T;
            if j = 2
                B(1 : j_e, :) = B(1 : j_e, :) − βv(B(1 : j_e, :)^T v)^T;
            end
        end
        i_b = i_b − N(j);
    end
    for i = N(1) : −1 : 1
        j_e = i + m − N(1);
        x = fliplr(B(i, 1 : j_e));
        [u, β] = house(x^T);
        v = flipud(u);
        B(1 : i, 1 : j_e) = B(1 : i, 1 : j_e) − βB(1 : i, 1 : j_e)vv^T;
        V(:, 1 : j_e) = V(:, 1 : j_e) − βV(:, 1 : j_e)vv^T;
    end
```

In this algorithm the MATLAB functions **fliplr** and **flipud** are used. **fliplr**(A) returns A with columns flippen in the left-right direction; **flipud**(A) returns A with rows flippen in the up-down direction.

3.7.2 The Basic Idea

In what follows we assume that (A, B) is controllable, and assume that we have reduced it to the staircase canonical form as in (3.7.1) and (3.7.2). Next we describe how to use QR-like techniques to compute an $m \times n$ matrix F_1 so that $\lambda(A_1 + B_1 F_1) = \mathcal{L}$, where $\mathcal{L} = \{\lambda_1, \dots, \lambda_n\}$ is defined as in Problem PA. As has done in Section 3.6.1, without loss of generality, we may assume that $n_1 = m$.

In order to show the basic idea of QR-like method, we first consider the single-input case, i.e., $m = 1$. In this case, A_1 is an unreduced upper Hessenberg matrix, and B_1 is a multiple of the unit vector e_1, written as b_1, that is, A_1 and b_1 have the following forms

$$A_1 = \begin{bmatrix} \alpha_{11} & \cdots & \alpha_{1,n-1} & \alpha_{1n} \\ \alpha_{21} & \cdots & \alpha_{2,n-1} & \alpha_{2n} \\ & \ddots & \vdots & \vdots \\ & & \alpha_{n,n-1} & \alpha_{nn} \end{bmatrix}, \quad b_1 = \begin{bmatrix} \beta_1 \\ 0 \\ \vdots \\ 0 \end{bmatrix},$$

with $\beta_1\alpha_{21}\cdots\alpha_{n,n-1}\neq 0$.

Suppose we want to assign a real pole λ_1 to the system (A_1, b_1), i.e., we wish to find some $f_1 \in \mathbf{R}^n$ such that λ_1 is an eigenvalue of $A_1 + b_1 f_1^T$. To do this, we may compute an orthogonal matrix Q_1 and the first component of $f_1^T Q_1$ so that

$$Q_1^T(A_1 + b_1 f_1^T)Q_1 = \begin{bmatrix} \lambda_1 & * \\ 0 & A_2 + b_2 f_2^T \end{bmatrix}, \tag{3.7.3}$$

where

$$Q_1^T A_1 Q_1 = \begin{array}{c} \begin{bmatrix} * & * \\ * & A_2 \end{bmatrix} \begin{array}{l} 1 \\ n-1 \end{array} \\ \begin{array}{cc} 1 & n-1 \end{array} \end{array}, \quad Q_1^T b_1 = \begin{bmatrix} * \\ b_2 \end{bmatrix} \begin{array}{l} 1 \\ n-1 \end{array}, \quad f_1^T Q_1 = \begin{array}{c} [f_{11},\ f_2^T\ ]. \\ \begin{array}{cc} 1 & n-1 \end{array} \end{array}$$

Of course, in order to be able to assign the remaining poles to (A_2, b_2) in the same way we should require that (A_2, b_2) be still a staircase canonical form. This can be done as follows.

Note that the elements $\alpha_{21}, \alpha_{32}, \ldots, \alpha_{n,n-1}$ are not affected by the term $b_1 f_1^T$, we can compute $n-1$ Givens rotations $G_n, G_{n-1}, \ldots, G_2$ such that

$$(A_1 - \lambda_1 I)G_n G_{n-1}\cdots G_2 = R = \begin{bmatrix} r_{11} & r_{12} & \cdots & r_{1n} \\ & r_{22} & \cdots & r_{2n} \\ & & \ddots & \vdots \\ & & & r_{nn} \end{bmatrix}$$

is upper triangular. Observe that $\alpha_{i,i-1}\neq 0$ implies that G_i is nontrivial and $r_{ii}\neq 0$ for each $i = 2, \ldots, n$. Now let $Q_1 = G_n G_{n-1}\cdots G_2$ and $(f_{11},\ f_2^T) = f_1^T Q_1^T$. Note that if λ_1 is an eigenvalue of $A_1 + b_1 f_1^T$, then $A_1 - \lambda_1 I + b_1 f_1^T$ must be singular. Hence, in order to ensure λ_1 is an eigenvalue of $A_1 + b_1 f_1^T$, we may take f_{11} to be $-r_{11}/\beta_1$, i.e., f_{11} satisfies $\beta_1 f_{11} + r_{11} = 0$. Thus we have

$$(A_1 - \lambda_1 I + b_1 f_1^T)Q_1 = \begin{bmatrix} 0 & r_{12} + \beta_1 f_{12} & \cdots & r_{1n} + \beta_1 f_{1n} \\ & r_{22} & \cdots & r_{2n} \\ & & \ddots & \vdots \\ & & & r_{nn} \end{bmatrix},$$

where $f_2 = (f_{12}, \ldots, f_{1n})^T$.

We next complete the equivalence transformation

$$Q_1^T A_1 Q_1 = Q_1^T(A_1 - \lambda_1 I)Q_1 + \lambda_1 I = Q_1^T R + \lambda_1 I = \begin{array}{c} \begin{bmatrix} * & * \\ a_1 & A_2 \end{bmatrix} \begin{array}{l} 1 \\ n-1 \end{array} \\ \begin{array}{cc} 1 & n-1 \end{array} \end{array},$$

$$Q_1^T b_1 = \begin{bmatrix} * \\ b_2 \end{bmatrix} \begin{array}{l} 1 \\ n-1 \end{array}.$$

Because the G_i are nontrivial Givens rotations and $\beta_1 r_{22}\cdots r_{nn} \neq 0$, it follows that $(A_2,\, b_2)$ is still a staircase canonical form, i.e., A_2 is still an unreduced upper Hessenberg matrix, and b_2 a nonzero multiple of the unit vector e_1. Moreover, it is easy to see that

$$Q_1^T(A_1 + b_1 f_1^T)Q_1 = Q_1^T(A_1 - \lambda_1 I + b_1 f_1^T)Q_1 + \lambda_1 I = \begin{bmatrix} \lambda_1 & * \\ 0 & A_2 + b_2 f_2^T \end{bmatrix},$$

and we can continue the process.

It is interesting to note that the above process for assigning a pole is just as the process for finding an eigenvalue by applying the well-known QR algorithm with shifts, the only difference here being that the eigenvalue is known.

In the remainder of this section, we will devote ourselves to extending this idea to general case. The real poles and the complex poles will be separately treated.

3.7.3 The Techniques for Assigning Real Poles

First of all we show how to assign a real pole λ_1 to the system $(A_1,\, B_1)$ without assuming $m = 1$.

If $n_1 > n_2$, we can immediately determine the first column f_1 of F_1 so that $A_1 + B_1F_1$ has λ_1 as its eigenvalue by solving

$$B_{11}f_1 + A_{11}(:,\, 1) = \lambda_1 e_1$$

for f_1. Once this has been done, we have

$$A_1 + B_1F_1 = \begin{bmatrix} \lambda_1 & * \\ 0 & A_2 + B_2F_2 \end{bmatrix},$$

where

$$A_2 = A_1(2:n,\, 2:n), \quad B_2 = B_1(2:n,\, 2:m), \quad F_2 = F_1(2:m,\, 2:n).$$

Clearly, $(A_2,\, B_2)$ is still a staircase canonical form as in (3.7.1) and (3.7.2); moreover, it is still controllable by Corollary 3.2.3. Thus we can continue to assign the remaining poles to $(A_2,\, B_2)$ in the same way or by the techniques to be developed in what follows.

Note that $F_1(1,\, 2:m)$ does not affect the eigenvalues of $A_1 + B_1F$, so it can be arbitrarily chosen. As has done in the previous two sections, we here simply set it to be zero.

If $n_1 = n_2 = \cdots = n_r > n_{r+1} \geq \cdots \geq n_k > 0$, $1 < r < k$, we can use a technique analogous to that used in the single-input case to assign λ_1 to $(A_1,\, B_1)$.

In this case $A_1 - \lambda_1 I$ has the following form

$$A_1 - \lambda_1 I = \begin{bmatrix} * & * \\ H & * \\ 0 & * \end{bmatrix} \begin{matrix} n_1 - 1 \\ s \\ n - rn_1 \end{matrix}$$
$$\quad s \quad n-s$$

with $s = (r-1)n_1 + 1$ and H unreduced upper Hessenberg. Thus we can compute an orthogonal matrix $\widetilde{Q}_1 \in \mathbf{R}^{s \times s}$ as a product of $(r-1)n_1$ nontrivial Givens rotations such that $H\widetilde{Q}_1 = R$ is nonsingular and upper triangular. Let

$$a_1 = (A_1 - \lambda_1 I)Q_1 e_1 \quad \text{and} \quad [f_1,\, F_2] = F_1 Q_1,$$

where $f_1 \in \mathbf{R}^n$ and $Q_1 = \operatorname{diag}(\widetilde{Q}_1,\, I) \in \mathbf{R}^{n \times n}$. Then a_1 must have the form $a_1 = (\alpha_1, \ldots, \alpha_{n_1}, 0, \ldots, 0)^T$. To ensure that $(A_1 - \lambda_1 I) + B_1 F_1$ is singular, we may take f_1 to be the solution of the linear system

$$B_{11} f_1 = -(\alpha_1, \ldots, \alpha_{n_1})^T.$$

Letting

$$Q_1^T A_1 Q_1 = \begin{bmatrix} * & * \\ * & A_2 \end{bmatrix} \begin{matrix} 1 \\ n-1 \end{matrix} \quad \text{and} \quad Q_1^T B_1 = \begin{bmatrix} * \\ B_2 \end{bmatrix} \begin{matrix} 1 \\ n-1 \end{matrix},$$
$$1 \quad n-1$$

it is easy to verify that

$$\begin{aligned} Q_1^T(A_1 + B_1 F_1)Q_1 &= Q_1^T\left[(A_1 - \lambda_1 I)Q_1 + B_1 F_1 Q_1\right] + \lambda_1 I \\ &= \begin{bmatrix} \lambda_1 & * \\ 0 & A_2 + B_2 F_2 \end{bmatrix}, \end{aligned}$$

which means that we have assigned λ_1 to $(A_1,\, B_1)$. Moreover, it follows from the construction of Q_1 that $(A_2,\, B_2)$ is controllable and has the following form

$$B_2 = \begin{bmatrix} B_{22} \\ 0 \end{bmatrix}, \quad A_2 = \begin{bmatrix} A_{11}^{(2)} & A_{12}^{(2)} & \cdots & A_{1,k-1}^{(2)} & A_{1k}^{(2)} \\ A_{21}^{(2)} & A_{22}^{(2)} & \cdots & A_{2,k-1}^{(2)} & A_{2k}^{(2)} \\ & A_{32}^{(2)} & \cdots & A_{3,k-1}^{(2)} & A_{3k}^{(2)} \\ & & \ddots & \vdots & \vdots \\ & & & A_{k,k-1}^{(2)} & A_{kk}^{(2)} \end{bmatrix},$$

with $A_{ij}^{(2)} \in \mathbf{R}^{m_i \times m_j}$, where $m_i = n_i$ for all i except for $i = r$; $m_r = n_r - 1$. The matrices $A_{21}^{(2)}, \ldots, A_{r-1,r-2}^{(2)}$, and B_{22} are nonsingular and upper triangular; the matrix $A_{r,r-1}^{(2)}$ is $(n_1 - 1) \times n_1$ and has the form $[0,\, R]$ with R nonsingular and upper triangular, $A_{r+1,r} = [0,\, A_{r+1,r}^{(2)}]$, and $A_{j,j-1}^{(2)} = A_{j,j-1}$ for $j = r+2, \ldots, k$.

That is, (A_2, B_2) is still a staircase canonical form as in (3.7.1) and (3.7.2). This ensures that we can continue the process.

Thus, in either case we can assign a real pole λ_1 to the system (A_1, B_1); moreover, this process can be continued. Consequently, repeatedly applying the two techniques described above, we can assign all the real poles in $\mathcal{L}$ to (A_1, B_1).

3.7.4 The Techniques for Assigning Complex Poles

Now we show how to assign a pair of complex conjugate poles $\lambda_1 = \overline{\lambda}_2 = \zeta + i\eta$ with $\eta \neq 0$ to the system (A_1, B_1). Just as QR algorithm with double shifts, to avoid complex arithmetic, we will assign two poles at time instead of just one as above.

First, we describe the main technique. Suppose that $n_2 = n_3 = \cdots = n_r = q > n_{r+1}$, $r > 2$, and $n_1 = q$ or $n_1 = q + 1$. That is, A_1 in (3.7.2) has the form

$$A_1 = \begin{bmatrix} A_{11} & \cdots & A_{1,r-1} & A_{1r} & \cdots & A_{1,k-1} & A_{1k} \\ A_{21} & \cdots & A_{2,r-1} & A_{2r} & \cdots & A_{2,k-1} & A_{2k} \\ & \ddots & \vdots & \vdots & \vdots & \vdots & \vdots \\ & & A_{r,r-1} & A_{rr} & \cdots & A_{r,k-1} & A_{rk} \\ & & & (0,\ R) & \cdots & A_{r+1,k-1} & A_{r+1k} \\ & & & & \ddots & \vdots & \vdots \\ & & & & & A_{k,k-1} & A_{kk} \end{bmatrix} \tag{3.7.4}$$

with $A_{31}, \dots A_{r,r-1}$ nonsingular and upper triangular, $A_{r+1,r} \equiv [0,\ R]$ being nonsquare with R nonsingular and upper triangular, and A_{21} being $n_1 \times n_2$ with $n_1 = n_2$ or $n_1 = n_2 + 1$. In this case we may assign λ_1, λ_2 to the system (A_1, B_1) by computing an orthogonal matrix Q and the first two columns of $F_1 Q$ such that

$$Q^T(A_1 + B_1F_1)Q = \begin{bmatrix} S & * \\ 0 & A_2 + B_2F_2 \end{bmatrix}, \tag{3.7.5}$$

where $S \in \mathbf{R}^{2\times 2}$ with λ_1, λ_2 as its eigenvalues, and

$$Q^T A_1 Q = \underset{\displaystyle 2 \quad n-2}{\begin{bmatrix} * & * \\ * & A_2 \end{bmatrix}} \begin{matrix} 2 \\ n-2 \end{matrix}, \qquad Q^T B_1 = \begin{bmatrix} * \\ B_2 \end{bmatrix} \begin{matrix} 2 \\ \end{matrix},$$

with (A_2, B_2) being a staircase canonical form as in (3.7.1) and (3.7.2), which ensures that we can continue to assign the remaining poles to the system (A_2, B_2) in the same way. To do this we need the following result.

Lemma 3.7.1 *Let*

$$C = \begin{bmatrix} C_{11} & C_{12} & \cdots & C_{1l} \\ C_{21} & C_{22} & \cdots & C_{2l} \\ & \ddots & & \vdots \\ & & C_{l,l-1} & C_{ll} \end{bmatrix} \in \mathbf{R}^{n\times n}, \qquad l \geq 3, \tag{3.7.6}$$

with $C_{ij} \in \mathbf{R}^{n_i \times n_j}$, where $n_2 = n_3 = \cdots = n_l = q$, $n_1 = q$ or $q+1$, $C_{i,i-1}$, $i = 3, \ldots, l$, are nonsingular, C_{21} nonsingular if $n_1 = q$ and $C_{21} = [0,\ R]$ with R nonsingular if $n_1 = q+1$. Let Q is an orthogonal matrix with the properties:

(1) *The first two columns of the matrix*

$$M \equiv (C - \lambda_1 I)(C - \lambda_2 I)Q \tag{3.7.7}$$

has the form

$$M(:, 1:2) = \begin{bmatrix} M_{10} \\ \vdots \\ M_{l-1,0} \\ 0 \end{bmatrix} \tag{3.7.8}$$

with $M_{i0} \in \mathbf{R}^{n_i \times 2}$, $i = 1, \ldots, l-1$, where $\lambda_1 = \overline{\lambda}_2 = \zeta + i\eta$ with $\eta \neq 0$.

(2) *The matrix $H \equiv Q^T C Q$ may be partitioned as*

$$H = \begin{bmatrix} S & X \\ 0 & Y \end{bmatrix}, \tag{3.7.9}$$

with $S \in \mathbf{R}^{2\times 2}$.

(3) *In the case $n_1 = q+1$, there is at least one $i \geq 3$ such that $e_1^T Q e_i \neq 0$.*

Then S has λ_1, λ_2 as its eigenvalues.

Proof First we have

$$CM = C(C - \lambda_1 I)(C - \lambda_2 I)Q = (C - \lambda_1 I)(C - \lambda_1 I)CQ = MQ^T CQ = MH. \tag{3.7.10}$$

Equating the first two columns of the two sides of (3.7.10) we get

$$\begin{bmatrix} C_{11} & C_{12} & \cdots & C_{1,l-1} & C_{1l} \\ C_{21} & C_{22} & \cdots & C_{2,l-1} & C_{2l} \\ & \ddots & & \vdots & \vdots \\ & & \ddots & C_{l-1,l-1} & C_{l-1,l} \\ & & & C_{l,l-1} & C_{ll} \end{bmatrix} \begin{bmatrix} M_{10} \\ M_{20} \\ \vdots \\ M_{l-1,0} \\ 0 \end{bmatrix} = \begin{bmatrix} M_{10} \\ M_{20} \\ \vdots \\ M_{l-1,0} \\ 0 \end{bmatrix} S,$$

and so

$$C_{l,l-1}M_{l-1,0} = 0, \qquad \sum_{j=i-1}^{l-1} C_{ij}M_{j0} = M_{i0}, \qquad i = 2, \dots, l-1.$$

Hence, it follows that $M_{i0} = 0$ since $C_{i+1,i}$ is nonsingular for $i = 2, \dots, l-1$, and that

$$C_{21}M_{10} = 0. \tag{3.7.11}$$

When $n_1 = q$, C_{21} is also nonsingular, and so $M_{12} = 0$, proving that first two columns of M are zero. If $n_1 = q+1$, because of the form of C_{21}, (3.7.11) implies that M_{10} must have the form

$$M_{10} = \begin{bmatrix} u^T \\ 0 \end{bmatrix}, \qquad \text{with } u \in \mathbf{R}^2.$$

Let

$$Q^T e_1 = (\alpha_1, \dots, \alpha_n)^T.$$

Then

$$Q^T M = Q^T \begin{bmatrix} u^T & \times & \cdots & \times \\ 0 & \times & \cdots & \times \\ 0 & \times & \cdots & \times \\ \vdots & \vdots & \vdots & \vdots \\ 0 & \times & \cdots & \times \end{bmatrix} = \begin{bmatrix} \alpha_1 u^T & \times & \cdots & \times \\ \alpha_2 u^T & \times & \cdots & \times \\ \alpha_3 u^T & \times & \cdots & \times \\ \vdots & \vdots & \vdots & \vdots \\ \alpha_n u^T & \times & \cdots & \times \end{bmatrix}. \tag{3.7.12}$$

On the other hand, we have

$$Q^T M = Q^T(C - \lambda_1 I)(C - \lambda_2 I)Q = (H - \lambda_1 I)(H - \lambda_2 I)$$

$$= \begin{bmatrix} (S - \lambda_1 I)(S - \lambda_2 I) & * \\ 0 & * \end{bmatrix}. \tag{3.7.13}$$

Comparing (3.7.12) and (3.7.13), we get $u = 0$ since condition (3) implies that there is at least one $i \geq 3$ such that $\alpha_i \neq 0$. Thus, in the case $n_1 = q+1$, we also prove that the first two columns of M are zero. In view of this, it follows from (3.7.13) that

$$(S - \lambda_1 I)(S - \lambda_2 I) = 0,$$

which implies that S must have λ_1, λ_2 as its eigenvalues, since $\lambda_1 = \overline{\lambda}_2 = \zeta + i\eta$ with $\eta \neq 0$ and S is real. □

Remark 3.7.1 It is worthwhile to point out that if λ_1, λ_2 are real, then the condition

$$(S - \lambda_1 I)(S - \lambda_2 I) = 0$$

does not ensure that both of them are the eigenvalues of S. For example, if $S = \operatorname{diag}(\lambda_1, \lambda_1)$, $\lambda_1 \in \mathbf{R}$, then

$$(S - \lambda_1 I)(S - \lambda_2 I) = 0$$

holds for any $\lambda_2 \in \mathbf{R}$.

Let now

$$A_1^{(1)} = \begin{bmatrix} A_{11} & A_{12} & \cdots & A_{1,r-1} & A_{1r} \\ A_{21} & A_{22} & \cdots & A_{2,r-1} & A_{2r} \\ & A_{32} & \cdots & A_{3,r-1} & A_{3r} \\ & & \ddots & \vdots & \vdots \\ & & & A_{r,r-1} & A_{rr} \end{bmatrix}, \qquad B_1^{(1)} = \begin{bmatrix} B_{11} \\ 0 \end{bmatrix} \begin{matrix} n_1 \\ (r-1)q \end{matrix},$$

and let

$$N = (A_1^{(1)} + B_1^{(1)} F_1^{(1)} - \lambda_1 I)(A_1^{(1)} + B_1^{(1)} F_1^{(1)} - \lambda_2 I),$$

where $F_1^{(1)} = F_1(:, 1 : (r-1)q + n_1)$. Clearly, the last block of q rows of N is not affected by $F_1^{(1)}$ and has the form

$$N(n_1 + (r-2)q + 1 : n_1 + (r-1)q, :) = [0,\ N_{r,r-2}, N_{r,r-1}, N_{rr}] \equiv N_r, \quad (3.7.14)$$

with $N_{rj} \in \mathbf{R}^{q \times q}$, $j = r-2, r-1, r$.

In view of Lemma 3.7.1, we hope to be able to compute an orthogonal matrix $\widetilde{Q} \in \mathbf{R}^{p \times p}$ with $p = n_1 + (r-2)q + 1$, written $Q_1 = \operatorname{diag}(\widetilde{Q},\ I_{q-1})$, so that

(I) $Q_1^T A_1^{(1)} Q_1$ and $Q^T B_1^{(1)}$ have the forms

$$Q_1^T A_1^{(1)} Q_1 = \left[\begin{array}{c|cccc} A_{00}^{(1)} & A_{01}^{(1)} & \cdots & A_{0,r-1}^{(1)} & A_{0r}^{(1)} \\ \hline A_{10}^{(1)} & A_{11}^{(1)} & \cdots & A_{1,r-1}^{(1)} & A_{1r}^{(1)} \\ & A_{21}^{(1)} & \cdots & A_{2,r-1}^{(1)} & A_{2r}^{(1)} \\ & & \ddots & \vdots & \vdots \\ & & & A_{r,r-1}^{(1)} & A_{rr}^{(1)} \end{array}\right] \begin{matrix} 2 \\ \\ \\ \\ \\ \end{matrix},$$

$$\underbrace{\hphantom{A_{00}^{(1)}}}_{2}$$

$$Q_1^T B_1^{(1)} = \begin{bmatrix} B_{10} \\ B_{22} \\ 0 \end{bmatrix} \begin{matrix} 2 \\ n_1 \\ (r-1)q-2 \end{matrix}$$

with $A_{ij}^{(1)} \in \mathbf{R}^{n_i' \times n_j'}$, where $n_i' = n_i$ for $i = 1, \ldots, r-2$ and $n_{r-1}' = n_r' = q-1$. The matrices B_{22} and $A_{j,j-1}^{(1)}$ are nonsingular and upper triangular for all j except for $j = 2,\ r-1$; $A_{r-1,r-2}^{(1)}$ is a $(q-1) \times q$ matrix of the form $[0, R]$ with R nonsingular and upper triangular; $A_{21}^{(1)}$ is nonsingular and upper triangular if $n_1 = q$ and $A_{21}^{(1)}$ is a $q \times (q+1)$ matrix of the form $[0, R]$ with R nonsingular and upper triangular if $n_1 = q+1$.

(II) The first two columns of $N_r Q_1$ are zero.

(III) In the case $n_1 = q+1$, there is at least one $i \geq 3$ such that $e_1^T Q_1 e_i \neq 0$.

If such a transforming matrix Q_1 has been computed, then we may solve

$$B_{22} F_{11} = -A_{10}^{(1)}$$

for F_{11}, where F_{11} is the first two columns of $F_1^{(1)} Q_1^T$. Thus we have

$$Q_1^T (A_1^{(1)} + B_1^{(1)} F_1^{(1)}) Q_1 = \begin{bmatrix} A_{00}^{(1)} + B_{10} F_{11} & * \\ 0 & * \end{bmatrix} \begin{matrix} 2 \\ \\ \end{matrix} .$$
$$\qquad\qquad\qquad\qquad 2$$

Now, applying Lemma 3.7.1 to matrix $C = A_1^{(1)} + B_1^{(1)} F_1^{(1)}$ and orthogonal matrix Q_1, we know that $S = A_{00}^{(1)} + B_{10} F_{11}$ has λ_1, λ_2 as its eigenvalues. Therefore, letting $Q = \operatorname{diag}(Q_1, I_s)$ with $s = n - (r-1)q - n_1$, it is easy to see that F_{11} is also the first two columns of $F_1 Q^T$ and such Q and F_{11} satisfy the requirements of (3.7.5). However, by Remark 3.7.1, if λ_1, λ_2 are real, we can't assert that both of λ_1 and λ_2 are the eigenvalues of $A_{00}^{(1)} + B_{10} F_{11}$. Consequently, the problem of whether the following technique can be used to assign a pair of real poles is not solved now and remains a topic of further research.

It remains for us to show how to compute an orthogonal matrix Q_1 that satisfies requirements (I), (II), and (III). The calculation is somewhat complicated. The transforming matrix Q_1 is computed as a product of four orthogonal matrices $Q_1 = P_1 P_2 P_3 P_4$ as follows. The matrix P_1 is a product of $q-1$ Givens rotations $P_1 = G_q G_{q-1} \ldots G_2$. The role of G_i is to zero the ith diagonal element of $A_{r,r-1}$ by applying it on the right side of $A_1^{(1)} G_q \ldots G_{i+1}$. P_1^T is then applied to the left side of $A_1^{(1)} P_1$, introducing $q-1$ nonzero elements in positions $(s+j+1, s-q+j)$, $j = 2, \ldots, q$, where $s = n_1 + (r-3)q$. In the case of $r = 4$, $n_1 = 3$, $n_2 = n_3 =$

$n_4 = 2$, we have

$$\left[\begin{array}{ccc|cc|cc|cc}
\times & \times & \times & \times & \times & \times & \times & \times & \times \\
\times & \times & \times & \times & \times & \times & \times & \times & \times \\
\times & \times & \times & \times & \times & \times & \times & \times & \times \\
\hline
 & + & \times & \times & \times & \times & \times & \times & \times \\
 & & + & \times & \times & \times & \times & \times & \times \\
\hline
 & & & + & \times & \times & \times & \times & \times \\
 & & & & + & \times & \times & \times & \times \\
\hline
 & & & & & + & \times & \times & \times \\
 & & & & & & + & \times & \times
\end{array}\right] \xrightarrow{P_1}$$

$$\left[\begin{array}{ccc|cc|cc|cc}
\times & \times & \times & \times & \times & \times & \times & \times & \times \\
\times & \times & \times & \times & \times & \times & \times & \times & \times \\
\times & \times & \times & \times & \times & \times & \times & \times & \times \\
\hline
 & + & \times & \times & \times & \times & \times & \times & \times \\
 & & + & \times & \times & \times & \times & \times & \times \\
\hline
 & & & + & \times & \times & \times & \times & \times \\
 & & & & \times & \times & \times & \times & \times \\
\hline
 & & & & + & \times & \times & \times & \times \\
 & & & & & & & + & \times
\end{array}\right],$$

as before, where $+$ denotes nonzero element. The matrix P_2 is now formed as a product of $q-1$ Householder transformations $P_2 = H_q H_{q-1} \dots H_2$. The role of H_j is to create zeros in positions $(s+j+1, s-q+j)$ and $(s+j+1, s-q+j+1)$ of $A_1^{(1)} P_1 H_q \dots H_{j+1}$, by applying H_j on its right side. P_2^T is then applied to the left side of $P_1^T A_1^{(1)} P_1 P_2$, introducing new nonzero elements as follows:

$$\xrightarrow{P_2} \left[\begin{array}{ccc|cc|cc|cc}
\times & \times & \times & \times & \times & \times & \times & \times & \times \\
\times & \times & \times & \times & \times & \times & \times & \times & \times \\
\times & \times & \times & \times & \times & \times & \times & \times & \times \\
\hline
 & + & \times & \times & \times & \times & \times & \times & \times \\
 & & \times & \times & \times & \times & \times & \times & \times \\
\hline
 & & \times & \times & \times & \times & \times & \times & \times \\
 & & + & \times & \times & \times & \times & \times & \times \\
\hline
 & & & & & & + & \times & \times \\
 & & & & & & & + & \times
\end{array}\right].$$

Up to now, we has not touched the matrix $B_1^{(1)}$, this is because in the case $r > 2$ we have $P_2^T P_1^T B_1^{(1)} = B_1^{(1)}$, that is, P_1 and P_2 do not affect $B_1^{(1)}$.

Observe that $P_1 P_2$ partly satisfies requirement (I). In order to satisfy requirement (II), we observe last q rows of $N P_1 P_2$, that is, $N_r P_1 P_2$. Note that

$$N P_1 P_2 = [(A_1^{(1)} + B_1^{(1)} F_1^{(1)} - \lambda_1 I) P_1 P_2][P_2^T P_1^T (A_1^{(1)} + B_1^{(1)} F_1^{(1)} - \lambda_2 I) P_1 P_2].$$

Because $P_2^T P_1^T B_1^{(1)} F_1^{(1)} P_1 P_2 = B_1^{(1)} F_1^{(1)} P_1 P_2$, it follows that

$$P_2^T P_1^T (A_1^{(1)} + B_1^{(1)} F_1^{(1)} - \lambda_2 I) P_1 P_2$$

has the same last $2(q-1)$ rows as $P_2^T P_1^T A_1^{(1)} P_1 P_2 - \lambda_2 I$, having the form

$$\begin{array}{c} \left[\begin{array}{cccc} 0 & \widetilde{A}_{r-1,r-2} & \widetilde{A}_{r-1,r-1} & \widetilde{A}_{r-1,r} \\ 0 & 0 & \widetilde{A}_{r,r-1} & \widetilde{A}_{r,r} \end{array}\right] \begin{array}{l} q-1 \\ q-1 \end{array} \\ \begin{array}{cccc} s & q-1 & q-1 & q-1 \end{array} \end{array} \tag{3.7.15}$$

where $s = n_1 + (r-1)q - 3(q-1)$ and $\widetilde{A}_{r-1,r-2}$ are nonsingular and upper triangular. Similarly, $(A_1^{(1)} + B_1^{(1)} F_1^{(1)} - \lambda_1 I) P_1 P_2$ has the same last $q-1$ rows as $A_1^{(1)} P_1 P_2 - \lambda_1 I$, having the form

$$\begin{array}{c} \left[\begin{array}{ccc} 0 & \widehat{A}_{r,r-1} & \widehat{A}_{rr} \end{array}\right] q-1, \\ \begin{array}{ccc} s_1 & q-1 & q-1 \end{array} \end{array} \tag{3.7.16}$$

with $\widetilde{A}_{r,r-1}$ nonsingular and upper triangular, where $s_1 = n_1 + (r-1)q - 2(q-1)$. Multiplying the necessary parts of (3.7.15) and (3.7.16), we see that the last $q-1$ rows of NP_1P_2 has the form

$$\begin{array}{c} \left[\begin{array}{cccc} 0 & \widetilde{M}_{r,r-2} & \widetilde{M}_{r,r-1} & \widetilde{M}_{rr} \end{array}\right] q-1 \\ \begin{array}{cccc} s & q-1 & q-1 & q-1 \end{array} \end{array}$$

where $\widetilde{M}_{r,r-2} = \widehat{A}_{r,r-1}\widetilde{A}_{r-1,r-2}$ is nonsingular and upper triangular.

Let w denote the first row of N_r defined as in (3.7.14). Then a short calculation gives that

$$w^T = (0, \ldots, 0, x^T, y^T, z^T) \tag{3.7.17}$$

where

$$\begin{aligned} x^T &= u^T A_{r-1,r-2} \\ y^T &= u^T A_{r-1,r-1} + v^T A_{r,r-1} - (2\zeta)u^T, \\ z^T &= u^T A_{r-1,r} + v^T A_{rr} - (2\zeta)v^T + (\zeta^2 + \eta^2)e_1^T, \end{aligned} \tag{3.7.18}$$

with $u^T = A_{r,r-1}(1,:)$, $v^T = A_{rr}(1,:)$, and $e_1 \in \mathbf{R}^q$. Because of the form of P_1P_2, it follows that $w^T P_1 P_2$ still has the form of (3.7.17), and component $n_1+(r-4)q+1$ of $w^T P_1 P_2$ (i.e., the first component of x) is $A_{r,r-1}(1,1)A_{r-1,r-2}(1,1)$, which is nonzero. Thus, $N_r P_1 P_2$ has the form

$$\begin{array}{c} \left[\begin{array}{c|ccc|ccc} 0 & + & \times & \times & * & * & * \\ 0 & 0 & 0 & 0 & \widetilde{M}_{r,r-2} & \widetilde{M}_{r,r-1} & \widetilde{M}_{rr} \end{array}\right] \begin{array}{l} 1 \\ q-1 \end{array} . \\ \begin{array}{ccccc} s-3 & & q-1 & q-1 & q-1 \end{array} \end{array}$$

To ensure that the resulting orthogonal matrix Q_1 satisfies requirement (II), we compute P_3 as a Householder transformation so that

$$w^T P_1 P_2 P_3 = (0, \ldots, 0, \widetilde{w}^T)^T, \qquad \text{with } \widetilde{w} \in \mathbf{R}^{3q-2},$$

that is, P_3 is to create zeros in positions $n_1 + (r-4)q + 1$ and $n_1 + (r-4)q + 2$ of $w^T P_1 P_2$. Thus, $N_r P_1 P_2 P_3$ (i.e., the last q rows of $N P_1 P_2 P_3$) has the form

$$N_r P_1 P_2 P_3 = \underset{s}{[0,}\ \underset{q}{M_{r,r-2},}\ \underset{q-1}{M_{r,r-1},}\ \underset{q-1}{M_{rr}]}\ q, \tag{3.7.19}$$

which will ensure that the resulting matrix Q_1 satisfies requirement (II). P_3 is then applied as $P_3^T P_2^T P_1^T A_1^{(1)} P_1 P_2 P_3$, with the following effect:

$$\xrightarrow{P_3} \left[\begin{array}{ccc|cc|cc|cc}
\times & \times & \times & \times & \times & \times & \times & \times & \times \\
\times & \times & \times & \times & \times & \times & \times & \times & \times \\
\times & \times & \times & \times & \times & \times & \times & \times & \times \\
\hline
 & \times & \times & \times & \times & \times & \times & \times & \times \\
 & \times & \times & \times & \times & \times & \times & \times & \times \\
\hline
 & + & \times & \times & \times & \times & \times & \times & \times \\
 & & + & \times & \times & \times & \times & \times & \times \\
\hline
 & & & & & & + & \times & \times \\
 & & & & & & & + & \times
\end{array}\right].$$

It is easy to see that P_3 does not zero any element of $P_2^T P_1^T A_1^{(1)} P_1 P_2$; it simply combines the appropriate columns and rows.

To satisfy the requirement (I), we need to reduce $P_3^T P_2^T P_1^T A_1^{(1)} P_1 P_2 P_3$ into the staircase as in (I). Hence, we compute $\widetilde{P}_4$ as a product of $(r-3)q$ Householder transformations

$$\widetilde{P}_4 = H_s H_{s-1} \cdots H_{n_1+3},$$

where $s = n_1 + (r-3)q + 2$. The role of H_j is to zero the first two nonzero elements of row j of

$$H_{j+1} \cdots H_s P_3^T P_2^T P_1^T A_1^{(1)} P_1 P_2 P_3 H_s \cdots H_{j+1}.$$

Thus, $\widetilde{P}_4^T P_3^T P_2^T P_1^T A_1^{(1)} P_1 P_2 P_3 \widetilde{P}_4$ has the desired form as in (I). In our example, it has the form

$$\xrightarrow{\widetilde{P}_4} \left[\begin{array}{ccc|cc|cc|cc}
\times & \times & \times & \times & \times & \times & \times & \times & \times \\
\times & \times & \times & \times & \times & \times & \times & \times & \times \\
\times & \times & \times & \times & \times & \times & \times & \times & \times \\
\hline
\times & \times & \times & \times & \times & \times & \times & \times & \times \\
\times & \times & \times & \times & \times & \times & \times & \times & \times \\
\hline
 & & & + & \times & \times & \times & \times & \times \\
 & & & & + & \times & \times & \times & \times \\
\hline
 & & & & & & + & \times & \times \\
 & & & & & & & + & \times
\end{array}\right].$$

If $n_1 = q$, then $P_4 = \widetilde{P}_4$. Otherwise if $n_1 = q+1$, to satisfy requirement (III), we take P_4 to be $\widetilde{P}_4 P$, where P is the identity matrix with rows 1 and 3 interchanged. Note that P dose not change the form of $\widetilde{P}_4^T P_3^T P_2^T P_1^T A_1^{(1)} P_1 P_2 P_3 \widetilde{P}_4$. Thus, we have completed the calculation of Q_1. Because of the form of Q_1, it follows that if $n_1 = q+1$, then

$$Q_1^T e_1 = P_4^T P_3^T P_2^T P_1^T e_1 = P_4^T e_1 = P \widetilde{P}_4^T e_1 = P e_1,$$

which means the third component of $Q_1^T e_1$ is 1, and so Q_1 satisfies the requirement (III). Moreover, it is clear that

$$N_r Q_1 = N_r P_1 P_2 P_3 P_4 = N_r P_1 P_2 P_3$$

which, together with (3.7.19), shows that Q_1 satisfies requirement (II). In addition, it is easy to see that Q_1 has the form

$$Q_1 = \mathrm{diag}\,(\widetilde{Q}, I_{q-1}).$$

To complete the equivalence transformation, we need to apply Q_1^T on the left side of $B_1^{(1)}$:

$$Q_1^T B_1^{(1)} = P_4^T P_3^t P_2^T P_1^T B_1^{(1)} = P_4^T B_1^{(1)} = P H_{n_1+3} \cdots H_s B_1^{(1)}$$

where $P = I$ or the identity matrix with rows 1 and 3 interchanged, which implies that $Q_1^T B_1^{(1)}$ has the desired form as in (I).

Observe that the algorithm we just described is to use implicitly shifted double steps to assign a pair of complex conjugate poles λ_1, λ_2 to the system (A_1, B_1). It only needs to compute row $n_1 + (r-2)q + 1$ of N explicitly, which is the only part of the process that λ_1, λ_2 explicitly take part in the computation. In this way the computation uses only real arithmetic, assigning complex conjugate poles in one double-step. However, the technique just developed can only treat the case of $n_2 = n_3 = \cdots = n_r > n_{r+1}$, $r > 2$, $n_1 = n_2$ or $n_2 + 1$. Therefore, we need to develop other techniques to treat the following cases:

(1) $n_1 = n_2 > n_3$;
(2) $n_1 = n_2 + 1 \geq n_3$;
(3) $n_1 > n_2 + 1$.

We first consider the case $q = n_1 = n_2 > n_3$. In this case, as before, to facilitate the pole assignment, we compute an orthogonal matrix Q as a product of $q-1$ Givens rotations $Q = G_q G_{q-1} \cdots G_2$. The G_j is applied on the right side of $A_1 G_q \cdots G_{j+1}$ to zero the jth diagonal element of A_{21}. Q is then applied as $(Q^T A_1 Q,\ Q^T B)$. Now let

$$Q^T A_1 Q = \underset{\displaystyle 2 \quad n-2}{\left[\begin{array}{cc} A_{00} & A_{01} \\ A_{10} & A_2 \end{array}\right]} \begin{array}{l} 2 \\ n-2 \end{array}, \quad Q^T B = \underset{\displaystyle 1 \quad m-1}{\left[\begin{array}{cc} b & B_{01} \\ B_{10} & B_2 \end{array}\right]} \begin{array}{l} 2 \\ n-2 \end{array}. \tag{3.7.20}$$

Then because of the form of Q, it is easy to see that

$$A_{10} = \begin{bmatrix} A_{11}^{(0)} \\ 0 \end{bmatrix} \begin{matrix} q-1 \\ \\ \end{matrix}, \quad B_2 = \begin{bmatrix} B_{22} \\ 0 \end{bmatrix} \begin{matrix} q-1 \\ \\ \end{matrix}, \quad B_{10} = 0, \; b = \begin{bmatrix} \beta \\ 0 \end{bmatrix} \quad \text{with } \beta \neq 0.$$

and that $(A_2,\, B_2)$ is still a staircase canonical form as in (3.7.1) and (3.7.2) with B_{22} and all subdiagonal blocks having full row rank. Thus $(A_2,\, B_2)$ is controllable.

Next we give an example that clarifies the above points. Consider the system $n = 6$, $m = 3$, $k = 2$, $n_1 = n_2 = q = 3$. Then

$$(A_1,\; B_1) = \left(\begin{array}{ccc|ccc} \times & \times & \times & \times & \times & \times \\ \times & \times & \times & \times & \times & \times \\ \times & \times & \times & \times & \times & \times \\ \hline + & \times & \times & \times & \times & \times \\ & + & \times & \times & \times & \times \\ & & + & \times & \times & \times \end{array} \quad \begin{array}{ccc} + & \times & \times \\ & + & \times \\ & & + \\ \hline & & \\ & & \\ & & \end{array} \right) \xrightarrow{Q}$$

$$\left(\begin{array}{cc|cc|cc} \times & \times & \times & \times & \times & \times \\ \times & \times & \times & \times & \times & \times \\ \hline \times & \times & \times & \times & \times & \times \\ + & \times & \times & \times & \times & \times \\ \cline{3-6} & & + & \times & \times & \times \\ & & & + & \times & \times \end{array} \quad \begin{array}{c|cc} + & \times & \times \\ & \times & \times \\ \hline & + & \times \\ & & + \\ & & \\ & & \end{array} \right).$$

Let

$$F_1 Q = \begin{bmatrix} f_1^T & F_{12} \\ F_{21} & F_2 \end{bmatrix} \begin{matrix} 1 \\ m-1 \end{matrix} \; . \qquad \begin{matrix} 2 & n-2 \end{matrix}$$

Then we have

$$\begin{aligned} & Q^T A_1 Q + (Q^T B_1)(F_1 Q) \\ & \quad = \begin{bmatrix} A_{00} + b f_1^T + B_{01} F_{21} & A_{01} + b F_{12} + B_{01} F_2 \\ A_{10} + B_2 F_{21} & A_2 + b_2 F_2 \end{bmatrix}. \end{aligned} \tag{3.7.21}$$

Since B_{22} is nonsingular and upper triangular, we can determine F_{21} such that $A_{10} + B_2 F_{21} = 0$ by solving

$$A_{11}^{(0)} + B_{22} F_{21} = 0.$$

Once F_{21} has been determined, we may compute f_1 such that the 2×2 matrix

$$A_{00} + B_{01} F_{21} + b f_1^T$$

has the given eigenvalues

$$\lambda_1 = \zeta_1 + i\eta, \quad \lambda_2 = \zeta_2 - i\eta,$$

with $\eta \neq 0$ implying $\zeta_1 = \zeta_2$. Note that here the given λ_1 and λ_2 may be real. Now let

$$A_{00} + B_{01}F_{21} = \begin{bmatrix} \alpha_{11} & \alpha_{12} \\ \alpha_{21} & \alpha_{22} \end{bmatrix}.$$

To compute f_1^T, we need to show $\alpha_{21} \neq 0$. Note that Q has the form $Q = \mathrm{diag}\,(1, Q_1, I)$ with $Q_1 \in \mathbf{R}^{q\times q}$. Comparing the first column of each side of (3.7.21), we have

$$Q_1^T v = \alpha_{21} e_1, \tag{3.7.22}$$

provided F_{21} has been determined as above, where

$$v = (A_1 + B_1 F_1)(2 : q+1,\ 1).$$

Since

$$v(q) = A_1(q+1,\ 1) = A_{21}(1,\ 1) \neq 0,$$

and Q_1 is a product of $q-1$ nontrivial Givens rotations, it follows from (3.7.22) that $\alpha_{21} \neq 0$. Hence $(A_{00} + B_{01}F_{00},\ b)$ is a controllable two-dimensional single-input system. Then we may assign two poles $\lambda_1,\ \lambda_2$ to it by using Algorithm 3.5.1.

Observe that from (3.7.21) F_{12} does not affect the eigenvalues of $Q^T(A_1 + B_1F_1)Q$. Therefore it can be used to satisfy the other requirements. For example, we could compute F_{12} such that

$$A_{10} + bF_{12} + B_{01}F_2 = 0,$$

once F_2 is known. Here we simply choose $F_{12} = 0$. We then continue the pole assignment with the controllable system (A_2, B_2).

Next we consider the case $n_1 = n_2 + 1 \geq n_3$. In this case we need a slightly different technique. Here we compute $Q = P_1P_2$, with P_1 being a product of $n_2 - 1$ Givens rotations as in the case $n_1 = n_2$, and P_2 a permutation matrix that will interchange columns 1 and 3 of $A_1 + B_1F_1$. In the case of $n = 7$, $m = 4$, $n_1 = 4$, $n_2 = 3$, $k = 2$. We have

$$(A_1,\ B_1) = \left(\begin{array}{cccc|ccc c cccc}
\times & \times & \times & \times & \times & \times & \times & & + & \times & \times & \times \\
\times & \times & \times & \times & \times & \times & \times & & & + & \times & \times \\
\times & \times & \times & \times & \times & \times & \times & & & & + & \times \\
\times & \times & \times & \times & \times & \times & \times & & & & & + \\
\hline
 & + & \times & \times & \times & \times & \times & & & & & \\
 & & + & \times & \times & \times & \times & & & & & \\
 & & & + & \times & \times & \times & & & & &
\end{array}\right) \xrightarrow{Q}$$

$$\left(\begin{array}{cc|ccccc}
\times & \times & \times & \times & \times & \times & \times \\
\times & \times & \times & \times & \times & \times & \times \\
\hline
\times & \times & \times & \times & \times & \times & \times \\
\times & \times & \times & \times & \times & \times & \times \\
\times & \times & \times & \times & \times & \times & \times \\
\cline{3-7}
 & & & + & \times & \times & \times \\
 & & & & + & \times & \times
\end{array}
\quad
\begin{array}{cccc}
 & & + & \times \\
 & + & \times & \times \\
\hline
+ & \times & \times & \times \\
 & & + & \times \\
 & & & + \\
 & & & \\
 & & &
\end{array}\right).$$

Observe that row 3 of $Q^T B$ is not the desired form as in (3.7.1). Hence we must zero the nonzero element in position $(3,1)$ of $Q^T B$ by applying a Givens rotation G on its right side. In our example the role of G is as follows

$$Q^T B = \left[\begin{array}{cccc}
 & & + & \times \\
 & + & \times & \times \\
\hline
+ & \times & \times & \times \\
 & & + & \times \\
 & & & + \\
 & & & \\
 & & &
\end{array}\right] \xrightarrow{G} \left[\begin{array}{cccc}
 & & + & \times \\
+ & \times & \times & \times \\
\hline
 & + & \times & \times \\
 & & + & \times \\
 & & & \times \\
 & & & \\
 & & &
\end{array}\right].$$

In general, we now have

$$Q^T A_1 Q = \begin{bmatrix} A_{00} & A_{01} \\ A_{10} & A_2 \end{bmatrix} \begin{matrix} 2 \\ n-2 \end{matrix}, \qquad Q^T B_1 G = \begin{bmatrix} b & B_{01} \\ 0 & B_2 \end{bmatrix} \begin{matrix} 2 \\ n-2 \end{matrix},$$
$$\qquad\quad 2 \quad n-2 \qquad\qquad\qquad\qquad\qquad 2 \quad m-2$$

with $(A_2,\ B_2)$ controllable and in the form of (3.7.1) and (3.7.2), and

$$A_{10} = \begin{bmatrix} A_{11}^{(0)} \\ 0, \end{bmatrix} \begin{matrix} n_2 \\ \\ \end{matrix}, \quad B_2 = \begin{bmatrix} B_{11} \\ 0 \end{bmatrix} \begin{matrix} n_2 \\ \\ \end{matrix}, \quad b = \begin{bmatrix} 0 \\ \beta \end{bmatrix},$$

where B_{11} is nonsingular and upper triangular, and $\beta \neq 0$. Thus, letting

$$G^T F_1 Q = \begin{bmatrix} f_1^T & F_{12} \\ F_{21} & F_2 \end{bmatrix} \begin{matrix} 1 \\ m-1 \end{matrix},$$
$$2 \quad n-2$$

we have

$$\begin{aligned} & Q^T A_1 Q + (Q^T B_1 G)(G^T F_1 Q) \\ & = \begin{bmatrix} A_{00} + b f_1^T + B_{01} F_{21} & A_{01} + b F_{12} + B_{01} F_2 \\ A_{10} + B_2 F_{21} & A_2 + B_2 F_2 \end{bmatrix}, \end{aligned} \tag{3.7.23}$$

which is apparently similar to (3.7.20). As before, we first compute F_{21} such that $A_{10} + B_2 F_{21} = 0$ by solving

$$A_{11}^{(0)} + B_{22} F_{21} = 0,$$

then compute f_1 such that the 2×2 matrix

$$A_{00} + B_{01}F_{21} + bf_1^T$$

has the given eigenvalues λ_1, λ_2. Here λ_1, λ_2 may be real or a pair of complex conjugate numbers. Similarly, we can prove that $(1, 2)$ element of $A_{00} + B_{01}F_{21}$ is nonzero. In fact, letting α_{12} denote this element and noting

$$(A_1 + (B_1G)(G^TF_1))Qe_2 = (A_1 + (B_1G)(G^TF_1))e_2 = (v^T, 0)^T$$

with $v \in \mathbf{R}^{n_1+1}$ and $v(n_1 + 1) = A_1(n_1 + 1, 2) = A_{21}(1, 2) \neq 0$, it follows from (3.7.23) that

$$P_2^T P_1^T (v^T, 0)^T = Q^T(A_1 + (B_1G)(G^TF_1))Qe_2 = (\alpha_{12}, \times, 0, 0, \dots, 0)^T,$$

provided F_{21} has determined, and so

$$P_1^T \begin{bmatrix} v \\ 0 \end{bmatrix} = P_2 \begin{bmatrix} \alpha_{12} \\ \times \\ 0 \\ 0 \\ \vdots \\ 0 \end{bmatrix} = \begin{bmatrix} 0 \\ \times \\ \alpha_{12} \\ 0 \\ \vdots \\ 0 \end{bmatrix}.$$

This implies that $\alpha_{12} \neq 0$ since $v(n_1 + 1) \neq 0$ and P_1 is a product of $n_2 - 1$ nontrivial Givens rotations. Thus $(A_{00} + B_{01}F_{22}, b)$ is controllable, and so we can use Algorithm 3.5.1 to carry out the calculation of f_1^T. After this, we set $F_{12} = 0$ as in the $n_1 = n_2$ case, and continue to assign the remaining poles to system (A_2, B_2).

Finally, we consider the $n_1 > n_2 + 1$ case. In this case we may immediately assign $2[\frac{n_1-n_2}{2}]$ poles to system (A_2, B_2) by solving

$$A_1(1 : n_1, 1 : s) + B_{11}F_1(:, 1 : s) = \underset{s}{\begin{bmatrix} S \\ 0 \end{bmatrix}} \begin{matrix} s \\ n_1 - s \end{matrix} \tag{3.7.24}$$

where $s = 2[\frac{n_1-n_2}{2}]$ and S is any real $s \times s$ matrix with the given eigenvalues λ_1, $\lambda_2, \dots, \lambda_s$, satisfying $\lambda_{2k-1} = \overline{\lambda}_{2k}$ if $\mathrm{Im}(\lambda_{2k}) \neq 0$, $k = 1, 2, \dots, s/2$. Here, as has done in Section 3.6, we simply choose

$$S = \mathrm{diag}\,(S_1, \dots, S_{s/2})$$

with

$$S_k = \begin{bmatrix} \mathrm{Re}\,(\lambda_{2k-1}) & \mathrm{Im}\,(\lambda_{2k}) \\ -\mathrm{Im}\,(\lambda_{2k}) & \mathrm{Re}\,(\lambda_{2k}) \end{bmatrix}.$$

The reason for such a choice has been stated in Section 3.6. Notice that here the assigned poles may be real or complex.

After the solution of (3.7.23), we have

$$A_1 + B_1 F_1 = \begin{bmatrix} S & A_{01} + B_{00} F_{12} + B_{01} F_2 \\ 0 & A_2 + B_2 F_2 \end{bmatrix}, \tag{3.7.25}$$

where

$$A_1 = \begin{bmatrix} A_{00} & A_{01} \\ A_{10} & A_2 \end{bmatrix} \begin{matrix} s \\ n-s \end{matrix}, \qquad B_1 = \begin{bmatrix} B_{00} & B_{01} \\ B_{10} & B_2 \end{bmatrix} \begin{matrix} s \\ m-s \end{matrix},$$
$$\qquad\quad s \quad n-s \qquad\qquad\qquad\qquad s \quad m-s$$

and

$$F_1(:,\, s+1:n) = \begin{bmatrix} F_{12} \\ F_2 \end{bmatrix} \begin{matrix} s \\ m-s \end{matrix}.$$

Observe that F_{12} dose not affect the eigenvalues of $A_1 + B_1 F_1$, and thus can be chosen any. Here, again we set it to be zero. Thus we can continue to assign the remaining poles to the system $(A_2,\, B_2)$.

3.7.5 QR-like Algorithm

We have developed several techniques that treat various possibly occurred cases. Starting with $(A_1\ B_1)$ and using these techniques repeatedly, we can eventually assign all the desired poles to the system $(A_1,\, B_1)$. How this is done is illustrated in the following algorithm:

Algorithm 3.7.2 (QR-like Algorithm) Given an $n \times n$ matrix A, an $n \times m$ matrix B, and an n-dimensional vector E, with $E(2l-1) = \overline{E(2l)}$ if $\mathrm{Im}(E(2l-1))$ is not zero, this algorithm computes an $m \times n$ matrix F such that the eigenvalues of $A + BF$ are the values specified in E if $(A,\, B)$ is controllable, and returns an empty matrix otherwise.

$[n, m] = \mathbf{size}(B)$;
$[A, B, U, V, N, k] = \mathbf{stairf}(A, B)$;
if $N(k) = 0$
  $F = [\ ]$; **return**
end
$n_1 = N(1)$; $B = B(:, m - n_1 + 1 : m)$; $F = \mathbf{zeros}(n_1, n)$;
$N(k+1) = 0$; $m_2 = 0$; $l = 1$;
$m_1 = \left\lceil \frac{N(1)-N(2)}{2} \right\rceil$;
if $m_1 > 0$
  $S = \mathbf{diag}\big(\mathbf{Re}(E(1)), \dots, \mathbf{Re}(E(2m_1))\big)$;

for $j = 1 : m_1$
  $S(2j-1, 2j) = \mathbf{Im}(E(2j-1));$
  $S(2j, 2j-1) = -S(2j-1, 2j);$
end
$O = \mathbf{zeros}(n_1 - 2m_1, 2m_1);$
$C = \begin{bmatrix} S \\ O \end{bmatrix} - A(1 : n_1, 1 : 2m_1);$
Solve $B(1 : n_1, 1 : n_1)X = C$ for X;
$F(:, 1 : 2m_1) = X;$
$m_2 = m_2 + 2m_1;\; l = l + 2m_1;\; N(1) = N(1) - 2m_1;$
end
if $k > 1$
  $r = 2;$
  while $N(r) = N(r+1)$
    $r = r + 1;$
  end
end
while $k > 1$
  if $r = 1$
    if $N(1) - N(2) > 1$
      $S = \begin{bmatrix} \mathbf{Re}(E(l)) & \mathbf{Im}(E(l)) \\ -\mathbf{Im}(E(l)) & \mathbf{Re}(E(l+1)) \end{bmatrix};$
      $O = \mathbf{zeros}(N(1) - 2, 2);$
      $C = -A(l : l + N(1) - 1, l : l + 1) + \begin{bmatrix} S \\ O \end{bmatrix};$
      Solve $B(l : l + N(1) - 1, m_2 + 1 : n_1)X = C$ for X;
      $F(m_2 + 1 : n_1, l : l + 1) = X;$
      $N(1) = N(1) - 2;\; l = l + 2;\; m_2 = m_2 + 2;$
    end
    $r = 2;$
    while $N(r) = N(r+1)$
      $r = r + 1;$
    end
  elseif $r = 2$
    if $N(1) = 1$
      $\delta = E(l) + E(l+1) - A(n-1, n-1) - A(n, n);$
      $F(m_2 + 1, l) = \delta \,/ B(l, m_2 + 1);$
      $\delta = \dfrac{(A(n,n) - E(l))(A(n,n) - E(l+1))}{A(n, n-1)};$
      $\delta = \delta + A(n-1, n);$
      $F(m_2 + 1, l + 1) = \delta \,/ B(l, m_2 + 1);$
      $N(r) = N(r) - 1;\; N(r-1) = N(r-1) - 1;\; l = l + 2;$
    else
      $j_b = l + N(1);\; q = N(2);$

for $j = j_b : -1 : j_b - N(1) + 2$
$[c, s] = \textbf{givens}(A(j+q-1, j-1), A(j+q-1, j));$
$G = \begin{bmatrix} -s & c \\ c & s \end{bmatrix};$
$A(:, j-1:j) = A(:, j-1:j)G;$
$A(j-1:j, :) = GA(j-1:j, :);$
$U(j-1:j, :) = GU(j-1:j, :);$
$B(j-1:j, :) = GB(j-1:j, :);$
end
if $N(1) = N(2)$
$C = B(l+2:l+q, m_2+2:n_1);$
$D = -A(l+2:l+q, l:l+1);$
Solve $CX = D$ for X;
$F(m_2+2:n_1, l:l+1) = X;$
$X = B(l:l+1, m_2+2:n_1)X;$
$X = X + A(l:l+1, l:l+1);$
$\delta = E(l) + E(l+1) - X(1,1) - X(2,2);$
$F(m_2+1, l) = \delta\,/B(l, m_2+1);$
$\delta = \dfrac{(X(2,2) - E(l))(X(2,2) - E(l+1))}{X(2,1)};$
$\delta = \delta + X(1,2);$
$F(m_2+1, l+1) = -\delta\,/B(l, m_2+1);$
else
$x = A(l, :);\ A(l, :) = A(l+2, :);$
$A(l+2, :) = x;\ x = A(:, l);$
$A(:, l) = A(:, l+2);\ A(:, l+2) = x;$
$x = B(l, :);\ B(l, :) = B(l+2, :);$
$B(l+2, :) = x;\ x = U(l, :);$
$U(l, :) = U(l+2, :);\ U(l+2, :) = x;$
$\alpha = B(l+2, m_2+1);\ \beta = B(l+2, m_2+2);$
$[c, s] = \textbf{givens}(\alpha, \beta);$
$G = \begin{bmatrix} -s & c \\ c & s \end{bmatrix};$
$C = B(l:l+2, m_2+1:m_2+2)G;$
$B(l:l+2, m_2+1:m_2+2) = C;$
$V(:, m_2+1:m_2+2) = V(:, m_2+1:m_2+2)G;$
$C = -A(l+2:l+q, l:l+1);$
Solve $B(l+2:l+q, m_2+2:n_1)X = C$ for X;
$F(m_2+2:n_1, l:l+1) = X;$
$X = B(l:l+1, m_2+2:n_1)X;$
$X = X + A(l:l+1, l:l+1);$
$\delta = E(l) + E(l+1) - X(1,1) - X(2,2);$
$F(m_2+1, l+1) = \delta\,/B(l+1, m_2+1);$

```
                δ = (X(1,1) − E(l))(X(1,1) − E(l+1)) / X(1,2);
                δ = δ + X(2,1);
                F(m_2 + 1, l) = δ / B(l+1, m_2 + 1);
            end
            m_2 = m_2 + 1; l = l + 2; N(1) = N(1) − 1;
            N(2) = N(2) − 1; r = 1;
        end
else
    if Im(E(l)) = 0
        s_1 = 2;
        while s > 0
            if N(1) > N(2)
                b = −A(l : l + N(1) − 1, l);
                b(1) = b(1) + E(l);
                Solve B(l : l + N(1) − 1, m_2 + 1 : n_1)x = b for x;
                F(m_2 + 1 : n_1, l) = x;
                N(1) = N(1) − 1; l = l + 1; m_2 = m_2 + 1; s_1 = s_1 − 1;
            else
                A = A − E(l)I_n; P = I_n;
                for j = l − 1 + (r − 1) ∗ N(1) : −1 : l
                    i = j + N(1);
                    [c, s] = givens(A(i, j), A(i, j + 1));
                    G = [−s, c; c, s];
                    A(:, j : j + 1) = A(:, j : j + 1)G;
                    P(j : j + 1, :) = GP(j : j + 1, :);
                end
                b = −A(l : l + N(1) − 1, l);
                Solve B(l : l + N(1) − 1, m_2 + 1 : n_1)x = b for x;
                F(m_2 + 1 : n_1, l) = x;
                B = PB; A = PA + E(l)I_n; U = PU;
                N(r) = N(r) − 1; l = l + 1; s_1 = s_1 − 1; r = r − 1;
            end
        end
    else
        q = N(2); y = zeros(1, n); s_3 = l + N(1) + (r − 4)q;
        δ = −Re(E(l) + E(l + 1));
        x = A(s_3 + 2q, s_3 + q : s_3 + 3q − 1);
        z = δA(s_3 + 2q, s_3 : s_3 + 2q);
        C = A(s_3 + q : s_3 + 3q − 1, s_3 : s_3 + 2q);
        y(s_3 : s_3 + 2q) = xC + z;
        y(s_3 + 2q) = y(s_3 + 2q) + E(l)E(l + 1);
        j_b = l + N(1) + (r − 2)q;
        for j = j_b : −1 : j_b − q + 2
```

```
        [c, s] = givens(A(j + q - 1, j - 1), A(j + q - 1, j));
        G = [ -s  c
               c  s ];
        A(:, j - 1 : j) = A(:, j - 1 : j)G;
        A(j - 1 : j, :) = GA(j - 1 : j, :);
        U(j - 1 : j, :) = GU(j - 1 : j, :);
        y(j - 1 : j) = y(j - 1 : j)G;
    end
    for j = j_b - q + 1 : -1 : j_b - 2q + 3
        x = fliplr(A(j + q - 1, j - 2 : j));
        [u, β] = house(x^T); v = flipud(u);
        C = A(l : j + q - 1, j - 2 : j);
        A(l : j + q - 1, j - 2 : j) = C - βCvv^T;
        A(j - 2 : j, :) = A(j - 2 : j, :) - βv(A(j - 2 : j, :)^T v)^T;
        U(j - 2 : j, :) = U(j - 2 : j, :) - βv(U(j - 2 : j, :)^T v)^T;
        y(j - 2 : j) = y(j - 2 : j) - βy(j - 2 : j)vv^T;
        if j - 2 < l + N(1)
            B(j - 2 : j, :) = B(j - 2 : j, :) - βv(B(j - 2 : j, :)^T v)^T;
        end
    end
    x = fliplr(y(s_3 : s_3 + 2));
    [u, β] = house(x^T); v = flipud(u);
    C = A(l : n, s_3 : s_3 + 2);
    A(l : n, s_3 : s_3 + 2) = C - βCvv^T;
    A(s_3 : s_3 + 2, :) = A(s_3 : s_3 + 2, :) - βv(A(s_3 : s_3 + 2, :)^T v)^T;
    U(s_3 : s_3 + 2, :) = U(s_3 : s_3 + 2, :) - βv(U(s_3 : s_3 + 2, :)^T v)^T;
    if r = 3
        C = βv(B(s_3 : s_3 + 2, :)^T v)^T;
        B(s_3 : s_3 + 2, :) = B(s_3 : s_3 + 2, :) - C;
    end
    j_b = l + N(1) + (r - 4)q + 1;
    j_e = l + N(1) - q + 2;
    for j = j_b : -1 : j_e
        x = fliplr(A(j + q, j - 2 : j));
        [u, β] = house(x^T); v = flipud(u);
        C = A(l : j + q, j - 2 : j);
        A(l : j + q, j - 2 : j) = C - βCvv^T;
        A(j - 2 : j, :) = A(j - 2 : j, :) - βv(A(j - 2 : j, :)^T v)^T;
        U(j - 2 : j, :) = U(j - 2 : j, :) - βv(U(j - 2 : j, :)^T v)^T;
        if j - 2 < l + N(1)
            B(j - 2 : j, :) = B(j - 2 : j, :) - βv(B(j - 2 : j, :)^T v)^T;
        end
    end
    if N(1) > q
```

$x = A(l,:);\ A(l,:) = A(l+2,:);\ A(l+2,:) = x;$
$x = A(:,l);\ A(:,l) = A(:,l+2);\ A(:,l+2) = x;$
$x = B(l,:);\ B(l,:) = B(l+2,:);\ B(l+2,:) = x;$
$x = U(l,:);\ U(l,:) = U(l+2,:);\ U(l+2,:) = x;$
end
$C = -B(l+2 : l+N(1)+1, m_2+1 : n_1);$
$D = A(l+2 : l+N(1)+1, l : l+1);$
Solve $CX = D$ for X;
$F(m_2+1 : n_1, l : l+1) = X;$
$l = l+2;\ N(r) = N(r)-1;$
$N(r-1) = N(r-1)-1;\ r = r-2;$
end
end
if $N(k) = 0$
$k = k-1;$
end
if $N(k) = 0$
$k = k-1;$
end
end
if $l \leq n$
if $l = n$
$S = E(n);$
else

$$S = \begin{bmatrix} \mathbf{Re}(E(n-1)) & \mathbf{Im}(E(n)) \\ -\mathbf{Im}(E(n)) & \mathbf{Re}(E(n)) \end{bmatrix};$$

end
Solve $B(l:n, m_2+1:n_1)X = S - A(l:n, l:n)$ for X;
$F(m_2+1 : n_1, l : n) = X;$
end
$O = \mathbf{zeros}(m-n_1, n);$

$$F = \begin{bmatrix} O \\ F \end{bmatrix};\ F = VFU;$$

This algorithm has good numerical behavior because it uses orthogonal transformations whenever possible.

3.8 Numerical Examples

To illustrate the performance of the three algorithms described in the previous sections, in this section we give several interesting examples, which were carried out using MATLAB on a Sun Workstation, with machine epsilon $\varepsilon \approx 2.204 \times 10^{-16}$.

Example 3.8.1 Consider the single-input controllable system (A, b) with

$$A = \begin{bmatrix} 20 & & & & \\ 20 & 19 & & & \\ & 20 & 18 & & \\ & & \ddots & \ddots & \\ & & & 20 & 1 \end{bmatrix}, \qquad b = \begin{bmatrix} 1 \\ 0 \\ 0 \\ \vdots \\ 0 \end{bmatrix}.$$

It is well known that A has a very ill-conditioned eigenproblem. Assume that we want to assign the poles $20, 19, 18, \ldots, 1$ to the system. The correct solution to this problem is obviously $f = 0$. Using Algorithms 3.5.4, 3.6.2, and 3.7.2 to solve this problem, we find

$$\begin{aligned} f_1^T = 10^{-13}\times(&-1.49, -0.36, 1.42, 2.77, 2.63, 6.75,\\ &-10.44, 10.62, -1.63, -1.31 - 2.80, 4.97, -2.72,\\ &1.50, 0.32, 0.12, -0.28, 0.11, -0.05, -0.01), \end{aligned}$$

$$\begin{aligned} f_2^T = 10^{-13}\times(&-0.56, 0.18, 0.07, -0.04, -0.42, 1.01,\\ &-1.13, 1.32, -1.64, 2.50, -3.37, 2.38, 0.91,\\ &-4.76, 6.55, -5.56, 3.43, -1.57, 0.43, -0.04), \end{aligned}$$

and

$$\begin{aligned} f_3^T = 10^{-13}\times(&0.20, 0.01, -0.20, 0.35, -0.51, 0.67,\\ &-0.82, 0.10, -1.10, 1.16, -1.17, 1.16, -1.08,\\ &0.87, -0.58, 0.30, -0.11, 0.03, -0.00, 0.00), \end{aligned}$$

where f_1, f_2, and f_3 denote the computed solutions of Algorithms 3.5.4, 3.6.2, and 3.7.2, respectively. Clearly, they all agree up to more than 12 significant digits with the mathematically correct solution, and the QR-like solution f_3 and the invariant subspace solution f_2 are slightly better than the Schur solution f_1.

Example 3.8.2 Let (A_k, b_k) with $A_k \in \mathbf{R}^{k\times k}$, $b_k \in \mathbf{R}^k$ be a random system generated by the MATLAB function **randn**, and let the assigned poles be the computed eigenvalues of $A_k + b_k f_k^T$ with $f_k \in \mathbf{R}^k$ generated by **randn**, which were computed by using the MATLAB function **eig**. Applying Algorithms 3.5.4, 3.6.2, and 3.7.2 to regenerate the f_k for $k = 15:5:35$ and $k = 36:50$, we find the computed errors shown as in Table 3.8.1, where $f_k^{(1)}$, $f_k^{(2)}$, and $f_k^{(3)}$ denote the computed solutions of the three algorithms, respectively. Clearly, for this example QR-like method is the best one.

Table 3.8.1 Comparison of the three algorithms for Example 3.8.2.

k	$\|f_k - f_k^{(1)}\|_\infty$	$\|f_k - f_k^{(2)}\|_\infty$	$\|f_k - f_k^{(3)}\|_\infty$
15	4.534×10^{-13}	5.704×10^{-14}	4.520×10^{-14}
20	1.879×10^{-11}	5.730×10^{-14}	5.301×10^{-14}
25	2.447×10^{-11}	5.578×10^{-13}	1.479×10^{-13}
30	2.161×10^{-9}	3.850×10^{-13}	4.888×10^{-13}
35	1.266×10^{-10}	1.195×10^{-13}	1.893×10^{-13}
36	2.368×10^{-9}	1.979×10^{-8}	4.081×10^{-13}
37	1.472×10^{-10}	1.386×10^{-13}	9.784×10^{-14}
38	2.501×10^{-9}	5.463×10^{-2}	4.558×10^{-13}
39	5.210×10^{-10}	8.260×10^{-14}	9.415×10^{-14}
40	4.465×10^{-10}	6.282×10^{-13}	1.759×10^{-11}
41	3.415×10^{-8}	6.509×10^{-8}	4.689×10^{-13}
42	1.364×10^{-8}	2.079×10^{-11}	4.660×10^{-13}
43	9.291×10^{-8}	1.596×10^{-12}	4.126×10^{-13}
44	1.010×10^{-7}	9.314×10^{-9}	2.536×10^{-13}
45	7.600×10^{-8}	9.223×10^{-14}	2.184×10^{-13}
46	6.003×10^{-9}	2.981×10^{-12}	7.275×10^{-13}
47	1.303×10^{-8}	2.203×10^{-11}	1.250×10^{-13}
48	9.741×10^{-9}	2.151×10^{-13}	1.792×10^{-13}
49	1.584×10^{-6}	2.261×10^{-12}	1.414×10^{-13}
50	1.222×10^{-7}	5.941×10^{-13}	3.000×10^{-13}

Example 3.8.3 Consider the controllable multi-input system (A, B) with

$$A = \left[\begin{array}{ccc|cc|c} 2 & 4 & 6 & 7 & 8 & 9 \\ 1 & 3 & 2 & 6 & 4 & 7 \\ 1 & 4 & 6 & 7 & 8 & 9 \\ \hline 0 & 7 & 4 & 2 & 4 & 5 \\ 0 & 0 & 6 & 7 & 8 & 9 \\ \hline 0 & 0 & 0 & 0 & 5 & 8 \end{array}\right], \qquad B = \left[\begin{array}{ccc} 10^{-4} & 10^2 & 10^2 \\ & 10^{-4} & 10^2 \\ & & 10^{-4} \\ \hline & & \\ & & \\ & & \end{array}\right],$$

and assume that we want to assign the poles 1, 2, 4, 5, 7, 8. Using the three algorithms to this problem, we find

$$F_1 = \begin{bmatrix} 1.6 \times 10^{-6} & 0.00002 & 0.00002 & 0.00002 & 0.00004 & 0.00005 \\ 1.58421 & 17.8963 & 20.6091 & 20.0351 & 37.2397 & 47.4209 \\ -0.138731 & -1.46550 & -1.75432 & -1.75933 & -3.17504 & -4.06325 \end{bmatrix},$$

$$F_2 = \begin{bmatrix} 0.12985 & 1.57982 & 1.65139 & 1.52265 & 3.14495 & 3.88378 \\ 129848. & 1579817. & 1651393. & 1522653. & 3144947. & 3883774. \\ -9861.98 & -119987. & -125424. & -115646. & -238859. & -294974. \end{bmatrix},$$

and

$$F_3 = \begin{bmatrix} 0.13167 & 1.60193 & 0.87777 & 0.61444 & 2.72109 & 2.74303 \\ 131665.2 & 1601929. & 877768. & 614436. & 2721082. & 2743025. \\ -10000.0 & -121667. & -15372.6 & 13176.4 & 1.2\times10^{-8} & -4.3\times10^{-9} \end{bmatrix}.$$

Using the MATLAB function **eig**, we find that the computed eigenvalues of $A + BF_1$ are

$$5.00000000,\ 4.00000000,\ 1.00000000,\ 7.00000000,\ 2.00000000,\ 8.00000000,$$

which are in good agreement with the assigned poles, whereas those of $A + BF_2$ and $A + BF_3$ are

$$8.001097 \pm 5.536717i,\ -2.952341,\ 3.943276,\ 2.006871,\ 8.000001,$$

and

$$11.509930 \pm 4.378134i,\ -0.157926 \pm 6.500778i,\ 0.176318,\ 4.119673,$$

respectively, which are far from the assigned poles. In addition, it is easy to calculate that

$$\|F_1\|_2 = 6.942180 \times 10, \quad \|F_2\|_2 = 5.720167 \times 10^6, \quad \|F_3\|_2 = 4.319974 \times 10^6.$$

These facts show that for this example the Schur method is much better than the other two.

Example 3.8.4 Consider a random multiple-input system (A, B) generated by **randn** as follows:

$$A(:,1:5) = \begin{bmatrix} 1.16495 & -0.36003 & 0.37504 & -0.55709 & -0.50770 \\ 0.62684 & -0.13558 & 1.12516 & -0.33671 & 0.88530 \\ 0.07508 & -1.34934 & 0.72864 & 0.41523 & -0.24809 \\ 0.35161 & -1.27045 & -2.37745 & 1.55781 & -0.72625 \\ -0.69651 & 0.98457 & -0.27378 & -2.44430 & -0.44504 \\ 1.69614 & -0.04488 & -0.32294 & -1.09820 & -0.61291 \\ 0.05906 & -0.79894 & 0.31799 & 1.12265 & -0.20914 \\ 1.79707 & -0.76517 & -0.51117 & 0.58167 & 0.56215 \\ 0.26407 & 0.86173 & -0.00204 & -0.27135 & -1.06392 \\ 0.87167 & -0.05623 & 1.60651 & 0.41419 & 0.35159 \\ -1.44617 & 0.51348 & 0.84765 & -0.97781 & 1.13300 \\ -0.70117 & 0.39668 & 0.26810 & -1.02147 & 0.14999 \\ 1.24598 & 0.75622 & -0.92349 & 0.31769 & 0.70314 \\ -0.63898 & 0.40049 & -0.07050 & 1.51611 & -0.05241 \\ 0.57735 & -1.34138 & 0.14789 & 0.74943 & 2.01850 \end{bmatrix},$$

$$A(:,6:10) = \begin{bmatrix} 0.92416 & 0.03988 & -0.62021 & 0.43871 & 0.67029 \\ -1.81411 & -2.48284 & 0.23715 & -1.24734 & 0.42015 \\ 0.03497 & 1.15865 & -1.58685 & 0.32467 & -2.87275 \\ -1.80786 & -1.02628 & -0.40148 & 0.39007 & 1.68587 \\ 1.02819 & 1.15349 & -0.77069 & -0.40514 & 0.02792 \\ 0.39460 & -0.78646 & -0.26268 & 0.29231 & -0.90203 \\ 0.63941 & 0.63481 & 0.97649 & 2.56591 & -2.05326 \\ 0.87421 & 0.82041 & 0.97782 & -0.45782 & 0.08909 \\ 1.75240 & -0.17603 & 1.17002 & -1.61083 & 2.08710 \\ -0.32005 & 0.56247 & 0.15931 & -2.66952 & 0.36512 \\ -0.13741 & -0.12744 & 0.49952 & -0.75970 & 0.84611 \\ 0.61577 & 0.55417 & -1.05538 & -0.67472 & -0.18454 \\ 0.97789 & -1.09734 & -0.45074 & -1.17169 & 1.03071 \\ -1.11535 & -0.73130 & 1.27038 & 2.03293 & -1.52762 \\ -0.55002 & 1.40473 & 0.89869 & 0.96848 & 0.96494 \end{bmatrix},$$

$$A(:,11:15) = \begin{bmatrix} 0.52616 & -1.40695 & -0.84441 & 0.78643 & -1.93113 \\ -0.18445 & 1.03081 & -0.31163 & -1.46164 & 0.66030 \\ 0.19878 & -0.75987 & 0.39781 & 1.55447 & -1.10251 \\ 1.59043 & 0.87413 & 1.04979 & -0.59754 & -0.10297 \\ 0.03219 & 0.76113 & -0.34080 & -1.21057 & -1.05980 \\ 0.88916 & -0.16592 & 0.33630 & -0.70267 & -1.23857 \\ -1.29915 & 0.30091 & -0.22136 & 0.35643 & -1.88924 \\ 1.18257 & -0.32247 & 0.01665 & 0.65264 & -0.97358 \\ 1.81747 & -0.36841 & -1.19236 & 0.21567 & 0.21212 \\ -0.58430 & 1.14790 & -0.13165 & -0.26390 & 0.49344 \\ -1.01067 & 0.04143 & 1.48752 & 1.80244 & 1.54718 \\ -0.96050 & -1.09805 & -0.83682 & -0.64298 & 0.64493 \\ 0.69116 & 1.56672 & -1.30098 & 0.10956 & -2.14836 \\ -0.75862 & -1.04842 & 1.57413 & -0.71904 & -1.02884 \\ -0.09697 & 0.42272 & 1.16604 & 0.42063 & -0.14158 \end{bmatrix},$$

and

$$B(1:9,:) = \begin{bmatrix} -2.52671 & -0.59452 & -0.03124 & 0.96899 & 1.15778 \\ -0.31298 & 0.13025 & 0.77821 & -0.74732 & 0.16191 \\ -0.59362 & 0.03501 & 2.18048 & -2.79602 & 1.55706 \\ 0.33232 & -0.62467 & 0.43781 & 0.69673 & -0.19354 \\ 0.55885 & -0.53978 & 1.33333 & 3.20691 & 1.65130 \\ 0.89988 & 1.87996 & 0.25108 & 0.53601 & -1.89878 \\ -0.20090 & -1.00385 & -0.31047 & 0.29845 & 1.82252 \\ -0.23373 & -0.49745 & -0.92300 & 0.28404 & -1.51842 \\ 1.44991 & -1.50440 & -0.38478 & 0.95966 & -1.05107 \end{bmatrix}$$

$$B(10:15,:) = \begin{bmatrix} 1.83613 & -0.09545 & 1.15818 & 2.08759 & 0.04993 \\ -0.38292 & 0.39673 & 0.86250 & 1.52468 & -1.45475 \\ 0.15508 & -0.52711 & -1.03471 & -0.19526 & 0.46655 \\ -0.96465 & 0.34457 & -0.19267 & 0.01726 & 0.54544 \\ 0.03876 & -0.72329 & -1.29972 & 0.24634 & 1.32032 \\ 0.76546 & 1.26819 & 0.30660 & -0.85448 & -0.40449 \end{bmatrix}.$$

Assume that we want to assign the poles

$$\lambda_1 = \overline{\lambda}_2 = -0.277775505972 + 1.226502341058i, \qquad \lambda_3 = -0.141404614026,$$
$$\lambda_4 = \overline{\lambda}_5 = -1.049022349784 + 0.396516210293i, \qquad \lambda_6 = -0.159447782443,$$
$$\lambda_7 = \overline{\lambda}_8 = -1.179059656670 + 0.302689036171i, \qquad \lambda_9 = -0.198499915965,$$
$$\lambda_{10} = \overline{\lambda}_{11} = -1.581053413802 + 0.069600095098i, \qquad \lambda_{12} = -0.247348749631,$$
$$\lambda_{13} = -0.411267926557, \quad \lambda_{14} = -0.418468509074, \quad \lambda_{15} = -2.704026048249,$$

which were also generated by **randn**. Applying the three algorithms to this problem, we find

$$F_1^T = \begin{bmatrix} 0.52262 & 0.50091 & 0.08333 & 0.00688 & 0.09148 \\ 0.16008 & 0.08079 & 0.07359 & -0.06649 & 0.35817 \\ -0.14842 & -0.40979 & -0.27026 & 0.10824 & 0.05070 \\ 0.25148 & 0.23049 & -0.16270 & -0.25305 & -0.03703 \\ 0.19321 & -0.71825 & -0.84792 & -0.12858 & -0.45656 \\ -0.72005 & 0.89418 & 1.12036 & 0.09387 & 0.13919 \\ -1.10258 & 0.07822 & 0.15359 & 0.10061 & -0.88045 \\ -0.10994 & 0.50147 & 0.32769 & -0.18994 & 0.36316 \\ -1.22293 & 0.59255 & 1.14119 & 0.16351 & 0.02659 \\ 0.43874 & -0.56853 & -0.60352 & -0.56737 & 0.07078 \\ 0.80483 & 0.71107 & 0.32982 & -0.06901 & 0.40817 \\ -0.06901 & -0.28130 & -0.75457 & -0.23220 & -0.27593 \\ 0.35195 & -0.39564 & -0.40554 & -0.41130 & 0.14891 \\ 0.34814 & 0.10096 & 0.22738 & -0.26183 & 0.21630 \\ -0.06203 & -1.20650 & -0.76451 & 0.26325 & -0.03284 \end{bmatrix},$$

$$F_2(:,1:10)^T = \begin{bmatrix} -0.25936 & 0.11290 & 0.52160 & 0.18720 & -0.39321 \\ 0.95179 & -1.16946 & 0.55691 & -0.85520 & 0.58136 \\ -0.76303 & 0.58994 & -2.57879 & 0.69636 & 0.18496 \\ -0.52644 & 2.64706 & -2.15516 & 0.95298 & 0.35205 \\ -0.11765 & -0.49138 & 0.22153 & -0.52973 & -0.53869 \\ -0.95753 & 0.19297 & 0.95987 & 0.68242 & -0.18240 \\ -1.76236 & 1.80434 & -1.61679 & 1.02993 & -0.94907 \\ -1.31950 & -0.42713 & 2.47761 & 0.41914 & -0.44064 \\ -1.63927 & -1.81640 & 5.12029 & 1.22800 & -1.59479 \\ 0.52250 & 2.24718 & -2.22121 & -0.76647 & 1.51510 \end{bmatrix}$$

$$F_2(:, 11:15)^T = \begin{bmatrix} 0.60727 & -0.19571 & 1.82991 & -0.25546 & 0.55623 \\ -0.08469 & -0.15198 & 0.28630 & -0.71661 & -0.48539 \\ 0.14992 & -0.52881 & 0.49200 & -0.12330 & -0.00183 \\ -0.70086 & 2.56414 & -2.87483 & 1.43194 & 0.95529 \\ 0.30510 & 1.35194 & -3.59333 & -0.19312 & 1.35646 \end{bmatrix},$$

and

$$F_3^T = \begin{bmatrix} 0.27366 & -1.56482 & 2.30851 & -0.26269 & -0.77522 \\ 0.03341 & -0.68896 & 0.73577 & -1.20659 & 0.37558 \\ -0.10219 & -2.30842 & 2.38111 & -0.32594 & -0.88279 \\ -0.55252 & -0.03243 & -3.58341 & 0.80905 & -0.12225 \\ 0.25667 & 2.53309 & -3.85262 & 0.12066 & 0.72052 \\ 0.70101 & 2.52407 & 3.35440 & 0.15714 & 0.97188 \\ 0.45261 & 2.05830 & -0.04068 & 0.33558 & -0.07442 \\ -0.42910 & -0.54869 & 0.79406 & -0.58297 & -0.16802 \\ 0.41365 & 2.66553 & 2.99061 & 1.90820 & 0.68455 \\ -0.63204 & 2.43394 & -8.61403 & -0.40748 & 1.80569 \\ -0.17766 & -0.75617 & 0.31795 & -0.80676 & 0.28789 \\ -0.60219 & -1.31571 & -3.54702 & -1.64405 & -0.48701 \\ -0.28229 & -1.99287 & -0.48545 & -0.46986 & -0.29061 \\ 0.17916 & 1.57315 & -1.31252 & 0.93871 & 0.75706 \\ -0.45053 & 0.75741 & -4.93252 & 0.32721 & 0.92203 \end{bmatrix},$$

where F_1^T, F_2^T, and F_3^T denote the computed solutions of Algorithms 3.5.4, 3.6.2, and 3.7.2. The absolute values of the differences of the assigned poles and the computed eigenvalues of $A + BF_j$ for j=1,2,3 are shown in the following table, where $\lambda_k^{(j)}$ denotes the kth computed eigenvalue of $A + BF_j$ that is closest to λ_k, which is computed using the MATLAB function **eig**.

Table 3.8.2 Comparison of the three algorithms for Example 3.8.3.

k	$\lvert\lambda_k - \lambda_k^{(1)}\rvert$	$\lvert\lambda_k - \lambda_k^{(2)}\rvert$	$\lvert\lambda_k - \lambda_k^{(3)}\rvert$
1	0.000000000000	0.000000000000	0.000000000032
3	0.000000001096	0.000000000004	0.000008699064
4	0.000000000050	0.000000000006	0.000000007192
6	0.000000001236	0.000000000003	0.000018466018
7	0.000000000030	0.000000000009	0.000000002217
9	0.000000000055	0.000000000036	0.000016165967
10	0.000000000000	0.000000000009	0.000000000229
12	0.000000000025	0.000000000040	0.000007195292
13	0.000000003962	0.000000000025	0.000006572449
14	0.000000004085	0.000000000034	0.000005774507
15	0.000000000000	0.000000000000	0.000000000001

This table clearly shows that all computed eigenvalues of $A + BF_2$ agree up to at least 10 significant digits with the assigned eigenvalues, all of those of $A + BF_1$ up to at least 8 significant digits, whereas some of those of $A + BF_2$ only up to 4 significant digits. This shows that for this example the invariant subspace method is the best one.

3.9 Robust Pole Assignment Problems

As has been pointed out in Section 3.3, in the multi-input case ($m > 1$) the solution to Problem PA is essentially undetermined, with many degrees of freedom. Therefore, the question arises as to how this freedom is to be parameterized and how it is to be exploited in practice. It is well known that for any system design a desirable property is that the eigenvalues of the closed-loop system matrix A_c should be insensitive to perturbations in matrix A_c. This leads to the robust pole assignment problem, which may be described as follows.

Problem RPA *Find a solution F to Problem* PA, *defined as in Section* 3.1, *such that the closed-loop system is robust, in the sense that the eigenvalues of $A + BF$ are as insensitive to perturbations in the matrix $A + BF$ as possible.*

This problem is important in the field of robust control, and has now attracted much attention. Therefore, in this section we will devote our attention to this problem.

3.9.1 Measures of Robustness

Clearly, Problem RPA is only a qualitative description for the robust pole assignment problem. In order to do further study, we must precisely formulate this problem in quantitative form. To do this, the key problem is how to choose a proper measure of robustness. This will be discussed in this subsection.

From the numerical analysis point of view the robustness problem of a closed-loop system is essentially the conditioning problem of the eigenproblem corresponding to the closed-loop matrix. It is known that the conditioning problem of eigenproblem is a problem that is both interesting and of fundamental importance, and so it has a rich theory. With the help of this theory we here introduce several basic measures of robustness.

Let $A \in \mathbf{R}^{n\times n}$, $B \in \mathbf{R}^{n\times m}$, $m > 1$, and $\mathcal{L} = \{\lambda_1, \ldots, \lambda_n\}$ a set of n complex numbers closed under complex conjugation. Assume that F is a solution to Problem PA with data A, B, and $\mathcal{L}$, i.e.,

$$\lambda(A_c) = \mathcal{L},$$

where $A_c = A + BF$ is the closed-loop system matrix. Now suppose A_c is non-defective, i.e., there exists a nonsingular matrix $X = [x_1, \dots, x_n] \in \mathbf{C}^{n\times n}$ such that

$$X^{-1}A_cX = \Lambda \equiv \operatorname{diag}(\lambda_1, \dots, \lambda_n). \tag{3.9.1}$$

Let

$$X^{-T} = Y \equiv [y_1, \dots, y_n]. \tag{3.9.2}$$

Then it follows from (3.9.1) that

$$A_cx_i = \lambda_i x_i, \quad y_i^T A_c = \lambda_i y_i^T, \quad y_i^T x_i = 1, \qquad i = 1, 2, \dots, n,$$

that is, y_i and x_i are the left and right eigenvectors of A_c corresponding to λ_i with $y_i^T x_i = 1$. Define

$$c_j = \|x_j\|_2 \, \|y_j\|_2, \quad j = 1, 2, \dots, n, \tag{3.9.3}$$

$$c = (c_1, c_2, \dots, c_n)^T. \tag{3.9.4}$$

It is worthwhile to point out that c_j is unique only if λ_i is simple. In addition, noting that

$$c_j^{-1} = \frac{1}{\|x_j\|_2 \, \|y_j\|_2} = \frac{y_j^T x_j}{\|y_j\|_2 \, \|x_j\|_2}$$

is just the cosine of the angle between the left and right eigenvectors associated with λ_j, it is clear that $c_j \geq 1$

Now assume that λ is an eigenvalue of multiplicity p, and without loss of generality we may assume that $\lambda = \lambda_1 = \dots = \lambda_p$. Let $B = [\beta_{ij}] \in \mathbf{R}^{n\times n}$ with $|\beta_{ij}| \leq 1$. Then it can be shown that there exist p eigenvalues $\mu_1(\varepsilon), \dots, \mu_p(\varepsilon)$ of $A_c + \varepsilon B$ $(0 < \varepsilon << 1)$ such that

$$|\mu_i(\varepsilon) - \lambda| \leq \big(\max_{1\leq j\leq p} c_j\big) np\varepsilon + O(\varepsilon^2)$$

for all $i = 1, 2, \dots, p$. See [221] or [195] for the details. This shows that the sensitivity of the multiple eigenvalue λ to perturbations in $A_c = A + BF$ depends upon the magnitude of $\max_{1\leq j\leq p} c_j$. Consequently, the quantity c_j is referred to as the **condition number** of λ_j. Thus every reasonable measure of the magnitude of the vector c is a reflection of the robustness of the closed-loop system. From the practical standpoint we here choose

$$\nu_1 = \nu_1(A_c) = \|c\|_\infty = \max_{1\leq j\leq n} c_j \tag{3.9.5}$$

as a measure of robustness.

On the other hand, Bauer and Fike (see [195]) showed that for any eigenvalue μ of $A_c + E$ with $E \in \mathbf{R}^{n\times n}$ there is an eigenvalue λ_j of A_c such that

$$|\lambda_j - \mu| \leq \kappa(X)\|E\|_2,$$

where $\kappa(X) = \|X\|_2\,\|X^{-1}\|_2$ is the **condition number** of the matrix X, which is defined by (3.9.1). Therefore, alternatively we may also take as a measure of robustness

$$\nu_2 = \nu_2(A_c) = \kappa(X) = \|X\|_2\,\|X^{-1}\|_2 = \|X\|_2\,\|Y\|_2. \tag{3.9.6}$$

Clearly, $\nu_2 \geq 1$.

Observe that when A_c is a normal matrix, that is $A_c^T A_c = A_c A_c^T$, the matrix X of eigenvectors can be taken to be unitary, and so X is perfectly conditioned with $\kappa_2(X) = 1$. In this case the condition numbers take minimum value $c_j = 1$, for all $j = 1, 2, \ldots, n$. Therefore, the aim of the robust pole assignment problem is essentially to select a solution F to Problem PA such that the resulting closed-loop matrix $A_c = A + BF$ is as "normal" as possible. This shows that we may also take the measure of nonnormality of A_c as a measure of robustness.

A straightforward measure of nonnormality of a matrix $M \in \mathbf{R}^{n\times n}$ is given by

$$\nu(M) = \|MM^T - M^T M\|_F.$$

It is clear that $\nu(A) = 0$ if and only if M is normal. We know also that a matrix M is normal if and only if there is a unitary matrix $Q \in \mathbf{C}^{n\times n}$ such that

$$Q^* M Q = \operatorname{diag}(\lambda_1, \ldots, \lambda_n).$$

Hence if $Q^* M Q = T = \operatorname{diag}(\lambda_1, \ldots, \lambda_n) + N$ is a Schur decomposition of a general matrix $M \in \mathbf{R}^{n\times n}$, then we may also take as a measure of nonnormality of M

$$\Delta(M) = \|N\|_F = \left(\|M\|_F^2 - \sum_{i=1}^{n} |\lambda_i|^2\right)^{\frac{1}{2}}.$$

This quantity is referred to as the **departure from normality** of M. The following inequality shows how $\Delta(M)$ and $\nu(M)$ are connected:

$$\frac{\nu(M)}{\sqrt{6}\|M\|_F} \leq \Delta(M) \leq \left(\frac{n^3 - n}{12}\right)^{\frac{1}{4}} \sqrt{\nu(M)}.$$

The left-hand inequality was proved by Eberlein and the right-hand inequality by Henrici. See [61] and [108] for the proofs. We can see that if $\nu(M)/\|M\|_F$ increases, then $\Delta(M)$ increases; reciprocally if $\Delta(M)$ is large, then $\nu(M)$ is large too.

In fact, the bad numerical behavior of highly nonnormal matrices has been known for long time. In 1962, Henrici [108] showed that for any eigenvalue μ of an $n \times n$ matrix $M + E$ with M nonnormal, there is an eigenvalue λ of M such that

$$|\mu - \lambda| \leq \frac{\eta}{g(\eta)} ||E||_F, \quad \eta = \frac{\Delta(M)}{||E||_F},$$

where for $x \geq 0$ defines as $g(x)$ the unique nonnegative root of

$$g + g^2 + \cdots + g^n = x.$$

From this result it easy to show that when $||E||_F \leq \frac{1}{n}\Delta(M)$, for any eigenvalue λ of $M + E$, there is an eigenvalue μ of M such that

$$|\mu - \lambda| \leq n^{\frac{1}{n}} (\Delta(M))^{1-\frac{1}{n}} ||E||_F^{\frac{1}{n}}.$$

See [195] for the details.

In addition, note that for the robust pole assignment problem the eigenvalues are known and unchanged. Consequently, we here take

$$\nu_3 = \nu_3(A_c) = \Delta(A_c) = \left(||A + BF||_F^2 - \sum_{i=1}^{n} |\lambda_i|^2 \right)^{\frac{1}{2}} \tag{3.9.7}$$

as a measure of the robustness of the closed-loop system.

So far we have defined three measures of robustness. The following theorem shows how these measures are connected.

Theorem 3.9.1 *Let ν_1, ν_2, and ν_3 be defined by* (3.9.5), (3.9.6), *and* (3.9.7), *respectively. Then*

(1) $\nu_2 \leq n\nu_1$, *assuming* $||x_j||_2 = 1$ *for all* j;

(2) $\nu_1 + \sqrt{\nu_1^2 - 1} \leq \nu_2$;

(3) $\nu_2 \geq \left(1 + \nu_3^2 / ||A_c||_F^2\right)^{\frac{1}{2}}$;

(4) $\nu_1 \leq \left(1 + \frac{\nu_3^2}{(n-1)\delta^2}\right)^{\frac{n-1}{2}}$, *where* $\delta = \min_{1 \leq j \leq n} \min_{i \neq j} |\lambda_i - \lambda_j| > 0$.

The inequality (2) was proved by Bauer and Fike, the inequality (3) by Loizou, and the inequality (4) by Smith. See [192] and [140] for the details. The inequality (1) can be proved as follows.

Proof of (1)

$$\nu_2 = \|X\|_2 \, \|X^{-1}\|_2 \le \|X\|_F \, \|X^{-1}\|_F = \left(\sum_{j=1}^{n} \|x_j\|_2^2\right)^{\frac{1}{2}} \left(\sum_{j=1}^{n} \|y_j\|_2^2\right)^{\frac{1}{2}}$$

$$= \sqrt{n}\left(\sum_{j=1}^{n} c_j^2\right)^{\frac{1}{2}} \le \sqrt{n}\cdot\sqrt{n} \max_{1\le j\le n} c_j = n\nu_1,$$

which is the desired result. □

This theorem shows that the measures ν_1, ν_2, and ν_3 defined as above are mathematically equivalent under some conditions, and the three measures all attain their minimal values simultaneously when A_c is normal.

Finally, it is worth mentioning that the three measures defined as above are invariant under orthogonal similarity transformations. More precisely, it is easy to prove the following theorem.

Theorem 3.9.2 *For any orthogonal matrix $Q \in \mathbf{R}^{n\times n}$ we have*

$$\nu_j(A_c) = \nu_j(Q^T A_c Q), \qquad j = 1, 2, 3.$$

3.9.2 Formulation I and Numerical Method I

3.9.2a Formulation I

In view of the fact that the eigenvalue sensitivity depends upon the condition number of the matrix of eigenvectors, and noticing that nondefective matrices (i.e., matrices that have n linearly independent eigenvectors) possess better robustness than defective ones, it is nature to formulate the robust pole assignment problem in quantitative form as follows.

Problem RPA1 *Given A, B and $\mathcal{L}$ as in Problem* PA, *find real matrix F and nonsingular matrix X satisfying*

$$(A + BF)X = X\Lambda, \tag{3.9.8}$$

where $\Lambda = \mathrm{diag}(\lambda_1, \cdots, \lambda_n)$, such that some measure ν (e.g., $\nu = \nu_1$ or ν_2) of the conditioning of the above eigenproblem is optimized.

Remark 3.9.1 Briefly, Problem RPA1 is to find a feedback matrix F so that:

(1) $A + BF$ is nondefective;

(2) $\lambda(A + BF) = \mathcal{L} = \{\lambda_1, \dots, \lambda_n\}$;

(3) $\nu_1(A+BF)$ or $\nu_2(A+BF)$ is as small as possible.

By Theorem 3.9.1, it is easy to see that if (A_1, B_1) is orthogonally equivalent to (A, B), i.e., if there exist orthogonal matrices $U \in \mathbf{R}^{n\times n}$, $V \in \mathbf{R}^{m\times m}$ such that

$$A_1 = UAU^T, \qquad B_1 = UBV,$$

then F_1 is a solution to Problem RPA1 with data A_1, B_1, and $\mathcal{L}$ if and only if $F = VF_1U$ is a solution to Problem RPA1 with data A, B, and $\mathcal{L}$. Therefore, without loss of generality, in what follows we may assume that

$$B = \begin{bmatrix} B_{11} \\ 0 \end{bmatrix} \begin{matrix} m \\ n-m \end{matrix} \tag{3.9.9}$$

with B_{11} nonsingular and upper triangular.

We observe that once the matrix X in Problem RPA1 has been determined, then the F is uniquely determined by

$$F = B_{11}^{-1}(X_1\Lambda X^{-1} - A_1), \tag{3.9.10}$$

where $X_1 = X(1:m,:)$, $A_1 = A(1:m,:)$. Consequently, the degrees of freedom available in the choice of the feedback F are reflected precisely by the degrees of freedom available in the selection of the matrix X of eigenvectors. Thus, the key problem now is how to determine the nonsingular matrix X. This needs to find under what conditions a nonsingular matrix X can be assigned to Problem RPA1. To this end, we first prove a preparation lemma.

Lemma 3.9.3 *Let nonsingular matrix $X = [x_1, \ldots, x_n] \in \mathbf{C}^{n\times n}$ satisfy that*

$$x_j \in \mathbf{R}^n, \qquad j = 1, 2, \ldots, r,$$
$$\overline{x}_{r+2k-1} = x_{r+2k} \in \mathbf{C}^n, \qquad k = 1, 2, \ldots, s,$$

where $r + 2s = n$. Then $X^{-T} = [y_1, \ldots, y_n]$ also satisfies that

$$y_j \in \mathbf{R}^n, \qquad j = 1, 2, \ldots, r,$$
$$\overline{y}_{r+2k-1} = y_{r+2k} \in \mathbf{C}^n, \qquad k = 1, 2, \ldots, s.$$

Proof By the assumptions we have

$$y_k^T x_j = \delta_{kj}, \quad k, j = 1, 2, \ldots, n. \tag{3.9.11}$$

Let $1 \le k \le r$, and $y_k = u_k + iv_k$, where u_k, $v_k \in \mathbf{R}^n$. Then (3.9.11) implies that

$$v_k^T x_j = 0, \qquad j = 1, 2, \ldots, n,$$

since x_k is real, and so v_k must be zero, hence $y_k \in \mathbf{R}^n$.

Next we consider the case $k = r + 2l - 1$, where $1 \leq l \leq s$. In this case it follows from (3.9.11) that

$$\overline{y}_k^T \overline{x}_j = \delta_{kj}, \qquad y_{k+1}^T x_j = \delta_{k+1,j}.$$

This, together with $\overline{x}_k = x_{k+1}$, implies that

$$(\overline{y}_k - y_{k+1})^T x_j = 0, \qquad j = 1, 2, \ldots, n,$$

and so $\overline{y}_k = y_{k+1}$, which is the desired result. □

For the sake of being clear and definite, in what follows we assume the given pole set $\mathcal{L} = \{\lambda_1, \ldots, \lambda_n\}$ is ordered as $\lambda_j \in \mathbf{R}$, $j = 1, 2, \ldots, r$, $\overline{\lambda}_{r+2j-1} = \lambda_{r+2j} = v_j + i\mu_j$ with $v_j, \mu_j \in \mathbf{R}$ and $\mu_j \neq 0$, $j = 1, 2, \ldots, s$, where $r + 2s = n$. For any $\lambda \in \mathbf{C}$ we define

$$\mathcal{S}(\lambda) = \mathcal{N}([A_{21}, A_{22} - \lambda I]), \tag{3.9.12}$$

where

$$A = \begin{bmatrix} A_{11} & A_{12} \\ A_{21} & A_{22} \end{bmatrix} \begin{matrix} m \\ n-m \end{matrix} . \tag{3.9.13}$$
$$\qquad\qquad m \quad n-m$$

The following theorem shows under what conditions a nonsingular matrix X can be assigned to Problem RPA1.

Theorem 3.9.4 *There exists $F \in \mathbf{R}^{m \times n}$ such that $A + BF$ is nondefective and $\lambda(A + BF) = \mathcal{L}$ if and only if there exist $x_j \in \mathcal{S}(\lambda_j)$ for $j = 1, 2, \ldots, n$, satisfying*

$$\begin{aligned} &x_j \in \mathbf{R}^n, && j = 1, 2, \ldots, r, \\ &\overline{x}_{r+2k-1} = x_{r+2k} \in \mathbf{C}^n, && k = 1, 2, \ldots, s, \end{aligned} \tag{3.9.14}$$

such that $X = [x_1, \ldots, x_n]$ is nonsingular.

Proof (Only if). Because $A + BF$ is real and nondefective and $\lambda(A + BF) = \mathcal{L}$, there is nonsingular matrix $X = [x_1, \ldots, x_n]$ satisfying (3.9.14) such that

$$(A + BF)x_j = \lambda_j x_j.$$

This implies that

$$[A_{21}, A_{22} - \lambda_j I]x_j = 0,$$

and so $x_j \in \mathcal{S}(\lambda_j)$.

(If). For given nonsingular matrix X, define F by (3.9.10). Then X satisfying (3.9.14) implies that F is real by lemma 4.3.1. Clearly, for such an F we have

$$\left([A_{11} - \lambda_j I, A_{12}] + B_{11}F\right)x_j = 0, \qquad j = 1, 2, \ldots, n. \tag{3.9.15}$$

On the other hand, $x_j \in \mathcal{S}(\lambda_j)$ implies that

$$[A_{21}, A_{22} - \lambda_j I]x_j = 0, \qquad j = 1, 2, \ldots, n. \tag{3.9.16}$$

Combining (3.9.15) and (3.9.16), we have

$$(A + BF)x_j = \lambda_j x_j, \qquad j = 1, 2, \ldots, n,$$

which implies that $A + BF$ is nondefective and $\lambda(A + BF) = \mathcal{L}$. □

Consequently, Problem RPA1 is now reduced to the problem of selecting vector $x_j \in \mathcal{S}(\lambda_j)$, $j = 1, 2, \ldots, n$, satisfying the conditions of Theorem 3.9.4, such that eigenproblem (3.9.8) is as well-conditioned as possible. Thus, it follows that the multiplicity of the eigenvalue $\lambda_j \in \mathcal{L}$ to be assigned must be less than or equal to $\dim(\mathcal{S}(\lambda_j))$, since the maximum number of independent eigenvectors which can be chosen to correspond to λ_j is equal to $\dim(\mathcal{S}(\lambda_j))$. It is easy to show that

$$\dim\left(\mathcal{S}(\lambda_j)\right) \geq m$$

with equality if and only if λ_j is controllable.

It is worth mentioning that minimizing the conditioning of X also leads to other desirable properties of the closed-loop system. For instance, it follows from (3.9.10) that

$$\|F\|_2 \leq \left(\|A\|_2 + \max_{1 \leq j \leq n} |\lambda_j| \kappa_2(X)\right) / \sigma_{\min}(B)\,. \tag{3.9.17}$$

This result shows that given $(A,\ B)$ and eigenvalue set $\mathcal{L}$, minimizing the conditioning $\kappa(X)$ of the assigned eigensystem also minimizes a bound on the feedback matrices.

The following Theorem shows that for given data the minimal conditioning that can be achieved is limited.

Theorem 3.9.5 *Let $X = SD$, where $S = [S_1, \ldots, S_k]$ with $S_j \in \mathbf{C}^{n \times m_j}$ having orthogonal columns (i.e., $S_j^* S_j = I$) for $j = 1, \ldots, k$, and $D = \operatorname{diag}(D_1, \ldots, D_k)$ with $D_j \in \mathbf{C}^{m_j \times r_j}$, $j = 1, 2, \ldots, k$, $r_1 + \cdots + r_k = n$. If X is nonsingular, then*

$$\kappa_2(S) = \|S\|_2 \|S^\dagger\|_2 \leq \sqrt{1 + (k-1)\delta}\,\kappa_2(X) \leq \sqrt{k}\,\kappa_2(X), \tag{3.9.18}$$

where

$$\delta = \max_{1 \leq j \leq k} \max_{i \neq j} \|S_j^* S_i\|_2 < 1. \tag{3.9.19}$$

This theorem gives a lower bound on the minimal conditioning of X satisfying the conditions of Theorem 3.9.4. Let S_{λ_j} is an $n \times m_j$ matrix with orthogonal columns spanning the space $\mathcal{S}(\lambda_j)$, $m_j = \dim(\mathcal{S}(\lambda_j))$, and let

$$S_{\mathcal{L}} = [S_{\lambda_1}, S_{\lambda_2}, \ldots, S_{\lambda_n}].$$

Then, by Theorem 3.9.5, for any X satisfying the conditions of Theorem 3.9.4 we must have

$$\kappa_2(X) \geq \kappa_2(S_{\mathcal{L}}) / \sqrt{n}. \tag{3.9.20}$$

This shows that if $S_{\mathcal{L}}$ is seriously ill-conditioned, i.e.,

$$\kappa_2(S_{\mathcal{L}}) = \|S_{\mathcal{L}}\|_2 \, \|S_{\mathcal{L}}^{\dagger}\|_2 = \frac{\sigma_{\max}(S_{\mathcal{L}})}{\sigma_{\min}(S_{\mathcal{L}})}$$

is very large, then the solution X must be badly conditioned.

In order to establish Theorem 3.9.5 we need the following lemma.

Lemma 3.9.6 *Let S and D be as in Theorem* 3.9.5. *Then*

$$\|D(SD)^{-1}\|_2 \geq \|S^{\dagger}\|_2. \tag{3.9.21}$$

Proof Since SD is invertible, rank (S) = rank $(D) = n$, and S and D have the following decompositions

$$S = [R,\ 0]U, \qquad D = V \begin{bmatrix} T \\ 0 \end{bmatrix},$$

where R, $T \in \mathbf{C}^{n \times n}$ are nonsingular, U, $V \in \mathbf{C}^{l \times l}$ unitary, $l = m_1 + \cdots + m_k$. Let

$$UV = \begin{bmatrix} U_{11} & U_{12} \\ U_{21} & U_{22} \end{bmatrix} \begin{matrix} n \\ l-n \end{matrix}, \qquad \begin{matrix} n & \ l-n \end{matrix}$$

Then

$$D(SD)^{-1} = V \begin{bmatrix} T \\ 0 \end{bmatrix} (RU_{11}T)^{-1} = V \begin{bmatrix} U_{11}^{-1}R^{-1} \\ 0 \end{bmatrix},$$

and so

$$\|D(SD)^{-1}\|_2 = \|(RU_{11})^{-1}\|_2 = 1 / \sigma_{\min}(RU_{11}) \,. \tag{3.9.22}$$

Noting U_{11} is the leading $n \times n$ principal submatrix of the unitary matrix UV, it is not hard to show that

$$\sigma_{\min}(RU_{11}) \leq \sigma_{\min}(R). \tag{3.9.23}$$

Substituting (3.9.23) into (3.9.22) gives the inequality (3.9.21). □

Using this lemma we may now give the proof of Theorem 3.9.5.

Proof of Theorem 3.9.5 Because of the form, we have

$$\|X\|_2 = \|SD\|_2 \geq \max_{1\leq j\leq k} \|S_j D_j\|_2 = \|D\|_2,$$

and so, by Lemma 3.9.6,

$$\kappa_2(X) = \|X\|_2 \, \|X^{-1}\|_2 \geq \|D\|_2 \, \|(SD)^{-1}\|_2 \geq \|D(SD)^{-1}\|_2 \geq \|S^\dagger\|_2. \quad (3.9.24)$$

On the other hand, for any $x = (x_1^T, \dots, x_k^T)^T$ with $\|x\|_2 = 1$ and $x_j \in \mathbf{C}^{m_j}$, we have

$$\begin{aligned} \|Sx\|_2^2 &= \sum_{j=1}^{k}\sum_{i=1}^{k} x_j^* S_j^* S_i x_i \leq \sum_{j=1}^{k} \|x_j\|_2^2 + \sum_{j=1}^{k}\sum_{\substack{i=1\\ i\neq j}}^{k} \|x_j\|_2 \, \|x_i\|_2 \delta \\ &= 1 + \left(\left(\sum_{j=1}^{k} \|x_j\|_2\right)^2 - 1\right)\delta \leq 1 + \left(k\sum_{j=1}^{k} \|x_j\|_2^2 - 1\right)\delta \\ &= 1 + (k-1)\delta, \end{aligned}$$

and therefore

$$\|S\|_2 \leq \sqrt{1 + (k-1)\delta}. \quad (3.9.25)$$

Combining (3.9.24) and (3.9.25) gives (3.9.18). □

3.9.2b Numerical Method I

We now show how to compute X and F to solve Problem RPA1 in practice. In fact, Theorem 3.9.5 has given us a recipe for doing this. In summary, the computation of the solution to Problem RPA1 requires three major steps:

Step 1. Construct an orthogonal basis, comprised by the columns of matrix S_j for the space $\mathcal{S}(\lambda_j)$ for $\lambda_j \in \mathcal{L}$, $j = 1, 2, \dots, n$.

Setp 2. Select vectors $x_j = S_j W_j \in \mathcal{S}(\lambda_j)$ with $\|x_j\|_2 = 1$, satisfying

$$x_j \in \mathbf{R}^n, \qquad j = 1, 2, \dots, r,$$

and

$$x_{r+2k-1} = \overline{x}_{r+2k} \in \mathbf{C}^n, \qquad k = 1, 2, \dots, s,$$

such that $X = [x_1, \dots, x_j]$ is as well-conditioned as possible.

Step 3. Find the $M = X_1\Lambda X^{-1}$ by solving $X^T M^T = (X_1\Lambda)^T$ and determine F by solving $B_{11}F = M - A_1$.

Step 1 can be carried out using either QR or SVD decomposition. Since the QR decomposition is computationally less expensive, we here use QR decomposition for this purpose. If the QR decomposition of $[A_{21}, A_{22} - \lambda_j I]^T$ is

$$\begin{bmatrix} A_{21}^T \\ A_{22}^T - \lambda_j I \end{bmatrix} = [\widetilde{S}_j, S_j] \begin{bmatrix} R_j \\ 0 \end{bmatrix} \tag{3.9.26}$$

with R_j of full row rank, then S_j is the required matrix. Setp 2 is the key step of this method. There are several methods for accomplishing this step. The objective of the method here is to choose vectors $x_j \in \mathcal{S}(\lambda_j)$, $j = 1, 2, \ldots, n$, satisfying

$$x_j \in \mathbf{R}^n,\ j = 1, 2, \ldots, r \quad \text{and} \quad x_{r+2k-1} = \overline{x}_{r+2k} \in \mathbf{C}^n,\ k = 1, 2, \ldots, s, \tag{3.9.27}$$

such that each vector is as orthogonal as possible to the space spanned by the remaining vectors. Let

$$\mathcal{X}_j = \operatorname{span}\{x_1, \ldots, x_{j-1}, x_{j+1}, \ldots, x_n\},$$

and let y_j be a normalized vector orthogonal to $\mathcal{X}_j$. Then choosing $x_j \in \mathcal{S}(\lambda_j)$ to be as orthogonal as possible to $\mathcal{X}_j$ is equivalent to choosing $x_j \in \mathcal{S}(\lambda_j)$ to minimize the angle between x_j and y_i. Note that the vector y_j^T is just the normalized left eigenvector corresponding to λ_j. Therefore, this method is essentially to find $x_j \in \mathcal{S}(\lambda_j)$ such that condition number $c_j = 1 / |y_j^T x_j|$ is minimized i.e., such that the sensitivity of eigenvalue λ_j is minimized.

The solution X can be found by an iteration as follows. Take any set of independent vectors $x_j \in \mathcal{S}(\lambda_j)$, $j = 1, 2, \ldots, n$, as an initial matrix $X = [x_1, \ldots, x_n]$. The method is then to sweep through the matrix X replacing x_j in turn with $\widetilde{x}_j$ which is the normalized projection of the corresponding left eigenvector onto the space $\mathcal{S}(\lambda_j)$. This ensures that the angle between y_j and $\widetilde{x}_j$ is minimized. The vector y_j may be obtained by performing a QR factorization of

$$X_j = [x_1, \ldots, x_{j-1}, x_{j+1}, \ldots, x_n].$$

That is,

$$X_j = [Q_j,\ y_j] \begin{bmatrix} \widetilde{R}_j \\ 0 \end{bmatrix} \begin{matrix} n-1 \\ 1 \end{matrix}\ .$$

Note that by Lemma 3.9.3 the y_j can be chosen to be real if x_j is real, i.e., if $1 \le j \le r$. Then $\widetilde{x}_j$ is computed from

$$\widetilde{x}_j = S_j S_j^* y_j / \|S_j^* y_j\|_2\ .$$

Of course, when $j = r + 2k$ (i.e., λ_j is complex), to satisfy the requirements (3.9.27), $\widetilde{x}_{r+2k}$ must be taken to be the complex conjugate of $\widetilde{x}_{r+2k-1}$. Hence, both $\widetilde{x}_{r+2k}$ and $\widetilde{x}_{r+2k-1}$ are known once one of them has been computed. The sweeps through the matrix are continued until the convergence criterion is satisfied or the maximum number of allowed sweeps through the matrix has been completed. A reasonable criterion for convergence is that

$$\frac{1}{n}\left(\sum_{i=1}^{n} \omega_j \Delta c_j\right) < \varepsilon_1,$$

where $\omega_1, \ldots, \omega_n$ are weights, ε_1 is a tolerance, and Δc_j is the change in the condition number c_j of λ_j due to the replacement of x_j with $\widetilde{x}_j$. Close to the imaginary axis are most likely to become unstable, a suitable set of weights is given by $\omega_j^{-1} = -Re(\lambda_j)$. It is worthwhile to point out that when $\widetilde{x}_j$ replaces x_j, the sensitivities of the other eigenvalues change. Thus the procedure is not guaranteed to reduce the overall sensitivity and the method may not converge.

Step 3 can be carried out using standard solvers of linear system. $M = X_1 \Lambda X^{-1}$ can be obtained by solving $X^T M^T = (X_1\Lambda)^T$ for M^T using Gaussian elimination with partial pivoting. F is then obtained by directly using back substitution to solve the upper triangular system $B_{11}F = M - A_1$.

Although this method does not guarantee convergence, it is simple to implement, and often gives good solutions for very small systems in practice.

Example 3.9.1 Let

$$A = \begin{bmatrix} 1.3800 & -0.2077 & 6.7150 & -5.6760 \\ -5.8140 & -4.2900 & 0.0000 & 0.6750 \\ 1.0670 & 4.2730 & -6.6540 & 5.8930 \\ 0.0480 & 4.2730 & 1.3430 & -2.1040 \end{bmatrix},$$

$$B^T = \begin{bmatrix} 0.0000 & 5.6790 & 1.1360 & 1.1360 \\ 0.0000 & 0.0000 & -3.1460 & 0.0000 \end{bmatrix},$$

$$\mathcal{L} = \begin{bmatrix} -0.2000 & -0.5000 & -5.0566 & -8.6659 \end{bmatrix}.$$

Applying the above method to this test problem, after only one sweep we obtain the computed feed back matrix as follows:

$$\widehat{F} = \begin{bmatrix} 0.94652320 & -0.17250577 & 0.17400794 & -0.13929285 \\ 1.09904963 & -0.29461930 & 0.57669297 & -0.14263732 \end{bmatrix}.$$

The computed eigenvalues of the resulting matrix $A + B\widehat{F}$ agree up to ten significant digits with the assigned poles, and $A + B\widehat{F}$ has a very good-conditioned eigenproblem as shown in the following:

$$\nu_1 = 1.78768584, \qquad \nu_2 = 3.59273619,$$

$$\nu_3 = 12.88540069, \qquad \|F\|_2 = 1.60267914.$$

However, it is worthwhile to point out that the method described above is numerically suspicious since it depends on the eigenvectors of the closed-loop system matrix and the eigenvectors of a matrix frequently suffer from being ill-conditioned. Experimental results also show that the performance of this method deteriorates rapidly with increasing system order.

3.9.3 Formulation II and Numerical Method II

3.9.3a Formulation II

In what follows we assume that the given set $\mathcal{L}$ in Problem RPA has l_1 distinct real numbers μ_j with multiplicity r_j, $j = 1, \ldots, l_1$, and l_2 distinct complex numbers λ_j with multiplicity s_j and $\mathrm{Im}(\lambda_j) > 0$, $j = 1, \ldots, l_2$, where $r_1 + \cdots + r_{l_1} + 2(s_1 + \cdots + s_{l_2}) = n$.

In order to give a proper formulation for the robust pole assignment problem, we need to find a numerically reliable and quantitative criterion to determine if a matrix has $\mathcal{L}$ as its spectrum. To determine whether a number λ is an eigenvalue of a matrix $M \in \mathbf{R}^{n\times n}$ is essentially to determine if the rank of the matrix $M - \lambda I$ is less than n. In fact, it is well known that λ is an eigenvalue of M with geometric multiplicity r if and only if the rank of the matrix $M - \lambda I$ is $n - r$. From the numerical point of view, the best way for determining the rank of a matrix is to compute the singular value decomposition of the matrix. Hence, we should use the singular values of $M - \lambda I$ to determine if λ is an eigenvalue of M. It is easy to prove that λ is an eigenvalue of M with geometric multiplicity r if and only if $M - \lambda I$ has just r zero singular values, that is,

$$\dim(\mathcal{N}(M - \lambda I)) = r$$

if and only if

$$\sigma_n(M - \lambda I) = \cdots = \sigma_{n-r+1}(M - \lambda I) = 0 < \sigma_{n-r}(M - \lambda I). \tag{3.9.28}$$

However, in practice this requires complex arithmetic if λ is a complex number. To avoid complex arithmetic, we first prove the following lemma.

Lemma 3.9.7 *Let $M \in \mathbf{R}^{n\times n}$, and $\lambda = \mu + i\nu$, μ, $\nu \in \mathbf{R}$, with $\nu \neq 0$. Then*

$$\dim(\mathcal{N}(M - \lambda I)) = r \tag{3.9.29}$$

if and only if

$$\dim(\mathcal{N}([M - \bar{\lambda} I][M - \lambda I])) = 2r. \tag{3.9.30}$$

Proof It is trivial if $r = 0$. In the case of $r \geq 1$, since both of (3.9.29) and (3.9.30) imply that λ is an eigenvalue of M, and they are invariant under a similarity transformation, it is sufficient to prove Lemma 3.9.7 in the special case

$$M = \begin{bmatrix} M_1 & 0 & 0 \\ 0 & M_2 & 0 \\ 0 & 0 & M_3 \end{bmatrix},$$

where

$$M_1 = \begin{bmatrix} \lambda & & * \\ & \ddots & \\ 0 & & \lambda \end{bmatrix}, \quad M_2 = \overline{M}_1, \quad \det(M_3 - \lambda I) \neq 0.$$

Thus we have

$$(M-\bar{\lambda}I)(M - \lambda I)$$

$$= \begin{bmatrix} (M_1 - \bar{\lambda}I)(M_1 - \lambda I) & 0 & 0 \\ 0 & (M_2 - \bar{\lambda}I)(M_2 - \lambda I) & 0 \\ 0 & 0 & (M_3 - \bar{\lambda}I)(M_3 - \lambda I) \end{bmatrix}.$$

It follows that

$$\begin{aligned} \dim(\mathcal{N}([M - \bar{\lambda}I][M - \lambda I])) &= \sum_{j=1}^{3} \dim(\mathcal{N}([M_j - \bar{\lambda}I][M_j - \lambda I])) \\ &= \dim(\mathcal{N}(M_1 - \lambda I)) + \dim(\mathcal{N}(M_2 - \bar{\lambda}I)). \end{aligned}$$

This implies that (3.9.29) is equivalent to (3.9.30). □

From Lemma 3.9.7 and (3.9.28) it is easy to drive the following theorem.

Theorem 3.9.8 *Let $M \in \mathbf{R}^{n\times n}$. Then M is nondefective and satisfies $\lambda(M) = \mathcal{L}$ if and only if*

$$\begin{aligned} \sigma_k(M - \mu_j I) = 0, &\qquad k = n, n-1, \ldots, n - r_j + 1,\ j = 1, \cdots, l_1, \\ \sigma_k([M - \bar{\lambda}_j I][M - \lambda_j I]) = 0, &\qquad k = n, n-1, \ldots, n - 2s_j + 1,\ j = 1, 2, \cdots, l_2. \end{aligned}$$

This theorem gives us a criterion to determine whether a matrix M has $\mathcal{L}$ as its spectrum. An attractive property of the singular values is that they are not very sensitive to changes in the matrix. Hence, this criterion is numerically reliable. Moreover, even if $\mathcal{L}$ contains some complex numbers, in practical applications it does not require complex arithmetic.

Exploiting this criterion and taking ν_3, defined by (3.9.7), as the measure of the robustness, We may now formulate the robust pole assignment problem as follows.

Problem RPA2 *Find $F \in \mathbf{R}^{m\times n}$ to solve the optimization problem*

$$\begin{aligned} &\textit{minimize} && \widetilde{f}(F), \\ &\textit{subject to} && \widetilde{g}_j(F) = 0, \quad j = 1, 2, \cdots, l_1, \\ & && \widetilde{h}_k(F) = 0, \quad k = 1, 2, \cdots, l_2. \end{aligned} \tag{3.9.31}$$

where

$$\begin{aligned} \widetilde{f}(F) &= \left(\|A + BF\|_F^2 - \sum_{j=1}^{l_1} r_j|\mu_j|^2 - 2\sum_{j=1}^{l_2} s_j|\lambda_j|^2 \right)^{\frac{1}{2}}, \\ \widetilde{g}_j(F) &= \sum_{l=n-r_j+1}^{n} [\sigma_l(A + BF - \mu_j I)]^2, \\ \widetilde{h}_k(F) &= \sum_{l=n-2s_k+1}^{n} [\sigma_l(H_k(F))]^2, \\ H_k(F) &= (A + BF - \bar{\lambda}_k I)(A + BF - \lambda_k I). \end{aligned} \tag{3.9.32}$$

Remark 3.9.2 Briefly, Problem RPA2 is to find a matrix $F \in \mathbf{R}^{m\times n}$ such that:

(1) The spectrum of $A + BF$ is $\mathcal{L}$;

(2) $A + BF$ is nondefective;

(3) The departure from normality of $A + BF$ is optimized.

Remark 3.9.3 Obviously there are other choices for the $\widetilde{g}_j$, $\widetilde{h}_k$, say,

$$\widetilde{g}_j(F) = \sigma_n(A + BF - \mu_j I) + \cdots + \sigma_{n-r_j+1}(A + BF - \mu_j I)$$

or

$$\widetilde{g}_j(F) = \left(\sigma_n(A + BF - \mu_j I), \cdots, \sigma_{n-r_j+1}(A + BF - \mu_j I)\right).$$

But the ones adopted here can guarantee that they are analytic. See Theorem 3.9.11 and Remarks 3.9.5 and 3.9.6.

Remark 3.9.4 It is easy to see that such a formulation for robust pole assignment problem possesses at least three advantages. The first one is that it is directly dependent on the elements of $A + BF$ rather than its eigenvectors. The second one is that it does not require complex arithmetic even if $\mathcal{L}$ contains some complex numbers. The third one is that all of those functions defined by (3.9.32) are not very sensitive to perturbations in the matrix F. These properties make it possible to develop some numerically reliable algorithms for solving it.

3.9.3b Numerical Method II

We now consider how to solve Problem RPA2 numerically. We begin by reducing Problem RPA2 into more simple form. For given A, B, $\mathcal{L}$ as before, assume that B has the following decomposition

$$B = U \begin{bmatrix} B_{11} \\ 0 \end{bmatrix} V, \tag{3.9.33}$$

where $U \in \mathbf{R}^{n\times n}$ and $V \in \mathbf{R}^{m\times m}$ are orthogonal, B_{11} is of the form $B_{11} = [0, R]$ with $R \in \mathbf{R}^{r\times r}$ nonsingular and upper triangular. This decomposition can be found by either the SVD or QR method. Let

$$A_1 = U^T AU = \begin{bmatrix} A_{11} \\ A_{21} \end{bmatrix}, \quad \text{with } A_{11} \in \mathbf{R}^{r\times n}, \tag{3.9.34}$$

$$A(X) = \begin{bmatrix} X \\ A_{21} \end{bmatrix}, \quad \text{for } X = [x_{il}] \in \mathbf{R}^{r\times n}, \tag{3.9.35}$$

$$H_j(X) = A(X)^2 - (\bar{\lambda}_j + \lambda_j)A(X) + |\lambda_j|^2 I. \quad j = 1, 2, \cdots l_2. \tag{3.9.36}$$

Consider the following optimization problem:

Find $X \in \mathbf{R}^{r\times n}$ to solve the optimization problem

$$\begin{array}{lll} \textit{minimize} & f(X), & \\ \textit{subject to} & g_j(X) = 0, & j = 1, 2, \cdots, l_1, \\ & h_k(X) = 0, & k = 1, 2, \cdots, l_2. \end{array} \tag{3.9.37}$$

where

$$\begin{aligned} f(X) &= \|X\|_F^2, \\ g_j(X) &= \sum_{l=n-r_j+1}^{n} [\sigma_l(A(X) - \mu_j I)]^2, \\ h_k(X) &= \sum_{l=n-2s_k+1}^{n} [\sigma_l(H_k(X))]^2, \end{aligned} \tag{3.9.38}$$

It is easy to prove that optimization Problem (3.9.37) is equivalent to Problem RPA2, that is, F is a solution to Problem RPA2 if and only if $X = A_{11} + RF_{21}$ is a solution to Problem (3.9.37), where

$$\begin{matrix} m-r \\ r \end{matrix} \begin{bmatrix} F_{11} \\ F_{21} \end{bmatrix} = VFU.$$

Notice that F_{11} does not affect the matrix

$$A + BF = U \begin{bmatrix} A_{11} + RF_{21} \\ A_{21} \end{bmatrix} U^T.$$

We will set it to zero, since it decreases the magnitude of F, which may make it cheaper to apply the control.

We will now show that $g_j(X)$, $h_k(X)$ defined by (3.9.38) are analytic.

Lemma 3.9.9 *Let $x \in \mathbf{R}^p$, and $S(x) = S(x)^T \in \mathbf{R}^{n\times n}$ be a real analytic function of x in some neighborhood $\mathcal{B}$ of the origin of $\mathbf{R}^p$. Suppose that $\lambda_1, \lambda_2, \cdots, \lambda_n$ are the eigenvalues of $S(0)$ with $\{\lambda_1, \cdots, \lambda_l\} \cap \{\lambda_{l+1}, \cdots, \lambda_n\} = \emptyset$, and $V = [v_1, \dots, v_n]$ is an orthogonal matrix of eigenvectors, i.e.,*

$$V^T S(0) V = \operatorname{diag}(\lambda_1, \cdots \lambda_n).$$

Let $\lambda_1(x), \cdots, \lambda_n(x)$ are the eigenvalues of $S(x)$ with $\lambda_j(0) = \lambda_j$, $j = 1, 2, \cdots n$. Then

$$h(x) = \lambda_1(x) + \cdots + \lambda_l(x)$$

is a real analytic function in some neighborhood of the origin, and

$$\frac{\partial h(0)}{\partial \xi_j} = v_1^T \frac{\partial S(0)}{\partial \xi_j} v_1 + \cdots + v_l^T \frac{\partial S(0)}{\partial \xi_j} v_l \tag{3.9.39}$$

for $j = 1, 2, \cdots, p$, where $x = (\xi_1, \dots, \xi_p)^T$.

Proof Let

$$\widetilde{S}(x) = V^T S(x) V = \begin{bmatrix} S_{11}(x) & S_{12}(x) \\ S_{21}(x) & S_{22}(x) \end{bmatrix},$$

where $S_{11}(x)$ is an $l \times l$ matrix, $S_{22}(x)$ an $(n-l) \times (n-l)$ matrix. Clearly,

$$S_{12}(0) = 0,\ S_{21}(0) = 0,\ S_{11}(0) = \operatorname{diag}(\lambda_1, \dots, \lambda_l),\ S_{22}(0) = \operatorname{diag}(\lambda_{l+1}, \dots, \lambda_n).$$

In a manner similar to the proof of Lemma 1.9.1, we may prove that there exists an analytic solution $Y(x) \in \mathbf{R}^{(n-1)\times l}$ in some neighborhood $\mathcal{B}_0 \subseteq \mathcal{B}$ of the origin of $\mathbf{R}^m$ such that $Y(0) = 0$ and

$$\begin{bmatrix} I^{(l)} & 0 \\ Y(x) & I^{(n-l)} \end{bmatrix}^{-1} \begin{bmatrix} S_{11}(x) & S_{12}(x) \\ S_{21}(x) & S_{22}(x) \end{bmatrix} \begin{bmatrix} I^{(l)} & 0 \\ Y(x) & I^{(n-l)} \end{bmatrix} = \begin{bmatrix} S_1(x) & 0 \\ 0 & S_2(0) \end{bmatrix}$$

with $S_1(0) = S_{11}(0) = \operatorname{diag}(\lambda_1, \cdots, \lambda_l)$. This implies that

$$h(x) = \operatorname{tr}(S_1(x))$$

in some neighborhood of the origin of $\mathbf{R}^m$. Note that

$$S_1(x) = S_{11}(x) + S_{12}(x)Y(x).$$

Then it follows that $h(x)$ is an analytic function in some neighborhood of the origin, and

$$\begin{aligned}\frac{\partial h(0)}{\partial \xi_j} &= \operatorname{tr}\left(\frac{\partial S_1(0)}{\partial \xi_j}\right) = \operatorname{tr}\left(\frac{\partial S_{11}(0)}{\partial \xi_j} + \frac{\partial S_{12}(0)}{\partial \xi_j}Y(0) + S_{12}(0)\frac{\partial Y(0)}{\partial \xi_j}\right)\\ &= \operatorname{tr}\left(\frac{\partial S_{11}(0)}{\partial \xi_j}\right) = v_1^T\frac{\partial S(0)}{\partial \xi_j}v_1 + \cdots + v_l^T\frac{\partial S(0)}{\partial \xi_j}v_l.\end{aligned}$$

This completes the proof of Lemma 3.9.9. □

Remark 3.9.5 It is worthwhile to point out that the eigenvalues $\lambda_j(x)$ of $S(x)$ in Lemma 3.9.9 are not analytic in general. For example, consider a simple example:

$$S(x) = \begin{bmatrix} 1+2\xi_1+2\xi_2 & \xi_2 \\ \xi_2 & 1+2\xi_2 \end{bmatrix}, \quad x = (\xi_1, \xi_2)^T \in \mathbf{R}^2.$$

$S(0)$ has eigenvalue 1 with multiplicity 2, and the eigenvalues of $S(x)$ are

$$\lambda_1(x) = 1+\xi_1+2\xi_2+\sqrt{\xi_1^2+\xi_2^2}, \quad \lambda_2(x) = 1+\xi_1+2\xi_2-\sqrt{\xi_1^2+\xi_2^2}.$$

It is easy to verify that $\lambda_1(x)$ and $\lambda_2(x)$ are not differentiable at $\xi_1 = \xi_2 = 0$, but $\lambda_1(x) + \lambda_2(x) = 2(1+\xi_1+2\xi_2)$ is analytic.

Corollary 3.9.10 *Let $x \in \mathbf{R}^p$, and $M(x) \in \mathbf{R}^{n\times n}$ be a real analytic function of x in some neighborhood $\mathcal{B}(0)$ of the origin of $\mathbf{R}^p$. Suppose that $M(0)$ has singular value decomposition*

$$M(0) = U\Sigma V^T, \quad \Sigma = \operatorname{diag}(\sigma_1, \cdots, \sigma_n),$$

where $U = [u_1, \cdots, u_n]$ and $V = [v_1, \cdots, v_n]$ are two $n \times n$ orthogonal matrices, $\sigma_n \le \sigma_{n-1} \le \cdots \le \sigma_l < \sigma_{l-1} \le \cdots \le \sigma_1$. Then

$$h(x) = [\sigma_n(M(x))]^2 + \cdots + [\sigma_l(M(x)]^2$$

is a real analytic function in some neighborhood of the origin, and

$$\frac{\partial h(0)}{\partial \xi_j} = 2\sigma_n u_n^T\frac{\partial M(0)}{\partial \xi_j}v_n + \cdots + 2\sigma_l u_l^T\frac{\partial M(0)}{\partial \xi_j}v_l \tag{3.9.40}$$

for $j = 1, 2, \ldots, p$, where $x = (\xi_1, \ldots, \xi_p)^T$.

Proof Applying Lemma 3.9.9 to symmetric matrix $M(x)^TM(x)$ and noting that

$$\begin{aligned} v_j^T \left.\frac{\partial(M(x)^TM(x))}{\partial \xi_j}\right|_{x=0} v_j &= v_j^T\frac{\partial M(0)^T}{\partial \xi_j}M(0)v_j + v_j^T M(0)^T\frac{\partial M(0)}{\partial \xi_j}v_j \\ &= \sigma_j v_j^T\frac{\partial M(0)^T}{\partial \xi_j}u_j + \sigma_j u_j^T\frac{\partial M(0)}{\partial \xi_j}v_j \\ &= 2\sigma_j u_j^T\frac{\partial M(0)}{\partial \xi_j}v_j, \end{aligned}$$

we immediately obtain the results of Corollary 3.9.10. □

Remark 3.9.6 Notice that $\sigma_n(M(x)) + \cdots + \sigma_l(M(x))$ is certainly not analytic. For example, let

$$M(x) = \begin{bmatrix} 1+\xi_1 & 0 \\ 0 & \xi_2 \end{bmatrix}, \quad x = (\xi_1, \xi_2)^T \in \mathbf{R}^2.$$

$M(0)$ has singular values $\sigma_2 = 0 < \sigma_1 = 1$. It easy to see that $\sigma_2(M(x)) = |\xi_2|$ is not differentiable at $\xi_1 = \xi_2 = 0$, but $[\sigma_2(M(x))]^2 = \xi_2^2$ is analytic.

Using Corollary 3.9.10 it is easy to prove the following theorem.

Theorem 3.9.11 *Let $X_0 \in \mathbf{R}^{r\times n}$ satisfy*

$$g_j(X_0) = 0, \quad j = 1, 2, \cdots l_1,$$

and

$$h_k(X_0) = 0, \quad k = 1, 2, \cdots, l_2,$$

where g_j, h_k defined by (3.9.38). Then $g_j(X)$ and $h_k(X)$ are real analytic functions in some neighborhood of X_0, and

$$\frac{\partial g_j(X)}{\partial x_{il}} = 2\sum_{t=n-r_j+1}^{n} \sigma_t(A(X) - \mu_j I)[u_t^{(j)}(X)]^T e_i e_l^T v_t^{(j)}(X), \tag{3.9.41}$$

$$\begin{aligned} \frac{\partial h_k(X)}{\partial x_{il}} = 2\sum_{t=n-2s_k+1}^{n} & \sigma_t(H_k(X))[\widetilde{u}_t^{(k)}(X)]^T \left(e_i e_l^T A(X) + A(X) e_i e_l^T\right. \\ & \left. - (\bar{\lambda}+\lambda) e_i e_l^T\right)\widetilde{v}_t^{(k)}(X). \end{aligned} \tag{3.9.42}$$

Here we assume that the singular value decompositions of $A(X) - \mu_j I$ and $H_k(X)$ are, respectively,

$$A(X) - \mu_j I = U_j \Sigma_j V_j^T, \quad \Sigma = \operatorname{diag}\left(\sigma_1(A(X) - \mu_j I), \cdots, \sigma_n(A(X) - \mu_j I)\right)$$

and

$$H_k(X) = \widetilde{U}_k \widetilde{\Sigma}_k \widetilde{V}_k^T, \quad \widetilde{\Sigma}_k = \operatorname{diag}\big(\sigma_1(H_k(X)), \cdots, \sigma_n(H_k(X))\big),$$

with

$$V_j = [v_1^{(j)}(X), \cdots, v_n^{(j)}(X)], \quad U_j = [u_1^{(j)}(X), \cdots, u_n^{(j)}(X)],$$
$$\widetilde{V}_k = [\widetilde{v}_1^{(k)}(X), \cdots, \widetilde{v}_n^{(k)}(X)], \quad \widetilde{U}_k = [\widetilde{u}_1^{(k)}(X), \cdots, \widetilde{u}_n^{(k)}(X)]$$

being $n \times n$ orthogonal matrices.

Now we apply penalty function method to solve the constrained optimization Problem (3.9.37). Define the function

$$q(\eta, X) = f(X) + \eta\Big(\sum_{j=1}^{l_1} g_j(X) + \sum_{k=1}^{l_2} h_k(X)\Big).$$

The procedure for solving Problem (3.9.37) by the penalty function method is this:

Step 1. Let $\varepsilon > 0$ be the termination scalar. Choose an initial matrix $X_0 \in \mathbf{R}^{r\times n}$ and an initial penalty constant $\eta_0 > 0$. Set $k = 0$.

Step 2. Determine X_{k+1} to solve the unconstrained optimization problem

$$\text{minimize } q(\eta_k, X). \tag{3.9.43}$$

Step 3. If $\|X_k - X_{k+1}\|_\infty < \varepsilon$, then $X = X_{k+1}$ is an approximate optimal solution to Problem (3.9.37); otherwise $\eta_{k+1} = 5\eta_k$, replace k by $k+1$, go to Step 2.

Using the standard methods of optimization it may prove that any limit point of the sequence $\{X_k\}$ generated by Step 2 for any given positive number sequence $\{\eta_k\}$ with $\lim\limits_{k\to\infty} \eta_k = \infty$ is a solution to Problem (3.9.37). See [142] for the details.

Here we apply the Davidon-Flecher-Powell method to solve the unconstrained optimization problem (3.9.43):

(1) Let $j = 0$, $l = 0$, $Y_0 = X_k$, and $H_0 = I_N$, where $N = n^2$.
(2) Compute $G_0 = [\frac{\partial q(\eta_k, Y_0)}{\partial x_{il}}]$, $D_0 = -G_0$.
(3) Determine $Y_{j+1} = Y_j + \alpha_j D_j$ by means of an approximate minimization

$$q(\eta_k, Y_j + \alpha_j D_j) \approx \min_{\alpha > 0} q(\eta_k, Y_j + \alpha D_j).$$

(4) Set $\Delta Y = Y_{j+1} - Y_j$. If $\|\Delta Y\|_2 \le \varepsilon$, then $X_{k+1} = Y_{j+1}$ is an approximate optimal solution to Problem (3.9.37); otherwise go to the next step.

(5) Compute

$$G_{j+1} = \left[\frac{\partial q(\eta_k, Y_{j+1})}{\partial x_{il}} \right], \quad \Delta G = G_{j+1} - G_j,$$
$$\beta_1 = (\mathrm{vec}\,(\Delta Y))^T \mathrm{vec}\,(\Delta G), \quad \beta_2 = (\mathrm{vec}\,(\Delta G))^T H_j \mathrm{vec}\,(\Delta G).$$

(6) If $l = m_0 (m_0$ is a proper positive integer), then set

$$H_{j+1} = I^{(N)}, \quad l = 0;$$

else if $\beta_1 < \beta_2$, then compute

$$H_{j+1} = H_j + \frac{1}{\beta_1} \mathrm{vec}\,(\Delta Y)(\mathrm{vec}\,(\Delta Y))^T - \frac{1}{\beta_2} H_j \mathrm{vec}\,(\Delta G)(H_j \mathrm{vec}\,(\Delta G))^T;$$

else compute

$$H_{j+1} = H_j + \frac{1}{\beta_1} \Bigg(1 + \frac{\beta_2}{\beta_1}) \mathrm{vec}\,(\Delta Y)[\mathrm{vec}\,(\Delta Y)]^T$$
$$- \mathrm{vec}\,(\Delta Y)[\mathrm{vec}\,(\Delta G)]^T H_j - H_j \mathrm{vec}\,(\Delta G)[\mathrm{vec}\,(\Delta Y)]^T \Bigg).$$

(7) Compute

$$\mathrm{vec}\,(D_{j+1}) = -H_{j+1} \mathrm{vec}\,(G_{j+1}).$$

(8) Replace k, l by $k+1$, $l+1$, respectively, and go to Step 3.

Remark 3.9.7 In practice we may apply the line search techniques based on curve fitting, such as cubic fit and quadratic fit, to compute α_j at Step 3.

Remark 3.9.8 At Step 5 we may compute each element of G_j by formula

$$\frac{\partial (\eta_k, Y_{j+1})}{\partial x_{il}} = 2x_{il} + \eta_k \left(\sum_{t=1}^{l_1} \frac{\partial g_t(Y_{j+1})}{\partial x_{il}} + \sum_{t=1}^{l_2} \frac{\partial h_t(Y_{j+1})}{\partial x_{il}} \right),$$

where $\frac{\partial g_t(Y_(j+1)}{\partial x_{il}}$ and $\frac{\partial h_t(Y_{j+1})}{\partial x_{il}}$ can be computed by the formula (3.9.41) and (3.9.42), in which the singular value decompositions can be obtained using standard routines. But observe that for each j or k the formula in (3.9.41) or (3.9.42) requires only a few singular vectors if the multiplicity r_j or s_k is not too large. Hence in such case it is much more efficient to use the inverse power method to compute the desired singular vectors.

Maybe the reader has noticed that this method is somewhat complicated to implement. But in practice it often gives better solutions than the method described in the previous subsection.

Example 3.9.2 Let

$$A = \begin{bmatrix} -0.10940 & 0.06280 & 0.00000 & 0.00000 & 0.00000 \\ 1.30600 & -2.13200 & 0.98070 & 0.00000 & 0.00000 \\ 0.00000 & 1.59500 & -3.14900 & 1.54700 & 0.00000 \\ 0.00000 & 0.03550 & 2.63200 & -4.25700 & 1.85500 \\ 0.00000 & 0.00227 & 0.00000 & 0.16360 & -0.16250 \end{bmatrix},$$

$$B^T = \begin{bmatrix} 0.00000 & 0.06380 & 0.08380 & 0.10040 & 0.00630 \\ 0.00000 & 0.00000 & -0.13960 & -0.20600 & -0.01280 \end{bmatrix},$$

$$\mathcal{L} = \{-0.2, \quad -0.5, \quad -1+i, \quad -1-i, \quad -1\}.$$

Applying the method described in the previous subsection to the above data, we obtain the computed feedback matrix as follows:

$$F_1 = \begin{bmatrix} -16.49333 & 36.96135 & -36.66062 & 17.03187 & -9.14835 \\ -2.81023 & -8.43420 & 61.43848 & -66.01980 & 21.55015 \end{bmatrix}.$$

The computed eigenvalues of the resulting closed-loop system matrix $A+BF_1$ are -0.2, -0.5, $-1.71206792 \pm 1.29846474$, and 0, some of which are far from the assigned poles. But applying the above method we find

$$F_2 = \begin{bmatrix} -48.74817 & 105.96879 & -223.14617 & 189.73044 & -45.92917 \\ -23.32831 & 34.00577 & -55.35486 & 40.97710 & -0.22674 \end{bmatrix}.$$

The eigenvalues of $A + BF_2$ are found to agree up to ten significant digits with the assigned poles.

Perhaps the reader has found that some further researches are still needed to develop some numerically stable and efficient algorithm based on the methods introduced in this section. In fact, to the best of our knowledge, up to now there are no numerically stable and efficient algorithms for solving Problem RPA. Consequently, further progress in this significant field of control engineering is needed.

Notes and References

Definition 3.2.1 is taken from [223]. The equivalent statements on the controllability (i.e., Theorem 3.2.1) are taken from [223], [171], and [124]. Section 3.3 is mainly based on Chapter 2 of [223]; the first proof of Theorem 3.3.4 was given by Wonham [222]; some simpler proofs were given in [47] and [109].

Section 3.4 is mainly based on [209], but the discussion given here need not embed the given data in the complex field so that some theoretical difficulties are avoided; Example 3.4.1 is taken from [209]; other related references include [126], [127], [148], [171], and [106].

The content of sections 3.5–3.7 represents three typical methods for solving Problem PA: the Schur method, the invariant subspace method, and the QR-like methods. The Schur method is based on [215] with some significant changes. The invariant subspace method was developed in [177] and [179], but here some significant improvement has been made. The QR-like method is mainly based on [155], [157], and [158] with some significant changes. Additional references on numerical method for solving Problem PA include [24], [26], [28], [29], [36], [37], [44], [45], [144], [152]–[154], [156], [159], [160], [166], [175], [182], and [187]. Examples 3.8.1 and 3.8.4 are taken from [157].

For a discussion of the measures of robustness see [204] and [124]. The formula I and the numerical method I are mainly based on [124]. Some papers concerned with robust pole assignment include [29], [123], [132], [201], and [202].

The pole assignment problem with minimized norm controllers and the pole assignment problem using output feedback are not included in this book. This is because, to the best of our knowledge, although there is a large literature on these problems, the basic theories haven't been well established and numerically stable and efficient algorithms haven't been developed. Those readers who are interested in these problems are referred to [23], [33], [48], [60], [66], [67], [128], [206], and the references contained therein.

Chapter 4

Additive and Multiplicative Inverse Eigenvalue Problems

4.1 Introduction

Having discussed Jacobi matrix inverse eigenvalue problems and pole assignment problems, we now direct our attention to the third major problem area in inverse algebraic eigenvalue problems, additive and multiplicative inverse eigenvalue problems. These two problems can be stated as follows:

Problem A (Additive Inverse Eigenvalue Problem) *Given an $n \times n$ real symmetric matrix $A = [\alpha_{ij}]$ and n real numbers $\lambda_1 \leq \cdots \leq \lambda_n$, find a vector $c = (c_1, \dots, c_n)^T \in \mathbf{R}^n$ such that the matrix $A + \operatorname{diag}(c_1, \dots, c_n)$ has eigenvalues $\lambda_1, \dots, \lambda_n$.*

Problem M (Multiplicative Inverse Eigenvalue Problem) *Given an $n \times n$ positive definite real symmetric matrix $A = [\alpha_{ij}]$ and n positive numbers $0 < \lambda_1 \leq \cdots \leq \lambda_n$, find a vector $c = (c_1, \dots, c_n)^T \in \mathbf{R}^n$ with $c_i > 0$, $i = 1, \dots, n$, such that the eigenvalues of the matrix DA are precisely the given values $\lambda_1, \dots, \lambda_n$. Here $D = \operatorname{diag}(c_1, \dots, c_n)$.*

The additive and multiplicative inverse eigenvalue problems arise in various areas of application. One classical example of the additive inverse eigenvalue problem is the solution of inverse Sturm-Liouville problems. Consider for example the boundary value problem

$$\begin{cases} -u''(x) + p(x)u(x) = \lambda u(x), \\ u(0) = u(\pi) = 0. \end{cases} \tag{4.1.1}$$

Suppose that the potential $p(x)$ is unknown, but the spectrum $\{\lambda_i\}_1^\infty$ is given. Can we determine $p(x)$? Let us discretize (4.1.1): let $h = \pi/n+1$, $u_k = u(kh)$, $p_k = p(kh)$, $k = 1, \dots, n$, and assume that $\{\lambda_1, \dots, \lambda_n\}$ is given. Using finite differences to approximate u'' we obtain

$$\begin{cases} \dfrac{-u_{k+1} + 2u_k - u_{k-1}}{h^2} + p_k u_k = \lambda u_k, \quad k = 1, 2, \dots, n, \\ u_0 = u_{n+1} = 0. \end{cases} \tag{4.1.2}$$

In matrix form (4.1.2) can be written as

$$(A + D)u = \lambda u \tag{4.1.3}$$

where

$$A = \frac{1}{h^2} \begin{bmatrix} 2 & -1 & & & \\ -1 & 2 & -1 & & \\ & \ddots & \ddots & \ddots & \\ & & \ddots & \ddots & -1 \\ & & & -1 & 2 \end{bmatrix}, \tag{4.1.4}$$

$$D = \operatorname{diag}(p_1, \dots, p_n), \quad u = (u_1, \dots, u_n)^T.$$

Therefore, the problem is : can we determine a diagonal matrix D such that $A+D$ has the given spectrum $\{\lambda_1, \dots, \lambda_n\}$? This is just the additive inverse eigenvalue problem.

One interesting example of the multiplicative inverse eigenvalue problem arises in studying a vibrating string. Here the boundary value problem is

$$\begin{cases} -u''(x) = \lambda \rho(x) u(x), \\ u(0) = u(\pi) = 0. \end{cases} \tag{4.1.5}$$

The question is whether we can determine the density function $\rho(x) > 0$ from the given spectrum $\{\lambda_i\}_1^\infty$. Discretizing the Problem As before, we obtain

$$Au = \lambda D u$$

or equivalently

$$D^{-1} A u = \lambda u$$

where $D = \operatorname{diag}(\rho(h), \rho(2h), \dots, \rho(nh))$, A and u are defined as in (4.1.4). Now the problem is: find a positive diagonal matrix D such that the matrix DA has the given eigenvalues $\lambda_1, \dots, \lambda_n$. This is just the multiplicative inverse eigenvalue problem.

A more general form of the additive and multiplicative inverse eigenvalue problem is the following problem.

Problem G *Given $n+1$ real $n \times n$ matrices $A_k = [\alpha_{ij}^{(k)}]$, $k = 0, 1, \ldots, n$, and given n real numbers $\lambda_1 \leq \cdots \leq \lambda_n$, find a vector $c = (c_1, \ldots, c_n)^T \in \mathbf{R}^n$ such that the matrix*

$$A(c) = A_0 + \sum_{k=1}^{n} c_k A_k \tag{4.1.6}$$

has the given values $\lambda_1, \ldots, \lambda_n$ as its eigenvalues.

Mathematically, Problem G is only a generalization of additive and multiplicative inverse eigenvalue problems: Problem A is obtained when A_0 is symmetric and $A_k = e_k e_k^T$, $k = 1, 2, \ldots, n$; Problem M is obtained when $A_0 = 0$ and $A_k = e_k a_k^T$, $k = 1, 2, \ldots, n$, where a_k^T is the kth row of A. Actually, however, Problem G arises in various areas of application. For example, this kind of problem occurs in shell model computations in nuclear spectroscopy. There $A(c)$ is the Hamiltonian, the variables c_i are the interactions of one and two bodies, the matrices A_i represent the result of adding and symmetrizing the effect of many particles, and the set $\{\lambda_i\}$ is obtained by experimental measurements.

In practical it happens frequently that only a few eigenvalues are given. However, for the purpose of analysis it is convenient to consider the case where the number of parameters, eigenvalues and the order of the matrices is the same.

Another special case of Problem G is obtained when the matrices A_k are all symmetric. In this case we refer to Problem G as Problem SG, the symmetric general inverse eigenvalue problem. The rich mathematical structure of symmetric matrices makes Problem SG become one of the most pleasing inverse eigenvalue problems.

It is easy to construct examples that show that none of the above three problems is always solvable. Therefore, the question that arises immediately is : when does there exist a solution to the above problems? A complete answer to the 2-dimensional problems will be given in the next section. It seems to be quite difficult to give a complete answer to the general case. Up to now only some necessary and sufficient conditions for the solubility of Problem G in many special cases have been known. In Section 4.3 we will present some necessary conditions for the solvability of the symmetric inverse eigenvalue problems. Some sufficient conditions for the solvability of the four problems described in this section will be given in Section 4.4.

To do the sensitivity analysis of the solutions of Problem G (or its special cases) is another important task. In Section 4.5, we will use some tools in mathematical analysis to develop several perturbation bounds of some solutions of the symmetric problems.

In the last two sections of this chapter we will concentrate on how to find a numerical solution of Problem G or its special cases. Most of existing numerical methods for solving these problems consist of two major steps: first to construct an equivalent nonlinear system and then to apply Newton's method to it. Three such methods will be developed in Section 4.6, in which we also give convergence analysis under some nonsingular assumptions.

However, Newton-like methods inherit a usual shortcoming of local convergence. In many cases, it often leads to the failure of the algorithms since it is hard to select a valid initial point. One possibility to overcome this shortcoming is to use homotopy method. Such a method for solving the additive inverse eigenvalue problem will be introduced in Section 4.7.

Finally, it is important to note that we can make some convenient assumptions on the three problems described above.

Obviously, without loss of generality we may always assume that $\alpha_{ii} = 0$ for all $i = 1, 2, \ldots, n$ in Problem A.

Note that for any symmetric positive definite matrix $A = [\alpha_{ij}] \in \mathbf{R}^{n\times n}$ and any positive definite diagonal matrix $X \in \mathbf{R}^{n\times n}$. We have

$$D^{\frac{1}{2}}(XA)D^{-\frac{1}{2}} = (D^{\frac{1}{2}}XD^{\frac{1}{2}})(D^{-\frac{1}{2}}AD^{-\frac{1}{2}}),$$

where $D = \operatorname{diag}(\alpha_{11}, \ldots, \alpha_{nn})$. The diagonal elements of $D^{-\frac{1}{2}}AD^{-\frac{1}{2}}$ are all equal to 1, and $D^{\frac{1}{2}}XD^{\frac{1}{2}}$ is still a positive definite diagonal matrix. Therefore, without loss of generality we may always assume that $a_{ii} = 1$ for all $i = 1, 2, \ldots, n$ in Problem M.

Assume the diagonal element matrix of the matrices A_k

$$E = \begin{bmatrix} \alpha_{11}^{(1)} & \alpha_{22}^{(1)} & \cdots & \alpha_{nn}^{(1)} \\ \alpha_{11}^{(2)} & \alpha_{22}^{(2)} & \cdots & \alpha_{nn}^{(2)} \\ \vdots & \vdots & & \vdots \\ \alpha_{11}^{(n)} & \alpha_{22}^{(n)} & \cdots & \alpha_{nn}^{(n)} \end{bmatrix}$$

is nonsingular, and let $E^{-1} = [\xi_{ij}]$ and $\widetilde{c} = (\widetilde{c}_1, \ldots, \widetilde{c}_n)^T = Ec$. Then we have the following transformation:

$$\begin{aligned} A(c) &= A_0 + \sum_{i=1}^{n} c_i A_i = A_0 + \sum_{i=1}^{n} \left(\sum_{j=1}^{n} \xi_{ij}\widetilde{c}_j \right) A_i \\ &= A_0 + \sum_{j=1}^{n} \widetilde{c}_j \left(\sum_{i=1}^{n} \xi_{ij} A_i \right). \end{aligned}$$

The diagonal element vector of $\sum_{i=1}^{n} \xi_{ij} A_i$ is the jth unit vector. Thus, without loss of generality we may assume that $\alpha_{ii}^{(j)} = \delta_{ij}$ for $i, j = 1, 2, \ldots, n$ in Problem G.

4.2 The Solvability of Two-demensional Problems

In this section, we consider the solvability of Problem G for the case where $n = 2$. More precisely, let

$$A_k = \begin{bmatrix} \alpha_{11}^{(k)} & \alpha_{12}^{(k)} \\ \alpha_{21}^{(k)} & \alpha_{22}^{(k)} \end{bmatrix} \in \mathbf{R}^{2\times 2}, \quad k = 0, 1, 2, \text{ and } \lambda_1, \lambda_1 \in \mathbf{R}.$$

We discuss under what conditions there exists a vector $x = (\xi_1, \xi_2)^T \in \mathbf{R}^2$ such that the matrix

$$A(x) = A_0 + \xi_1 A_1 + \xi_2 A_2 = \begin{bmatrix} \alpha_{11}(x) & \alpha_{12}(x) \\ \alpha_{21}(x) & \alpha_{22}(x) \end{bmatrix} \tag{4.2.1}$$

has the given values λ_1 and λ_2 as its eigenvalues, where

$$\alpha_{ij}(x) = \alpha_{ij}^{(0)} + \xi_1 \alpha_{ij}^{(1)} + \xi_2 \alpha_{ij}^{(2)}, \quad i, j = 1, 2. \tag{4.2.2}$$

There are two reasons why we deal with the case of $n = 2$ separately. One is that the rich mathematical theory for this case allows us to develop a complete theory of the solvability. Another is to provide us an example that shows that how difficult it would be to give some general sufficient and necessary conditions for the solvability of Problem G when $n \geq 3$.

Using the theorem of Vieta on the relations between the roots and the coefficients of a quadratic equation in one variable, we know that there exists a vector $x \in \mathbf{R}^2$ such that the eigenvalues of $A(x)$ are λ_1 and λ_2 if and only if x is a solution to the following system of polynomials

$$\begin{cases} \alpha_{11}(x) + \alpha_{22}(x) = \lambda_1 + \lambda_2, \\ \alpha_{11}(x)\alpha_{22}(x) - \alpha_{12}(x)\alpha_{21}(x) = \lambda_1 \lambda_2. \end{cases} \tag{4.2.3}$$

Substituting (4.2.2) into (4.2.3), by a simple calculation, we obtain

$$\begin{cases} t_1\xi_1 + t_2\xi_2 + t_0 = \lambda_1 + \lambda_2, \\ d_1\xi_1^2 + d_{12}\xi_1\xi_2 + d_2\xi_2^2 + d_{01}\xi_1 + d_{02}\xi_2 + d_0 = \lambda_1\lambda_2, \end{cases} \tag{4.2.4}$$

where

$$t_k = \operatorname{tr} A_k, \quad d_k = \det A_k, \quad k = 0, 1, 2,$$

$$d_{ij} = \det A_{ij} + \det A_{ji}, \quad A_{ij} = \begin{bmatrix} \alpha_{11}^{(i)} & \alpha_{12}^{(j)} \\ \alpha_{21}^{(i)} & \alpha_{22}^{(j)} \end{bmatrix}, \quad i \neq j, \quad i, j = 0, 1, 2. \tag{4.2.5}$$

We first consider the special case where $t_1 = t_2 = 0$. In this case, if there is a x such that the eigenvalues of $A(x)$ are λ_1 and λ_2, then it follows from (4.2.4) that it must have $t_0 = \lambda_1 + \lambda_2$. When $t_0 = \lambda_1 + \lambda_2$, applying the classification of conics in a plane to the second equation of (4.2.4), we obtain the following theorem.

Theorem 4.2.1 *If $t_1 = t_2 = 0$ and $t_0 = \lambda_1 + \lambda_2$, then there exists a real vector $x = (\xi_1, \xi_2)^T$ such that $A(x)$ has λ_1 and λ_2 as its eigenvalues if and only if one of the following four conditions holds:*

(1) $\Delta = 0$ *and* $\delta \neq 0$;

(2) $\Delta = \delta = 0$ *and* $d_{01}^2 + d_{02}^2 - 4(d_1 + d_2)(d_0 - \lambda_1\lambda_2) \geq 0$;

(3) $\Delta \neq 0$ *and* $\delta \geq 0$;

(4) $\delta < 0$ *and* $(d_1 + d_2)\Delta < 0$.

Where

$$\Delta = \det \begin{bmatrix} 2d_1 & d_{12} & d_{01} \\ d_{12} & 2d_2 & d_{02} \\ d_{01} & d_{02} & 2(d_0 - \lambda_1\lambda_2) \end{bmatrix}, \quad \delta = d_{12}^2 - 4d_1d_2. \tag{4.2.6}$$

Moreover, when condition (3) or (4) holds, there exist infinitely many vectors x such that the eigenvalues of $A(x)$ are λ_1 and λ_2.

Example 4.2.1 Let

$$A_0 = \begin{bmatrix} -1 & 1 \\ 2 & -2 \end{bmatrix}, \quad A_1 = \begin{bmatrix} -1 & 2 \\ 3 & 1 \end{bmatrix}, \quad A_2 = \begin{bmatrix} 0 & 1 \\ -1 & 0 \end{bmatrix},$$
$$\lambda_1 = -1, \qquad \lambda_2 = -2.$$

It is easy to see that $t_1 = \operatorname{tr} A_1 = 0 = \operatorname{tr} A_2 = t_2$ and $t_0 = \operatorname{tr} A_0 = -3 = \lambda_1 + \lambda_2$. A straightforward calculation gives that

$$d_0 = \det A_0 = 0, \quad d_1 = \det A_1 = -7, \quad d_2 = \det A_2 = 1,$$
$$d_{12} = \det A_{12} + \det A_{21} = -3 + 2 = -1,$$
$$d_{01} = \det A_{01} + \det A_{10} = -5 - 1 = -6,$$
$$d_{02} = \det A_{02} + \det A_{20} = -2 + 1 = -1,$$
$$\Delta = \det \begin{bmatrix} -14 & -1 & -6 \\ -1 & 2 & -1 \\ -6 & -1 & -4 \end{bmatrix} = 46,$$
$$\delta = 1 + 28 = 29.$$

By Theorem 4.2.1, we know that Problem G with the above given data has infinitely many solutions. In fact, $x = (\xi_1, \xi_2)^T$ is a solution to Problem G if and only if x is a point of the hyperbola

$$-7\xi_1^2 - \xi_1\xi_2 + \xi_2^2 - 6\xi_1 - \xi_2 - 2 = 0.$$

Next we consider the case where $t_1 \neq 0$ or $t_2 \neq 0$. Without loss of generality we may assume that $t_1 \neq 0$. Then from the first equation of (4.2.4) we get

$$\xi_1 = (\lambda_1 + \lambda_2 - t_0 - t_2\xi_2)/t_1, \tag{4.2.7}$$

Substituting (4.2.7) into the second equation of (4.2.4) and rearranging the terms we obtain

$$a_1 x_2^2 + b_1\xi_2 + c_1 = 0, \tag{4.2.8}$$

where

$$\begin{aligned} a_1 &= d_1 t_2^2 - d_{12} t_1 t_2 + d_2 t_1^2, \\ b_1 &= (d_{12} t_1 - 2d_1 t_2)(\lambda_1 + \lambda_2 - t_0) - d_{01} t_1 t_2 + d_{02} t_1^2, \\ c_1 &= d_1(\lambda_1 + \lambda_2 - t_0)^2 + d_{01} t_1(\lambda_1 + \lambda_2 - t_0) + t_1^2(d_0 - \lambda_1\lambda_2). \end{aligned}$$

Thus, applying the elementary theory of real quadratic equations to (4.2.8), we can easily obtain the following result.

Theorem 4.2.2 *If there is $i \in \{1,2\}$ such that $t_i = \operatorname{tr} A_i \neq 0$, then there is a real vector $x = (\xi_1, \xi_2)$ such that the eigenvalues of $A(x)$ is λ_1 and λ_2 if and only if $b_i^2 - 4a_ic_i \geq 0$, where*

$$\begin{aligned} a_i &= d_1 t_2^2 - d_{12} t_1 t_2 + d_2 t_1^2, \\ b_i &= (d_{12} t_i - 2d_i t_{k_i})(\lambda_1 + \lambda_2 - t_0) - d_{0i} t_1 t_2 + d_{0k_i} t_i^2, \\ c_i &= d_i(\lambda_1 + \lambda_2 - t_0)^2 + d_{0i} t_i(\lambda_1 + \lambda_2 - t_0) + t_i^2(d_0 - \lambda_1\lambda_2), \\ k_i &= \begin{cases} 2, & i = 1, \\ 1, & i = 2. \end{cases} \end{aligned}$$

Moreover, when the condition holds, there are at most two vectors x such that $A(x)$ has λ_1 and λ_2 as its eigenvalues.

Applying Theorem 4.2.2 to the additive and multiplicative inverse eigenvalue problems for the case of $n = 2$, we immediately obtain the following well-known results.

Corollary 4.2.3 *Let $\lambda_1, \lambda_2 \in \mathbf{R}$ and $A = \begin{bmatrix} 0 & \alpha \\ \alpha & 0 \end{bmatrix}$ with $\alpha \in \mathbf{R}$. Then there is a real diagonal matrix $D = \operatorname{diag}(\xi_1, \xi_2)$ such that $A + D$ has λ_1 and λ_2 as its eigenvalues if and only if*

$$(\lambda_1 - \lambda_2)^2 \geq 4\alpha^2. \tag{4.2.9}$$

Corollary 4.2.4 *Let $\lambda_1, \lambda_2 \in \mathbf{R}$ with $\lambda_i > 0$ and $A = \begin{bmatrix} 1 & \beta \\ \beta & 1 \end{bmatrix}$ with $\beta \in \mathbf{R}$ and $|\beta| < 1$. Then there exists a positive diagonal matrix $D = \operatorname{diag}(\xi_1, \xi_2)$ such that the matrix DA has λ_1 and λ_2 as its eigenvalues if and only if*

$$(\lambda_1 - \lambda_2)^2 \geq \beta^2(\lambda_1 + \lambda_2)^2. \tag{4.2.10}$$

From conditions (4.2.9) and (4.2.10) we can see that to ensure that the 2-dimensional additive or multiplicative inverse eigenvalue problem has a solution, there must be a enough separation between the two given eigenvalues; that is, the distance between the two given eigenvalues must be large enough.

Example 4.2.2 Let

$$A = \begin{bmatrix} 1 & 0.5 \\ 0.5 & 1 \end{bmatrix}, \qquad \lambda_1 = 1, \quad \lambda_2 = 2.$$

Then

$$(\lambda_1 - \lambda_2)^2 = 1 < \frac{9}{4} = \beta^2(\lambda_1 + \lambda_2)^2.$$

Thus, Corollary 4.2.4 tells us that there is no positive diagonal matrix D such that DA has the given values as its eigenvalues.

4.3 Necessary Conditions for the Solvability

In the previous section we have given some criteria for the existence of solutions to a given 2-dimensional inverse eigenvalue problem. However, for the general case no such a criterion appears to be known; only some necessary conditions and sufficient conditions are known. In this section we will present some necessary conditions for a given symmetric inverse eigenvalue problem to have a solution . We begin with some notation and definitions.

Let k and n be integers, $1 \leq k \leq n$. We use $\mathcal{G}_{k,n}$ to denote the set of all increasing sequences of integers:

$$\pi = (j_1, j_2, \dots, j_k), \quad \text{with } 1 \leq j_1 < j_2 < \dots < j_k \leq n.$$

For any given $\pi = (j_1, \dots, j_k) \in \mathcal{G}_{k,n}$ and $A = [\alpha_{ij}] \in \mathbf{R}^n$, we use $A(\pi)$ to denote the $k \times k$ principal submatrix of A whose (i, l) element is $\alpha_{j_i j_l}$, $i, l = 1, 2, \dots, k$; and we define

$$A^{(0)} = A - \operatorname{diag}(\alpha_{11}, \dots, \alpha_{nn}).$$

To establish our main results, we need two elementary results.

Lemma 4.3.1 *Let $B = [\beta_{ij}]$ be a $k \times k$ real symmetric matrix with eigenvalues $\mu_1 \le \cdots \le \mu_k$. Then we have*

$$\sum_{1\le i<j\le k} (\mu_j - \mu_i)^2 \ge k\|B^{(0)}\|_F^2. \tag{4.3.1}$$

Proof Since

$$\begin{aligned}\sum_{i=1}^k \mu_i^2 &= \operatorname{tr}(B^2) = \|B^{(0)}\|_F^2 + \sum_{i=1}^k \beta_{ii}^2 \\ &\ge \|B^{(0)}\|_F^2 + \frac{1}{k}\left(\sum_{i=1}^k \beta_{ii}\right)^2 = \|B^{(0)}\|_F^2 + \frac{1}{k}\left(\sum_{i=1}^k \mu_i\right)^2,\end{aligned}$$

we have

$$\sum_{1\le i<j\le k} (\mu_j - \mu_i)^2 = k\sum_{i=1}^k \mu_i^2 - \left(\sum_{i=1}^k \mu_i\right)^2 \ge k\|B^{(0)}\|_F^2,$$

which is the desired result. □

Lemma 4.3.2 *Let $A_k \in \mathbf{R}^{n\times n}$, $k = 0, 1, 2, \ldots, m$, and let*

$$\mathcal{X} = \left\{C \in \mathbf{R}^{n\times n} \,:\, C = \sum_{k=1}^m \xi_k A_k,\ (\xi_1, \ldots, \xi_n)^T \in \mathbf{R}^n\right\}.$$

Then there exists a unique $C_0 \in \mathcal{X}$ such that

$$\|A_0 + C_0\|_F = \min\{\|A_0 + C\|_F \,:\, C \in \mathcal{X}\}. \tag{4.3.2}$$

Moreover, C_0 satisfies (4.3.2) if and only if

$$\operatorname{tr}(C_0 A_k^T) + \operatorname{tr}(A_0 A_k^T) = 0 \tag{4.3.3}$$

for all $k = 1, 2, \ldots, m$.

Proof Clearly, $\mathcal{X}$ is a subspace of $\mathbf{R}^{n\times n}$. Hence, there exists $B_k \in \mathbf{R}^{n\times n}$, $k = 1, 2, \ldots, l$, with $\operatorname{tr}(B_i B_j^T) = \delta_{ij}$, $i, j = 1, 2, \ldots, l$, such that

$$\begin{aligned}\mathcal{X} &= \operatorname{span}\{B_1, \ldots, B_l\} \\ &= \left\{C \,:\, C = \sum_{k=1}^l \xi_k B_k,\ (\xi_1, \ldots, \xi_l)^T \in \mathbf{R}^l\right\};\end{aligned}$$

that is, $\{B_1, \ldots, B_l\}$ is an orthonormal basis for $\mathcal{X}$.

Now let

$$C_0 = -\sum_{k=1}^{l} \operatorname{tr}(A_0 B_k^T) B_k. \tag{4.3.4}$$

Then $C_0 \in \mathcal{X}$ and satisfies

$$\begin{aligned}\operatorname{tr}\left((A_0 + C_0)\Big(\sum_{k=1}^{l} \xi_k B_k\Big)^T\right) &= \sum_{k=1}^{l} \xi_k \left(\operatorname{tr}(A_0 B_k^T) + \operatorname{tr}(C_0 B_k^T)\right) \\ &= \sum_{k=1}^{l} \xi_k \left(\operatorname{tr}(A_0 B_k^T) - \operatorname{tr}(A_0 B_k^T)\right) = 0\end{aligned}$$

for any $(\xi_1 \ldots, \xi_l)^T \in \mathbf{R}^l$; that is, C_0 satisfies

$$\operatorname{tr}\left(A_0 + C_0)C^T\right) = 0$$

for any $C \in \mathcal{X}$. Consequently, for any $C \in \mathcal{X}$ we have

$$\begin{aligned}\|A_0 + C\|_F^2 &= \|(A_0 + C_0) + (C - C_0)\|_F^2 \\ &= \operatorname{tr}\left([(A_0 + C_0) + (C - C_0)][(A_0 + C_0) + (C - C_0)]^T\right) \\ &= \operatorname{tr}\left((A_0 + C_0)(A_0 + C_0)^T\right) + \operatorname{tr}\left((C - C_0)(C - C_0)^T\right) \\ &= \|A_0 + C_0\|_F^2 + \|C - C_0\|_F^2,\end{aligned}$$

and so

$$\|A_0 + C\|_F \geq \|A_0 + C_0\|_F$$

for any $C \in \mathcal{X}$ with equality if and only if $C = C_0$. This proves that there exists a unique $C_0 \in \mathcal{X}$ satisfies (4.3.2).

The foregoing proof also shows that $C_0 \in \mathcal{X}$ satisfies (4.3.2) if and only if C_0 can be expressed as (4.3.4). This is equivalent to

$$\operatorname{tr}\left((A_0 + C_0)C^T = 0\right. \tag{4.3.5}$$

for any $C \in \mathcal{X}$. It is not difficult to verify that (4.3.5) holds if and only if (4.3.3) holds. □

With the foregoing preparations we now establish the main results of this section.

Theorem 4.3.3 *If $\alpha_{ii} = 1$ for all $i = 1, \dots, n$ in Problem* **M**, *and if Problem* **M** *has a solution, then the given data must satisfy the following conditions*

$$\sum_{1 \le i < j \le k} (\lambda_i - \lambda_{j+n-k})^2 \ge \lambda_1^2 k \max \left\{ \|A^{(0)}(\pi)\|_F^2 \ : \ \pi \in \mathcal{G}_{k,n} \right\} \tag{4.3.6}$$

for all $k = 2, \dots, n$.

Proof Suppose that Problem **M** has a solution $c = (c_1, c_2, \dots c_n)^T \in \mathbf{R}^n$ with $c_i > 0$, $i = 1, 2, \dots, n$, that is, DA has $\lambda_1, \dots, \lambda_n$ as its eigenvalues, where $D = \operatorname{diag}(c_1, \dots, c_n)$. Let $\pi = (j_1, \dots, j_k) \in \mathcal{G}_{k,n}$ be given, and let $\mu_1 \le \dots \le \mu_k$ be eigenvalues of $(D(\pi))^{\frac{1}{2}} A(\pi)(D(\pi))^{\frac{1}{2}}$. By Lemma 4.3.1, we have

$$\sum_{1 \le i < j \le k} (\mu_j - \mu_i)^2 \ge k \left\| (D(\pi))^{\frac{1}{2}} A^{(0)}(\pi)(D(\pi))^{\frac{1}{2}} \right\|_F^2. \tag{4.3.7}$$

Notice that $D^{\frac{1}{2}} A D^{\frac{1}{2}}$ is similar to DA and $(D(\pi))^{\frac{1}{2}} A(\pi)(D(\pi))^{\frac{1}{2}}$ is a $k \times k$ principal submatrix of $D^{\frac{1}{2}} A D^{\frac{1}{2}}$. By Theorem 1.5.2, we know that

$$\lambda_i \le \mu_i \le \lambda_{i+n-k}, \quad i = 1, 2, ..., k, \tag{4.3.8}$$

and so

$$\lambda_{j+n-k} - \lambda_i \ge \mu_j - \mu_i \ge 0, \quad 1 \le i < j \le k. \tag{4.3.9}$$

Combining it with (4.3.7) gives

$$\sum_{1 \le i < j \le k} (\lambda_i - \lambda_{j+n-k})^2 \ge k \|[D(\pi)]^{\frac{1}{2}} A^{(0)}(\pi)[D(\pi)]^{\frac{1}{2}}\|_F^2. \tag{4.3.10}$$

By Theorem 1.5.3, we have

$$c_i \ge \lambda_1 > 0, \quad i = 1, 2, ..., n. \tag{4.3.11}$$

Let $A^{(0)}(\pi) = [\beta_{il}]$. It follows from (4.3.11) that

$$\begin{aligned} \left\| (D(\pi))^{\frac{1}{2}} A^{(0)}(\pi)(D(\pi))^{\frac{1}{2}} \right\|_F^2 &= \sum_{i,l=1}^{k} c_{j_i} c_{j_l} |\beta_{il}|^2 \ge \lambda_1^2 \sum_{i,l=1}^{k} |\beta_{il}|^2 \\ &= \lambda_1^2 \left\| A^{(0)}(\pi) \right\|_F^2. \end{aligned}$$

This, together with (4.3.10), gives

$$\sum_{1 \le i < j \le k} (\lambda_i - \lambda_{j+n-k})^2 \ge \lambda_1^2 \|A^{(0)}(\pi)\|_F^2. \tag{4.3.12}$$

Noting that $\pi \in \mathcal{G}_{k,n}$ is arbitrary, (4.3.6) follows immediately from (4.3.12). Thus the proof of Theorem 4.3.3 is completed. □

As a immediate consequence of Theorem 4.3.3, we have the following corollary, which is more convenient in practice.

Corollary 4.3.4 *If $\alpha_{ii} = 1$ for all $i = 1, \dots, n$ in Problem* M, *and if Problem* M *has a solution, then the give data must satisfy*

$$\sum_{1 \le i < j \le n} (\lambda_i - \lambda_j)^2 \ge 2n\lambda_1^2 \sum_{1 \le i < j \le n} |\alpha_{ij}|^2 \tag{4.3.13}$$

and

$$\lambda_n - \lambda_1 \ge 2\lambda_1 \max_{1 \le i < j \le n} |\alpha_{ij}|. \tag{4.3.14}$$

The next theorem gives some necessary conditions for the solvability of Problem SG.

Theorem 4.3.5 *Let $A_k \in \mathbf{R}^{n \times n}$, $k = 0, 1, \dots, n$, be symmetric with $\alpha_{ii}^{(k)} = \delta_{ik}$ for $i, k = 1, \dots, n$. If Problem* G *with these given data has a solution, then the data must satisfy the following conditions*

$$\sum_{1 \le i < j \le k} (\lambda_i - \lambda_{j+n-k})^2 \ge k \max\left\{ \left\| A^{(0)}(\pi) + \sum_{i=1}^{n} \xi_i^{(\pi)} A_i^{(0)}(\pi) \right\|_F^2 : \pi \in \mathcal{G}_{k,n} \right\} \tag{4.3.15}$$

for $k = 2, \dots, n$, where

$$(\xi_1^{(\pi)}, \dots, \xi_n^{(\pi)})^T = (S_\pi)^\dagger b_\pi,$$

$$S_\pi = \begin{bmatrix} \operatorname{tr}\left(A_1^{(0)}(\pi) A_1^{(0)}(\pi)\right) & \dots & \operatorname{tr}\left(A_1^{(0)}(\pi) A_n^{(0)}(\pi)\right) \\ \vdots & & \vdots \\ \operatorname{tr}\left(A_n^{(0)}(\pi) A_1^{(0)}(\pi)\right) & \dots & \operatorname{tr}\left(A_n^{(0)}(\pi) A_n^{(0)}(\pi)\right) \end{bmatrix}, \tag{4.3.16}$$

$$b_\pi = \left(-\operatorname{tr}\left(A_1^{(0)}(\pi) A_0^{(0)}(\pi)\right), \dots, -\operatorname{tr}\left(A_n^{(0)}(\pi) A_0^{(0)}(\pi)\right) \right)^T.$$

Let $A_k = e_k e_k^T$, $k = 1, 2, \dots, n$, in Theorem 4.3.5, we immediately obtain the necessary conditions for the solvability of the additive inverse eigenvalue problem as follows.

Theorem 4.3.6 *If $\alpha_{ii} = 0$ for all $i = 1, \dots, n$ in Problem A, and if Problem A has a solution, then the given data must satisfy the following conditions*

$$\sum_{1 \le i < j \le k} (\lambda_i - \lambda_{j+n-k})^2 \ge k \max \left\{ \|A(\pi)\|_F^2 \; : \; \pi \in \mathcal{G}_{k,n} \right\} \tag{4.3.17}$$

for $k = 2, \dots, n$.

Especially, when $k = 2$, n we obtain the following corollary, which is more convenient in practice.

Corollary 4.3.7 *If $\alpha_{ii} = 0$ for all $i = 1, \dots, n$ in Problem A, and if Problem A has a solution, then the given data must satisfy*

$$\sum_{1 \le i < j \le n} \left(\lambda_i - \lambda_j\right)^2 \ge 2n \sum_{1 \le i < j \le n} |\alpha_{ij}|^2 \tag{4.3.18}$$

and

$$\lambda_n - \lambda_1 \ge 2 \max_{1 \le i < j \le n} |\alpha_{ij}|. \tag{4.3.19}$$

Example 4.3.1. Let

$$A = \begin{bmatrix} 0 & 2.01 & 0 \\ 2.01 & 0 & 0.2 \\ 0 & 0.2 & 0 \end{bmatrix}, \quad \lambda_1 = 2, \; \lambda_2 = 5, \; \lambda_3 = 6.$$

It is easy to verify that the give data satisfies the condition (4.3.18), but does not satisfy the condition (4.3.19). Hence, by Corollary 4.3.7, we know that there is no diagonal matrix D such that $A + D$ has eigenvalues 2, 5, and 6.

We now give the proof of Theorem 4.3.5.

Proof of Theorem 4.3.5 Suppose that $c = (c_1, \dots, c_n)^T \in \mathbf{R}^n$ is a solution of Problem G with the given data. Let $\pi = (j_1, \dots, j_k) \in \mathcal{G}_{k,n}$ be given. Similar to the proof of (4.3.10), we can easily prove that

$$\sum_{1 \le i < j \le k} (\lambda_i - \lambda_{j+n-k})^2 \ge k \left\| A_0^{(0)}(\pi) + \sum_{i=1}^{n} c_i A_i^{(0)}(\pi) \right\|_F^2. \tag{4.3.20}$$

By Lemma 4.3.2, we know that there exists a vector $x_\pi = (\xi_1^{(\pi)}, \dots, \xi_n^{(\pi)})^T \in \mathbf{R}^n$ such that

$$\left\| A_0^{(0)}(\pi) + \sum_{i=1}^{n} \xi_i^{(\pi)} A_i^{(0)}(\pi) \right\|_F^2 = \min_{x = (\xi_i) \in \mathbf{R}^n} \left\| A_0^{(0)}(\pi) + \sum_{i=1}^{n} \xi_i A_i^{(0)}(\pi) \right\|_F^2; \tag{4.3.21}$$

moreover, x_π satisfies (4.3.21) if and only if

$$\sum_{i=1}^{n} \xi_i^{(\pi)} \operatorname{tr}\left(A_j^{(0)}(\pi) A_i^{(0)}(\pi)\right) = -\operatorname{tr}\left(A_j^{(0)}(\pi) A_0^{(0)}(\pi)\right) \tag{4.3.22}$$

for $j = 1, 2, \ldots, n$. In matrix-vector form (4.3.22) can be written as

$$S_\pi x_\pi = b_\pi, \tag{4.3.23}$$

where S_π and b_π are defined as in (4.3.16).

Since S_π is positive semidefinite, there exists an orthogonal matrix Q such that

$$S_\pi = Q^T \Sigma Q$$

where $\Sigma = \operatorname{diag}(\sigma_1, \sigma_2, \ldots, \sigma_m, 0, \ldots, 0)$ with $\sigma_i > 0$ for $i = 1, 2, \ldots, m$. Let

$$y = (\eta_1, \ldots, \eta_n)^T = Q x_\pi \quad \text{and} \quad d = (\delta_1, \ldots, \delta_n)^T = Q b_\pi.$$

Then (4.3.23) can be reduced to

$$\Sigma y = d. \tag{4.3.24}$$

Since (4.3.24) must have a solution, it follows that

$$\delta_i = 0, \qquad i = m+1, \ldots, n.$$

Therefore, if we set

$$\eta_i^{(0)} = \begin{cases} \dfrac{\delta_i}{\sigma_i}, & 1 \le i \le m, \\ 0, & m+1 \le i \le n, \end{cases}$$

then $y_0 = \left(\eta_1^{(0)}, \ldots, \eta_n^{(0)}\right)^T$ is a solution of system (4.3.24).

Now let $x_\pi = Q^T y_0$, that is, $x_\pi = (S_\pi)^\dagger b_\pi$. Then x_π satisfies (4.3.23), and hence x_π satisfies (4.3.21). Note that $\pi \in \mathcal{G}_{k,n}$ is arbitrary. Combining it with (4.3.20), we obtain (4.3.15) at once. This establishes Theorem 4.3.5. □

4.4 Sufficient Conditions for the Solvability

In this section we will establish several sufficient conditions for the solvability of Problem A, Problem M, and Problem G. Our discussion will be based on degree theory. Therefore, we first recall some basic definitions and results concerning the degree of a mapping (see[168] for further details).

In its original form, the degree of a mapping is a concept of combinatorial topology, but there are also purely analytic definitions for it. We will present here one such analytic approach to degree theory.

For given set $\Omega \subset \mathbf{R}^n$, we will use $\overline{\Omega}$ to denote the closure of Ω, and $\partial\Omega = \overline{\Omega} \setminus \Omega$ boundary of Ω. For given $\alpha > 0$, let $\mathcal{W}_\alpha$ be the set of all real functions $\varphi : [0, \infty) \subset \mathbf{R} \longrightarrow \mathbf{R}$ which are continuous on $[0, \infty)$ and for which there exists a $\delta \in (0, \alpha)$ such that $\varphi(t) = 0$ whenever $t \notin [\delta, \alpha]$. We call every $\varphi \in \mathcal{W}_\alpha$ a **weight function** of index α. Let

$$\mathcal{W}_\alpha^1 = \Big\{\varphi \in \mathcal{W}_\alpha : \int_{\mathbf{R}^n} \varphi(\|x\|_2)\mathrm{d}x = 1\Big\}.$$

Let $f : \mathcal{D} \subset \mathbf{R}^n \to \mathbf{R}^n$ be continuously differentiable on the open set $\mathcal{D}$, Ω an open, bounded set with $\overline{\Omega} \subset \mathcal{D}$, and $y \notin f(\partial\Omega)$ a given point. Then the **degree** of f at point y with respect to Ω is defined by

$$\deg(f, \Omega, y) = \int_\Omega \varphi\big(\|f(x) - y\|_2\big) \det f'(x)\mathrm{d}x, \tag{4.4.1}$$

where $\varphi \in \mathcal{W}'_\alpha$ is any weight function of index α such that $0 < \alpha < \min\{\|f(x) - y\|_2 : x \in \partial\Omega\}$.

It can be proved that the degree integral (4.4.1) is independent of the choice of the weight function. Moreover, if $\{x \in \Omega : f(x) = y\} = \{x_1, \cdots, x_m\}$, and $f'(x_i)$ is nonsingular for all $i = 1, 2, \cdots, m$, then we can prove that

$$\deg(f, \Omega, y) = \sum_{i=1}^m \operatorname{sgn}\big(\det f'(x_i)\big), \tag{4.4.2}$$

where

$$\operatorname{sgn}(t) = \begin{cases} 1, & \text{if } t > 0, \\ -1, & \text{if } t < 0. \end{cases}$$

This shows that, at least in this particular case, the degree introduced in (4.4.1) is integer-valued.

We can extend the definition of the degree to continuous mappings by using the standard Weierstrass approximation theorem.

Let $f : \overline{\Omega} \subset \mathbf{R}^n \to \mathbf{R}^n$ be continuous and Ω an open, bounded set. Then, for any $y \notin f(\partial\Omega)$, the **degree** of f at y with respect to Ω is defined by

$$\deg(f, \Omega, y) = \lim_{k\to\infty} \deg(f_k, \Omega, y), \tag{4.4.3}$$

where $f_k : \mathcal{D} \subset \mathbf{R}^n \to \mathbf{R}^n$ is any sequence of maps which are continuously differentiable on an open set $D \supset \overline{\Omega}$ and for which $\lim_{k\to\infty} \sup_{x\in\overline{\Omega}} \|f(x) - f_k(x)\|_2 = 0$

It can be proved that the limit in (4.4.3) is always reached after finitely many steps and is independent of the choice of the sequence $\{f_k\}$. Therefore the definition of (4.4.3) is meaningful.

The following result plays a central role in degree theory.

Lemma 4.4.1 (**Homotopy Invariance Theorem**) *Let Ω be open and bounded, and $h : \overline{\Omega} \times [0,1] \subset \mathbf{R}^{n+1} \to \mathbf{R}^n$ a continuous map from $\overline{\Omega} \times [0,1]$ into $\mathbf{R}^n$. Suppose further that $y \in \mathbf{R}^n$ satisfies $h(x,t) \neq y$ for all $(x,t) \in \partial\overline{\Omega} \times [0,1]$. Then $\deg\big(h(\cdot,t),\Omega,y\big)$ is constant for $t \in [0,1]$.*

The following theorem is a useful tool for the proof of other existence theorems.

Lemma 4.4.2 (**Kronecker Theorem**) *Let $f : \overline{\Omega} \subset \mathbf{R}^n \to \mathbf{R}^n$ be continuous and Ω an open, bounded set. If $y \notin f(\partial\Omega)$ and if $\deg(f,\Omega,y) \neq 0$, then the equation*

$$f(x) = y$$

has a solution in Ω.

Using Kronecker Theorem, we can prove the well-know existence theorem, Brouwer Fixed-Point Theorem, which is one of the most powerful tools in analysis.

Lemma 4.4.3 (**Brouwer Fixed-Point Theorem**) *Let $f : \overline{\Omega} \subset \mathbf{R}^n \to \mathbf{R}^n$ be continuous on the compact, convex set $\overline{\Omega}$, and suppose that $f(\overline{\Omega}) \subset \overline{\Omega}$. Then f has a fixed point in Ω, that is, there exists a point $x \in \Omega$ such that $f(x) = x$.*

Next we will use Kronecker theorem and Brouwer fixed-point theorem to establish some sufficient conditions for the solvability of the problems described in Section 4.1. It is well-known that symmetric matrices possess much better properties than nonsymmetric ones. Therefore we will here deal with the two different cases separately.

4.4.1 The Nonsymmetric Case

First we recall the well-known Gerschgorin disc theorem. For given $A = [\alpha_{ij}] \in \mathbf{R}^{n\times n}$, define

$$g_i(A) = \sum_{\substack{j=1 \\ j\neq i}}^{n} |\alpha_{ij}|, \quad i = 1,2,\ldots,n, \tag{4.4.4}$$

and

$$G_i(A) = \{z \in \mathcal{C} : |z - \alpha_{ii}| \leq g_i(A)\}, \quad i = 1, 2, \ldots, n, \tag{4.4.5}$$

which is known as the ith **Gerschgorin disc**. Then we have

Lemma 4.4.4 **(Gerschgorin Disc Theorem)** *Let $A \in \mathbf{R}^{n\times n}$. Then all the eigenvalues of A are located in the union of n Gerschgorin discs, that is,*

$$\lambda(A) \subset \bigcup_{i=1}^{n} G_i(A).$$

Furthermore, if a union of k of these n discs forms a connected region that is disjoint from all the remaining $n-k$ discs, then there are precisely k eigenvalues of A in this region.

The proof of this theorem can be found in many text books on matrix theory, for example, see [118]. Here we apply Gerschgorin disc theorem to prove a basic result, which will be repeatedly used in the sequel.

Lemma 4.4.5 *Let $A = [\alpha_{ij}] \in \mathbf{R}^n$ with $\alpha_{11} \leq \alpha_{22} \leq \cdots \leq \alpha_{nn}$. Suppose further that the eigenvalues $\lambda_1, \ldots, \lambda_n$ of A are all real and satisfy that*

$$\lambda_i - \lambda_{i-1} > 2\max\{g_i(A), g_{i-1}(A)\}, \quad i = 2, \ldots, n. \tag{4.4.6}$$

Then we have $\lambda_i \in G_i(A)$ for each $i = 1, 2, \ldots, n$.

Proof Note that (4.4.6) implies that

$$\lambda_1 < \lambda_2 < \cdots < \lambda_n.$$

Hence, by Lemma 4.4.4, each connected region of $\bigcup_{i=1}^{n} G_i(A)$ must have the form

$$\bigcup_{i=p}^{q} G_i(A), \quad p \leq q,$$

and the $q-p+1$ eigenvalues that are located in this region are just $\lambda_p, \ldots, \lambda_q$. Thus, if there exists an integer $1 \leq j \leq n$ so that $\lambda_j \notin G_j(A)$, then there must be two integers $p < q$ such that $\bigcup_{i=p}^{q} G_i(A)$ forms a connected region that is disjoint from all the remaining $n+p-q-1$ discs, and

$$\{\lambda_p, \ldots, \lambda_j, \ldots, \lambda_q\} \subset \bigcup_{i=p}^{q} G_i(A). \tag{4.4.7}$$

Without loss of generality we may now assume that

$$\lambda_j < \alpha_{jj} - g_j(A). \tag{4.4.8}$$

It follows from (4.4.7) and (4.4.8) that there are two possibilities:

(1) $\lambda_j, \lambda_p \in \bigcup_{i=p}^{j-1} G_i(A)$; or

(2) $\lambda_j \in \bigcup_{i=j+1}^{q} G_i(A)$ or $\lambda_p \in \bigcup_{i=j}^{q} G_i(A)$.

Now we prove that neither (1) nor (2) can be true. If (1) is true, then

$$\lambda_j - \lambda_p \le 2\big(g_p(A) + \cdots + g_{j-1}(A)\big).$$

Hence, (4.4.6) gives the contradiction that

$$\lambda_j - \lambda_p > 2\big(g_p(A) + \cdots + g_{j-1}(A)\big).$$

If (2) is true, say, $\lambda_p \in G_i(A), i \ge j$, then

$$\alpha_{ii} - g_i(A) \le \lambda_p < \lambda_j < \alpha_{jj} - g_j(A) \le \alpha_{jj} \le \alpha_{ii},$$

so that $G_j(A) \subset G_i(A)$, thus $\big(\bigcup_{i=p}^{j-1} G_i(A)\big) \bigcup \big(\bigcup_{i=j+1}^{q} G_i(A)\big)$ is still a connected region, and so

$$\lambda_q - \lambda_p \le 2\big(g_p(A) + \cdots + g_{j-1}(A) + g_{j+1}(A) + \cdots + g_q(A)\big).$$

Hence, (4.4.6) provides the contradiction that

$$\lambda_q - \lambda_p > 2\big(g_p(A) + \cdots + g_{j-1}(A) + g_{j+1}(A) + \cdots + g_q(A)\big).$$

This establishes the lemma. □

We now establish the main result of this subsection.

Theorem 4.4.6 *Let $\alpha_{ii}^{(k)} = \delta_{ik}\ (i,k = 1,2,\dots,n)$ and $\lambda_i \ne \lambda_j\ (i \ne j)$ in Problem G. Suppose further that there exists a permutation π of $\{1,2,\dots,n\}$ such that*

$$\max\Big\{\sum_{k=1}^{n} g_{\pi(1)}(A_k),\quad \sum_{k=1}^{n} g_{\pi(n)}(A_k)\Big\} < 1, \tag{4.4.9}$$

and

$$\lambda_i - \lambda_{i-1} \ge 2\max\{l_{\pi(i)}(\tau), l_{\pi(i-1)}(\tau)\}, \quad i = 2,\dots,n, \tag{4.4.10}$$

where

$$l_i(\rho) = \sum_{\substack{j=1\\ j\neq i}}^{n} \left|\alpha_{ij}^{(0)} - \sum_{k=1}^{n} \alpha_{kk}^{(0)}\alpha_{ij}^{(k)}\right| + (\lambda_0+\rho)\sum_{k+1}^{n} g_i(A_k), \quad \rho \geq 0, \tag{4.4.11}$$

$$\tau = \max\left\{\frac{l_{\pi(1)}(0)}{1-\sum_{k=1}^{n} g_{\pi(1)}(A_k)}, \frac{l_{\pi(n)}(0)}{1-\sum_{k=1}^{n} g_{\pi(n)}(A_k)}\right\}, \quad \lambda_0 = \max\{|\lambda_1|, |\lambda_n|\}. \tag{4.4.12}$$

Then there exists a vector $c = (c_1, \dots, c_n)^T \in \mathbf{R}^n$, *satisfying*

$$\lambda_1 - \tau \leq \alpha_{\pi(1),\pi(1)}^{(0)} + c_1 \leq \cdots \leq \alpha_{\pi(n),\pi(n)}^{(0)} + c_n \leq \lambda_n + \tau \tag{4.4.13}$$

and

$$\left|c_{\pi(i)} - \left(\lambda_i - \alpha_{\pi(i),\pi(i)}^{(0)}\right)\right| \leq l_{\pi(i)}(\tau), \quad i = 1, 2, \dots, n, \tag{4.4.14}$$

such that $A(c) = A_0 + \sum_{k=1}^{n} c_k A_k$ *has the given values* $\lambda_1, \dots, \lambda_n$ *as its eigenvalues, that is, Problem* G *has a solution satisfying* (4.4.13) *and* (4.4.14).

Proof The proof is in two parts.

(1) Assume that all the inequalities in (4.4.10) are strict, that is,

$$\lambda_i - \lambda_{i-1} > 2\max\{l_{\pi(i)}(\tau),\ l_{\pi(i-1)}(\tau)\}, \quad i = 2, \dots, n.$$

For the sake of simplicity, without loss of generality we may assume that $\pi(i) = i, \quad i = 1, 2, \dots, n$. Thus, there is a positive number ε_0 such that for any $0 < \varepsilon \leq \varepsilon_0$,

$$\lambda_i - \lambda_{i-1} > 2\max\{l_i(\tau+\varepsilon),\ l_{i-1}(\tau+\varepsilon)\}, \quad i = 2, \dots, n. \tag{4.4.15}$$

For fixed $\varepsilon \in (0, \varepsilon_0)$, define

$$\mathcal{D} = \big\{x = (\xi_1, \dots, \xi_n)^T \in \mathbf{R}^n \ : \ \lambda_1 - (\tau+\varepsilon) < \alpha_{11}^{(0)} + \xi_1 < \cdots < \alpha_{nn}^{(0)} + \xi_n < \lambda_n + (\tau+\varepsilon)\big\}.$$

Then it is easy to verify that $\mathcal{D}$ is open and bounded. Define the homotopy $h : \overline{\mathcal{D}} \times [0,1] \to \mathbf{R}^n$ by

$$h(x,t) = \Big(\mathrm{Re}\,\big(\lambda_1(x,t)\big), \dots, \mathrm{Re}\,\big(\lambda_n(x,t)\big)\Big)^T, \quad x \in \overline{\mathcal{D}},\ t \in [0,1], \tag{4.4.16}$$

where $\lambda_i(x,t)$ is the ith eigenvalue of the matrix

$$A(x,t) = A_0(t) + \sum_{k=1}^{n} \xi_k A_k(t), \quad \text{with } A_k(t) = (1-t)\mathrm{diag}\,(\alpha_{ii}^{(k)}) + tA_k \tag{4.4.17}$$

in the sense that $\mathrm{Re}(\lambda_1(x,t)) \le \mathrm{Re}(\lambda_2(x,t)) \le \cdots \le \mathrm{Re}(\lambda_n(x,t))$. Clearly, $h(x,t)$ is continuous by the continuity of eigenvalues.

Now assume that $c \in \overline{\mathcal{D}}$ satisfies $h(c,t) = \lambda$ for some $t \in [0,1]$, where $\lambda = (\lambda_1, \ldots, \lambda_n)^T$. In what follows we devote to the proof of $c \notin \partial\mathcal{D}$. Note that if a real matrix has complex eigenvalues, they must occur in complex conjugate pairs. Hence, (4.4.15), together with $h(c,t) = \lambda$, implies that the eigenvalues of $A(c,t)$ are all real and just the given values $\lambda_1, \ldots, \lambda_n$.

Notice that for any $x \in \overline{\mathcal{D}}$ and $t \in [0,1]$, we have

$$\begin{aligned} g_i(A(x,t)) &= \sum_{\substack{j=1 \\ j\neq i}}^{n} |\alpha_{ij}(x,t)| = \sum_{\substack{j=1 \\ j\neq i}}^{n} \Big| t\alpha_{ij}^{(0)} + t\sum_{k=1}^{n} \xi_k \alpha_{ij}^{(k)} \Big| \\ &\le t\sum_{\substack{j=1 \\ j\neq i}}^{n} \Big| \alpha_{ij}^{(0)} - \sum_{k=1}^{n} \alpha_{kk}^{(0)} \alpha_{ij}^{(k)} \Big| + t\sum_{\substack{j=1 \\ j\neq i}}^{n} \sum_{k=1}^{n} |\alpha_{kk}^{(0)} + \xi_k|\, |\alpha_{ij}^{(k)}| \\ &\le t\sum_{\substack{j=1 \\ j\neq i}}^{n} \Big| \alpha_{ij}^{(0)} - \sum_{k=1}^{n} \alpha_{kk}^{(0)} \alpha_{ij}^{(k)} \Big| + t(\tau + \varepsilon + \lambda_0) \sum_{k=1}^{n} g_i(A_k) \\ &\le t l_i(\tau+\varepsilon) \le l_i(\tau+\varepsilon). \end{aligned} \tag{4.4.18}$$

Hence, by Lemma 4.4.5, it follows from (4.4.15) that $\lambda_i \in G_i(A(c,t))$, i.e,

$$|\lambda_i - (\alpha_{ii}^{(0)} + c_i)| \le g_i(A(c,t)), \quad i = 1, 2, \ldots, n. \tag{4.4.19}$$

This, together with (4.4.9) and (4.4.12), implies

$$\lambda_1 - (\alpha_{11}^{(0)} + c_1) \le g_1(A(c,t) \le l_1(\tau+\varepsilon) \le l_1(0) + (\tau+\varepsilon)\sum_{k=1}^{n} g_1(A_k) < \tau + \varepsilon$$

and

$$(\alpha_{nn}^{(0)} + c_n) - \lambda_n \le g_n(A(c,t)) \le l_n(\tau+\varepsilon) \le l_n(0) + (\tau+\varepsilon)\sum_{k=1}^{n} g_n(A_k) < \tau + \varepsilon.$$

Thus in order to prove that $c \notin \partial\mathcal{D}$, it is sufficient to prove that

$$\alpha_{ii}^{(0)} + c_i \neq \alpha_{i+1,i+1}^{(0)} + c_{i+1}$$

for all $i = 1, 2, \ldots, n-1$. In fact, if there is an i such that $\alpha_{ii}^{(0)} + c_i = \alpha_{i+1,i+1}^{(0)} + c_{i+1}$, then from (4.4.18) and (4.4.19) we have

$$\begin{aligned} \lambda_{i+1} - \lambda_i &= \lambda_{i+1} - (\alpha_{i+1,i+1}^{(0)} + c_{i+1}) + (\alpha_{ii}^{(0)} + c_i) - \lambda_i \\ &\le l_{i+1}(\tau+\varepsilon) + l_i(\tau+\varepsilon) \\ &\le 2\max\{l_{i+1}(\tau+\varepsilon),\ l_i(\tau+\varepsilon)\}. \end{aligned}$$

Hence, (4.4.15) provides a contradiction. Thus we have proved that $\lambda \notin h(\partial D \times [0,1])$, and so, by Lemma 4.4.1, we have

$$\deg\big(h(\cdot,1),\mathcal{D},\lambda\big) = \deg\big(h(\cdot,0),\mathcal{D},\lambda\big).$$

Noting that $h(c,0) \doteq \lambda$ has a unique solution

$$c = (\lambda_1 - \alpha_{11}^{(0)}, \lambda_2 - \alpha_{22}^{(0)}, \dots, \lambda_n - \alpha_{nn}^{(0)})^T \in \mathcal{D} \quad \text{and} \quad \det h'(c,0) = 1,$$

it follows from (4.4.2) that

$$\deg\big(h(\cdot,1),\mathcal{D},\lambda\big) = \deg\big(h(\cdot,0),\mathcal{D},\lambda\big) = 1.$$

Hence, by Lemma 4.4.2, there is a vector $c_\varepsilon = (c_1^{(\varepsilon)}, \dots, c_n^{(\varepsilon)})^T \in \mathcal{D}$ such that $h(c,1) = \lambda$, that is, $A(c) = A_0 + \sum_{k=1}^n c_k^{(\varepsilon)} A_k$ has $\lambda_1, \dots, \lambda_n$ as its eigenvalues. Note that $\varepsilon > 0$ is arbitrary and c_ε satisfies (4.4.19). By the continuity of eigenvalues, we know that there must be a $c = (c_1, \dots, c_n)^T$ that satisfies that (4.4.13) and (4.4.14) such that the eigenvalues of $A(c)$ are the given values $\lambda_1 \dots, \lambda_n$.

(2) We now use the continuity of eigenvalues to extend the above result to general case. As before, for simplicity, we assume that $\pi(i) = i$, $i = 1, 2, \dots, n$. For any given t with $0 < t < 1$, consider the matrices $A_k(t)$ defined by (4.4.17). Since $\lambda_i \neq \lambda_j$, $i \neq j$, (4.4.10) implies that

$$\lambda_i - \lambda_{i-1} > 2\max\{tl_i(\tau),\ tl_{i-1}(\tau)\},$$

for $i = 2, \dots, n$. Notice that

$$\sum_{\substack{j=1\\ j\neq i}}^{n} \left|\alpha_{ij}^{(0)}(t) - \sum_{k=1}^{n} \alpha_{kk}^{(0)}(t)\alpha_{ij}^{(k)}(t)\right| + (\lambda_0 + \tau)\sum_{k=1}^{n} g_i(A_k(t)) = tl_i(\tau),$$

where $\alpha_{ij}^{(k)}(t)$ denotes the (i,j) element of $A_k(t)$. Hence, by the result that has been proved in (1), it follows that for any $t \in (0,\ 1)$ there is a vector $c(t) \in \mathbf{R}^n$, satisfying (4.4.13) and

$$|c_i(t) - (\lambda_i - \alpha_{ii})| \le tl_i(\tau), \quad i = 1, 2, \dots, n,$$

such that $A(c(t), t)$ has $\lambda_1, \dots, \lambda_n$ as its eigenvalues. Using the continuity of eigenvalues we can obtain a vector $c \in \mathbf{R}^n$ satisfying (4.4.13) and (4.4.14) such that $A(c) = A(c,1)$ has $\lambda_1, \dots, \lambda_n$ as its eigenvalues. This completes the proof.

□

Remark 4.4.1 The proof of Theorem 4.4.6 implies that under the hypotheses of Theorem 4.4.6, if all the inequalities in (4.4.10) are strict, then Problem G has a solution $c = (c_1, \dots, c_n)^T \in \mathbf{R}^n$ satisfying

$$\alpha_{\pi(1),\pi(1)}^{(0)} + c_1 < \alpha_{\pi(2),\pi(2)}^{(0)} + c_2 < \cdots < \alpha_{\pi(n),\pi(n)}^{(0)} + c_n.$$

By considering the homotopy $h(x,t)$ defined by (4.4.10) on the set $\mathcal{B} \times [0,1]$, where

$$\mathcal{B} = \Big\{ x = (\xi_1, \dots, \xi_n)^T \in \mathbf{R}^n \ : \ \alpha_{11}^{(0)} + \xi_1 < \cdots < \alpha_{nn}^{(0)} + \xi_n \quad \text{and} \quad |\xi_i - (\lambda_i - \alpha_{ii}^{(0)})| < \varepsilon, \quad i = 1, \dots, n \Big\}, \quad \varepsilon > 0,$$

in a similar manner to the proof of Theorem 4.4.6 we can prove the following theorem.

Theorem 4.4.7 *Let $\alpha_{ii}^{(k)} = \delta_{ik}$ $(i,k = 1,2,\dots,n)$ and $\lambda_i \neq \lambda_j$ $(i \neq j)$ in Problem G. Suppose that $\sum_{k=1}^n g_i(A_k) < 1$ $(i = 1,2,\dots,n)$ and there exists a permutation π of $\{1,2,\dots,n\}$ such that*

$$\lambda_i - \lambda_{i-1} \geq 2\max\{s_{\pi(i)}(\sigma), s_{\pi(i-1)}(\sigma)\}, \quad i = 2,3,\dots,n, \tag{4.4.20}$$

where

$$\begin{aligned} s_i(\rho) &= g_i(A_0) + (\delta + \rho)\sum_{k=1}^n g_i(A_k), \quad \rho \geq 0, \\ \sigma &= \max_{1\leq i\leq n} \frac{s_i(0)}{1 - \sum_{k=1}^n g_i(A_k)}, \\ \delta &= \max_{1\leq i\leq n} |\alpha_{\pi(i),\pi(i)}^{(0)} - \lambda_i|. \end{aligned} \tag{4.4.21}$$

Then there exists a vector $c = (c_1, \dots, c_n)^T \in \mathbf{R}^n$, satisfying

$$\alpha_{\pi(1),\pi(1)}^{(0)} + c_{\pi(1)} \leq \cdots \leq \alpha_{\pi(n),\pi(n)}^{(0)} + c_{\pi(n)}$$

and

$$\left| c_{\pi(i)} - (\lambda_i - \alpha_{\pi(i),\pi(i))}^{(0)} \right| \leq s_{\pi(i)}(\sigma), \quad i = 1,2,\dots,n,$$

such that the eigenvalues of $A(c) = A_0 + \sum_{k=1}^n c_k A_k$ are precisely the given values $\lambda_1, \dots, \lambda_n$.

Proof Left to the reader as an exercise. □

As a simple consequence of Theorem 4.4.7, we have the following more convenient result in practice.

Corollary 4.4.8 *Let $\alpha_{ij}^{(k)} = \delta_{ik}(i,k = 1,2,\dots,n)$, and $\lambda_1 < \cdots < \lambda_n$ in Problem G, and let*

$$d(\lambda) = \max_{2\leq i\leq n} (\lambda_i - \lambda_{i-1}), \quad g = \max_{1\leq i\leq n} g_i(A_0), \quad h = \max_{1\leq i\leq n} \sum_{k=1}^n g_i(A_k). \tag{4.4.22}$$

If $h < 1$ and there exists a permutation π of $\{1, \ldots, n\}$ such that

$$d(\lambda) \geq \frac{2(g + \|\lambda - a_\pi\|_\infty h)}{(1-h)}, \tag{4.4.23}$$

where

$$a_\pi = \left(\alpha^{(0)}_{\pi(1),\pi(1)}, \ldots, \alpha^{(0)}_{\pi(n),\pi(n)}\right)^T,$$

then Problem G *has a solution $c = (c_1, \ldots, c_n)^T \in \mathbf{R}^n$ satisfying*

$$\left|c_{\pi(i)} - \left(\lambda_i - \alpha^{(0)}_{\pi(i),\pi(i)}\right)\right| \leq \frac{g + \|\lambda - a_\pi\|_\infty h}{1-h}, \quad 1 \leq i \leq n.$$

4.4.2 The Symmetric Case

It is well-known that the symmetric eigenvalue problem has a theory that is both rich and elegant. This permits us to derive some better conditions for the solvability of Problem A, Problem M, and Problem SG.

First, let us see how to improve on the conditions of Theorem 4.4.6 when the given matrices are symmetric. From the proof of Theorem 4.4.6 we know that the condition (4.4.9) and the value τ defined as in (4.4.12) are only to guarantee that when $A(x,t)$ has the given values $\lambda_1, \ldots, \lambda_n$ as its eigenvalues for some $x = (\xi_1, \ldots, \xi_n)^T \in \overline{\mathcal{D}}$ and $t \in [0,1]$, we can derive that

$$\alpha^{(0)}_{11} + \xi_1 > \lambda_1 - (\tau + \varepsilon) \quad \text{and} \quad \alpha^{(0)}_{nn} + \xi_n < \lambda_n + (\tau + \varepsilon),$$

for any $\varepsilon > 0$. Whereas for any symmetric matrix $A = [\alpha_{ij}] \in \mathbf{R}^{n\times n}$ with eigenvalues $\lambda_1 \leq \cdots \leq \lambda_n$, by Theorem 1.5.3, we have

$$\lambda_1 \leq \alpha_{ii} \leq \lambda_n, \quad i = 1, 2, \ldots, n,$$

and so, for any $\varepsilon > 0$, we have

$$\lambda_1 - \varepsilon < \alpha_{ii} < \lambda_n + \varepsilon, \quad i = 1, 2, \ldots, n.$$

Therefore, for Problem SG neither the condition (4.4.9) nor the value τ is required. Thus we have the following theorem.

Theorem 4.4.9 *Let $A_k = [\alpha^{(k)}_{ij}] \in \mathbf{R}^{n\times n}$, $k = 0, 1, \ldots, n$, be symmetric with $\alpha^{(k)}_{ij} = \delta_{ik}$, $i, k = 1, 2, \ldots, n$, and let $\lambda_1 < \lambda_2 < \cdots < \lambda_n$. Suppose further that there exists a permutation π of $\{1, 2, \ldots, n\}$ such that*

$$\lambda_i - \lambda_{i-1} \geq 2\max\{l_{\pi(i)},\ l_{\pi(i-1)}\}, \quad i = 2, \ldots, n, \tag{4.4.24}$$

where

$$l_i = \sum_{\substack{j=1 \\ j\neq i}}^{n} \Big|\alpha_{ij}^{(0)} - \sum_{k=1}^{n} \alpha_{kk}^{(0)}\alpha_{ij}^{(k)}\Big| + \max\{|\lambda_1|,\ |\lambda_n|\} \sum_{k=1}^{n} g_i(A_k). \tag{4.4.25}$$

Then there exists a vector $c = (c_1, \dots, c_n)^T \in \mathbf{R}^n$, *satisfying*

$$\lambda_1 \le \alpha_{\pi(1),\pi(1)}^{(0)} + c_{\pi(1)} \le \cdots \le \alpha_{\pi(n),\pi(n)}^{(0)} + c_{\pi(n)} \le \lambda_n$$

and

$$\Big|c_{\pi(i)} - (\lambda_i - \alpha_{\pi(i),\pi(i)}^{(0)})\Big| \le l_{\pi(i)}, \quad i = 1, 2, \dots, n,$$

such that $A(c) = A_0 + \sum_{k=1}^{n} c_k A_k$ *has* $\lambda_1, \dots, \lambda_n$ *as its eigenvalues; that is, Problem* SG *is solvable.*

Proof Left to the reader. □

Applying this theorem to additive inverse eigenvalue problem, we immediately get the following result.

Corollary 4.4.10 *Let* $A = [\alpha_{ij}] \in \mathbf{R}^{n\times n}$ *be symmetric with* $\alpha_{ii} = 0$ *for all* $i = 1, 2, \dots, n$, *and let* $\lambda_1 < \lambda_2 < \cdots < \lambda_n$. *If there exists a permutation* π *of* $\{1, 2, \dots, n\}$ *so that*

$$\lambda_i - \lambda_{i-1} \ge 2\max\{g_{\pi(i)}(A),\ g_{\pi(i-1)}(A)\}, \quad i = 2, \dots, n, \tag{4.4.26}$$

then there is a vector $c = (c_1, \dots, c_n)^T \in \mathbf{R}^n$, *satisfying*

$$\lambda_1 \le c_{\pi(1)} \le \cdots \le c_{\pi(n)} \le \lambda_n$$

and

$$|\lambda_i - c_{\pi(i)}| \le g_{\pi(i)}(A), \quad i = 1, 2, \dots, n,$$

such that $A + \mathrm{diag}(c_1, \dots, c_n)$ *has* $\lambda_1, \dots, \lambda_n$ *as its eigenvalues.*

Similarly, we can derive a sufficient condition for the solvability of Problem M as follows.

Theorem 4.4.11 *Let* $A = [\alpha_{ij}] \in \mathbf{R}^{n\times n}$ *be symmetric and positive definite with* $\alpha_{ii} = 1$ *for all* $i = 1, 2, \dots, n$, *and let* $0 < \lambda_1 < \cdots < \lambda_n$. *If there exists a permutation* π *of* $\{1, 2, \dots, n\}$ *so that*

$$\lambda_i - \lambda_{i-1} \ge 2\lambda_n \max\{g_{\pi(i)}(A),\ g_{\pi(i-1)}(A)\}, \quad i = 2, \dots, n, \tag{4.4.27}$$

then there is a vector $c = (c_1, \ldots c_n)^T \in \mathbf{R}^n$ *satisfying*

$$\lambda_1 \leq c_{\pi(1)} \leq \cdots \leq c_{\pi(n)} \leq \lambda_n \tag{4.4.28}$$

and

$$|\lambda_i - c_{\pi(i)}| \leq \lambda_n g_{\pi(i)}(A), \quad i = 1, 2, \ldots, n, \tag{4.4.29}$$

such that DA *has eigenvalues* $\lambda_1, \ldots, \lambda_n$*, where* $D = \operatorname{diag}(c_1, \ldots, c_n)$*, which is positive definite; that is to say that Problem* **M** *has a solution that satisfies* (4.4.28) *and* (4.4.29).

Proof Left to the reader. □

Next, utilizing an excellent estimate of the differences between the main diagonal elements of a symmetric matrix and its eigenvalues, with the help of Brouwer's fixed-point theorem we develop other kind of sufficient condition for the solvability of symmetric problems. We begin with establishing the estimate.

Lemma 4.4.12 *Let* $B = [\beta_{ij}] \in \mathbf{R}^{n\times n}$ *be a symmetric matrix. Then for each* $i = 1, \ldots, n$*, there exists a* $\mu \in \lambda(B)$ *such that*

$$|\mu - \beta_{ii}| \leq k_i(B) \equiv \Big(\sum_{\substack{j=1 \\ j\neq i}}^{n} \beta_{ij}^2\Big)^{\frac{1}{2}}. \tag{4.4.30}$$

Furthermore, if there is precisely one eigenvalue μ *of* B *in* $(\beta_{ii} - \delta,\ \beta_{ii} + \delta)$ *for some* $\delta > 0$*, then*

$$|\mu - \beta_{ii}| \leq \frac{k_i(B)^2}{\delta}. \tag{4.4.31}$$

Proof Let $Q \in \mathbf{R}^{n\times n}$ be an orthogonal matrix such that

$$Q^T B Q = \Lambda \equiv \operatorname{diag}(\mu_1, \ldots, \mu_n). \tag{4.4.32}$$

For fixed $i \in \{1, \ldots, n\}$, without loss of generality we may assume that $\beta_{ii} \neq \mu_j$, $j = 1, 2, \ldots, n$. Thus it follows from (4.4.32) that

$$(\Lambda - \beta_{ii} I)^{-1} Q^T (B - \beta_{ii} I) = Q^T,$$

and so

$$\begin{aligned} 1 = \|Q^T e_i\|_2 &\leq \|(\Lambda - \beta_{ii} I)^{-1}\|_2 \|Q^T (B - \beta_{ii} I) e_i\|_2 \\ &= \big(\min_{1\leq j\leq n} |\mu_j - \beta_{ii}|\big)^{-1} k_i(B), \end{aligned}$$

from which (4.4.30) follows immediately.

Now assume that only one eigenvalue μ is located in $(\beta_{ii}-\delta,\ \beta_{ii}+\delta)$. Then it follows that

$$(B-\mu I)(B-(\beta_{ii}+\delta)I) = Q(\Lambda-\mu I)\big(\Lambda-(\beta_{ii}+\delta)I\big)Q^T$$

is a positive semidefinite matrix since the assumption of $(\mu,\ \beta_{ii}+\delta)\bigcap\lambda(B)=\emptyset$ implies that $(\mu_i-\mu)(\mu_i-\beta_{ii}-\delta)\ge 0$ for all $i=1,2,\dots,n$. Hence,

$$(\beta_{ii}-\mu)(-\delta)+k_i(B)^2 = e_i^T(B-\mu I)(B-(\beta_{ii}+\delta)I)e_i \ge 0$$

and so

$$\beta_{ii}-\mu \le \frac{k_i(B)^2}{\delta}. \tag{4.4.33}$$

Similarly, $(\beta_{ii}-\delta,\ \mu)\bigcap\lambda(B)=\emptyset$ implies that

$$\mu-\beta_{ii} \le \frac{k_i(B)^2}{\delta}.$$

This, together with (4.4.33), gives (4.4.31). □

Lemma 4.4.13 *Let $B=[\beta_{ij}]\in\mathbf{R}^{n\times n}$ be a symmetric matrix with $\beta_{11}\le\beta_{22}\le\beta_{33}\le\cdots\le\beta_{nn}$, and let*

$$k(B)=\max_{1\le i\le n} k_i(B)=\max_{1\le i\le n}\Bigg(\sum_{\substack{j=1\\ j\ne i}}^{n}\beta_{ij}^2\Bigg)^{\frac{1}{2}}>0. \tag{4.4.34}$$

Suppose further that there exists a positive number d such that

$$\min_{1\le j\le n}\min_{i\ne j}|\beta_{ii}-\beta_{jj}|\ge d\ge 2k(B). \tag{4.4.35}$$

Then we have

$$|\beta_{ii}-\mu_i|\le\frac{1}{2}\Big(d-\big(d^2-4k(B)^2\big)^{\frac{1}{2}}\Big),\quad i=1,2,\dots,n, \tag{4.4.36}$$

where $\mu_1\le\mu_2\le\cdots\le\mu_n$ are the ordered eigenvalues of B.

Proof By Lemma 4.4.12, it follows from the assumptions that

$$|\beta_{ii}-\mu_i|\le k_i(B),\quad i=1,2,\dots,n. \tag{4.4.37}$$

If $d = 2k(B)$, it is clear that (4.4.37) implies (4.4.36). Therefore, in what follows we assume that $d > 2k(B)$. It follows from (4.4.35) that

$$\beta_{i-1,i-1} + k_{i-1}(B) \leq \beta_{ii} - d + k(B) < \beta_{ii} + d - k(B) \leq \beta_{i+1,i+1} - k_{i+1}(B),$$

and hence only the eigenvalue μ_i is located in

$$\big(\beta_{ii} - d + k(B), \quad \beta_{ii} + d - k(B)\big),$$

so that, applying Lemma 4.4.12 with $\delta = d - k(B)$, we have

$$|\mu_i - \beta_{ii}| \leq \frac{k_i(B)^2}{d - k(B)} \leq \frac{k(B)^2}{d - k(B)}.$$

Now let $\alpha_0 = k(B)$, and define

$$\alpha_m = \frac{k(B)^2}{d - \alpha_{m-1}}, \qquad m = 1, 2, \ldots .$$

Then the previous proof shows that

$$|\mu - \beta_{ii}| \leq \alpha_m, \quad m = 0, 1;$$

moreover, it is easy to verify that

$$0 < \alpha_1 < \alpha_0 = k(B) < \frac{d}{2}.$$

Assume that we have proved that $|\mu - \beta_{ii}| \leq \alpha_m$ and $0 < \alpha_m < \alpha_{m-1} < \frac{d}{2}$ for some $m > 1$. Then we have

$$\beta_{i-1,i-1} + \alpha_m \leq \beta_{ii} - d + \alpha_m < \beta_{ii} + d - \alpha_m \leq \beta_{i+1,i+1} - \alpha_m,$$

and so μ_i is the unique eigenvalue in the interval $(\beta_{ii} - d + \alpha_m, \beta_{ii} + d - \alpha_m)$. Applying Lemma 4.4.12 again, we have

$$|\mu_i - \beta_{ii}| \leq \frac{k_i(B)^2}{d - \alpha_m} \leq \frac{k(B)^2}{d - \alpha_m} = \alpha_{m+1}.$$

Besides, noting that

$$\alpha_{m+1} - \alpha_m = \frac{k(B)^2(\alpha_m - \alpha_{m-1})}{(d - \alpha_m)(d - \alpha_{m-1})} < 0,$$

we have $0 < \alpha_{m+1} < \alpha_m < \dfrac{d}{2}$.

Consequently, by mathematical induction principle, it follows that

$$|\mu_i - \beta_{ii}| \leq \alpha_m, \quad m = 0, 1, \ldots \tag{4.4.38}$$

and the sequence $\{\alpha_m\}$ is monotonically decreasing and bounded. Thus the sequence $\{\alpha_m\}$ converges, and it is easy to derive from the definition formula of α_m that

$$\lim_{m\to\infty} \alpha_m = \frac{1}{2}\Big(d - \big(d^2 - 4k(B)^2\big)^{\frac{1}{2}}\Big).$$

This, together with (4.4.38), gives (4.4.36). The lemma is established. □

Now using Lemma 4.4.13, with the help of Brouwer's fixed-point theorem we can prove the following theorem.

Theorem 4.4.14 *Let* $A_k = [\alpha_{ij}^{(k)}] \in \mathbf{R}^{n\times n}$, $k = 0, \dots, n$, *be symmetric with* $\alpha_{ii}^{(k)} = \delta_{ik}$, $i, k = 1, 2, \dots, n$. *Assume that there exists a permutation* π *of* $\{1, 2, \dots, n\}$ *such that*

$$d(\lambda) \geq 2\delta_\pi\big(3 + k(S)^2\big)^{\frac{1}{2}} + 2\delta_\pi k(S), \tag{4.4.39}$$

where $d(\lambda)$ *is defined as in* (4.4.22), $k(\cdot)$ *is defined by* (4.4.34), *and*

$$\delta_\pi = k(S) \max_{1\leq i\leq n} \big|\lambda_i - \alpha_{\pi(i),\pi(i)}^{(0)}\big| + k(A_0), \quad S = [s_{ij}] \quad \text{with } s_{ij} = \sum_{k=1}^{n} |\alpha_{ij}^{(k)}|.$$

Then there exists a vector $c = (c_1, \dots, c_n)^T \in \mathbf{R}^n$ *such that*

$$A(c) = A_0 + \sum_{k=1}^{n} c_k A_k$$

has eigenvalues $\lambda_1, \dots, \lambda_n$ *; that is, Problem* SG *is solvable.*

Proof For the sake of simplicity of statement, without loss of generality we may assume that $\pi(i) = i$, $i = 1, 2, \dots, n$. Let

$$\mathcal{D}_\varepsilon = \Big\{x = (\xi_1, \dots, \xi_n)^T \in \mathbf{R}^n \;:\; \max_{1\leq i\leq n} |\xi_i + \alpha_{ii}^{(0)} - \lambda_i| \leq \varepsilon\Big\},$$

where

$$\varepsilon = \frac{d(\lambda) - 2\delta_\pi k(S) - \Big(\big(d(\lambda) - 2\delta_\pi k(S)\big)^2 - 4\delta_\pi^2\big(3 + k(S)^2\big)\Big)^{\frac{1}{2}}}{6 + 2k(S)^2},$$

which is a solution of the quadratic equation

$$\big(d(\lambda) - 4\varepsilon\big)^2 = \big(d(\lambda) - 2\varepsilon\big)^2 - 4\big(\delta_\pi + \varepsilon k(S)\big)^2.$$

Note that (4.4.39) implies

$$0 < \varepsilon \leq \frac{2\delta_\pi \left(3 + k(S)^2\right)^{\frac{1}{2}}}{6 + 2k(S)^2} \leq \delta_\pi,$$

and so the $\mathcal{D}_\varepsilon$ is a nonempty convex compact set. Define $f : \mathcal{D}_\varepsilon \to \mathbf{R}^n$ by

$$f(x) = \left(f_1(x), \dots, f_n(x)\right)^T,$$

with

$$f_i(x) = \xi_i + \lambda_i - \lambda_i(x), \quad x \in \mathcal{D}_\varepsilon, \quad i = 1, 2, \dots, n,$$

where $\lambda_1(x) \leq \lambda_2(x) \leq \cdots \leq \lambda_n(x)$ are the eigenvalue of $A(x) = A_0 + \sum_{k=1}^n \xi_k A_k$. Clearly, $f(x)$ is continuous by the continuity of eigenvalues. Furthermore, to prove that Problem SG has a solution in $\mathcal{D}_\varepsilon$ is equivalent to proving that $f(x)$ has a fixed-point in $\mathcal{D}_\varepsilon$. For this, by Brouwer's fixed-point theorem, it is sufficient to prove that $x \in \mathcal{D}_\varepsilon$ implies $f(x) \in \mathcal{D}_\varepsilon$. For any given $x = (\xi_1, \dots, \xi_n)^T \in \mathcal{D}_\varepsilon$, we have

$$(\alpha_{i+1,i+1}^{(0)} + \xi_{i+1}) - (\alpha_{ii}^{(0)} + \xi_i) \geq \lambda_{i+1} - \lambda_i - 2\varepsilon \geq d(\lambda) - 2\varepsilon,$$

and

$$\begin{aligned} k(A(x)) &\leq k(A_0) + \|x\|_\infty k(S) \\ &\leq k(A_0) + \left(\max_{1\leq i\leq n} |\xi_i + \alpha_{ii}^{(0)} - \lambda_i| + \max_{1\leq i\leq n} |\alpha_{ii}^{(0)} - \lambda_i|\right) k(S) \\ &\leq \delta_\pi + \varepsilon k(S), \end{aligned}$$

and hence

$$\begin{aligned} d(\lambda) &\geq 2\delta_\pi k(S) + 2\delta_\pi \left(3 + k(S)^2\right)^{\frac{1}{2}} \\ &\geq 2\varepsilon k(S) + 0.7\delta_\pi \left(3 + k(S)^2\right)^{\frac{1}{2}} + 1.3 \times \sqrt{3}\delta_\pi \\ &\geq 2\varepsilon k(S) + 0.7 \times 3\varepsilon + 2\delta_\pi \\ &\geq 2\varepsilon k(S) + 2\varepsilon + 2\delta_\pi \\ &\geq 2\left(k(A(x)) + \varepsilon\right). \end{aligned}$$

Here we use the fact that $\delta_\pi \left(3 + k(S)^2\right)^{\frac{1}{2}} \geq 3\varepsilon$. Thus, we have proved that $x \in \mathcal{D}_\varepsilon$ implies that

$$d(\lambda) - 2\varepsilon \geq 2k(A(x)).$$

Hence all the conditions of Lemma 4.4.13 are satisfied for the matrix $A(x)$, and so

$$\begin{aligned} \left|\lambda_i(x) - (\alpha_{ii}^{(0)} + \xi_i)\right| &\leq \frac{1}{2}\left(d(\lambda) - 2\varepsilon - \left((d(\lambda) - 2\varepsilon)^2 - 4k(A(x))^2\right)^{\frac{1}{2}}\right) \\ &\leq \frac{1}{2}\left(d(\lambda) - 2\varepsilon - \left((d(\lambda) - 2\varepsilon)^2 - 4(\varepsilon k(S) + \delta_\pi)^2\right)^{\frac{1}{2}}\right) \\ &= \varepsilon. \end{aligned}$$

Consequently, we have

$$|\alpha_{ii}^{(0)} + f_i(x) - \lambda_i| = |\lambda_i(x) - (\alpha_{ii}^{(0)} - \xi_i)| \le \varepsilon, \quad i = 1, 2, \dots, n,$$

which is to say $f(x) \in \mathcal{D}_\varepsilon$. This establishes the theorem. □

Applying Theorem 4.4.14 to Problem A, we obtain the following corollary.

Corollary 4.4.15 *If $\alpha_{ii} = 0$ for $i = 1, 2, \dots, n$ in Problem A and if*

$$d(\lambda) \ge 2\sqrt{3}k(A),$$

then Problem A is solvable.

In the remainder of this section we will devote to the establishment of the third kind of sufficient condition for the solvability of the symmetric problems. To do this, we need the precise relationship between the main diagonal elements of a real symmetric and its eigenvalues which is given by the notion of majorization. Let us begin with the definition of majorization.

Let $x = (\xi_1, \dots, \xi_n)^T \in \mathbf{R}^n$ and $y = (\eta_1, \dots, \eta_n)^T \in \mathbf{R}^n$ be given. The vector y is said to **majorize** the vector x if

$$\min\left\{\sum_{j=1}^{k} \eta_{i_j} \ : \ 1 \le i_1 < \cdots < i_k \le n\right\} \ge \min\left\{\sum_{j=1}^{k} \xi_{i_j} \ : \ 1 \le i_1 < \cdots < i_k \le n\right\}$$

for all $k = 1, 2, \dots, n$ with equality for $k = n$. If the vector y majorizes the vector x, we write as $y \succ x$.

Remark 4.4.2 If we arrange the entries of x and y in increasing order

$$\xi_{i_1} \le \cdots \le \xi_{i_n}, \qquad \eta_{m_1} \le \cdots \le \eta_{m_n},$$

the defining inequalities can be restated in the equivalent form

$$\sum_{i=1}^{k} \eta_{m_i} \ge \sum_{j=1}^{k} \xi_{i_j}$$

for all $k = 1, 2, \dots, n$ with equality for $k = n$.

The notion of majorization is an important one that arises in many places in matrix theory. One example is the following result.

Lemma 4.4.16 *Let $B = [\beta_{ij}]$ be an $n \times n$ symmetric matrix. The vector of diagonal elements of B majorizes the vector of eigenvalues of B.*

Proof The proof is by induction on n. For $n = 1$, there is nothing to show, so we suppose the result is valid for real symmetric matrices of dimension k for all $k \leq n-1$. Let $B \in \mathbf{R}^{n\times n}$ be a given symmetric matrix and let B_1 be the principal submatrix of B obtained by deleting the row and column corresponding to the largest diagonal element of B. Let $\lambda_1 \leq \cdots \leq \lambda_n$ be the ordered eigenvalues of B, let $\mu_1 \leq \cdots \leq \mu_{n-1}$ be the ordered eigenvalues of B_1, and let $\beta_{i_1 i_1} \leq \beta_{i_2 i_2} \leq \cdots \leq \beta_{i_n i_n}$ be a rearrangement of the diagonal elements into increasing order. By the induction hypothesis, we have

$$\sum_{j=1}^{k} \beta_{i_j i_j} \geq \sum_{j=1}^{k} \mu_j, \qquad \text{for all } k = 1, \dots, n-1.$$

By Theorem 1.5.2 (i.e., Cauchy's interlacing theorem), we also have

$$\lambda_1 \leq \mu_1 \leq \lambda_2 \leq \mu_2 \leq \cdots \leq \mu_{n-1} \leq \lambda_n,$$

and hence

$$\sum_{j=1}^{k} \mu_j \geq \sum_{j=1}^{k} \lambda_j, \qquad \text{for all } k = 1, \dots, n-1.$$

Thus

$$\sum_{j=1}^{k} \beta_{i_j i_j} \geq \sum_{j=1}^{k} \lambda_j, \quad k = 1, \dots, n-1,$$

and equality holds for $k = n$ because the trace is the sum of the eigenvalues. By induction principle, the lemma is established. □

We also need the following equivalent statement of the majorization.

Lemma 4.4.17 *Let $x = (\xi_i) \in \mathbf{R}^n$ and $y = (\eta_i) \in \mathbf{R}^n$. Then $y \succ x$ if and only if there exists a doubly stochastic matrix S such that $y = Sx$ (An $n \times n$* **doubly stochastic matrix** *is a matrix which has n^2 nonnegative elements such that the sum of elements in every row and every column is 1).*

The proof of this result depends on Birkhoff's theorem, and so we cannot give it here (see [118] for the proof).

Lemma 4.4.18 *Let $\lambda = (\lambda_1, \dots, \lambda_n)^T \in \mathbf{R}^n$, and let*

$$\mathcal{K}_\lambda = \{x \in \mathbf{R}^n \ : \ x \succ \lambda\}.$$

Then $\mathcal{K}_\lambda$ is a nonempty convex compact set in $\mathbf{R}^n$.

Proof (1) Since $\lambda \in \mathcal{K}_\lambda$, $\mathcal{K}_\lambda \neq \emptyset$.

(2) For any $x \in \mathcal{K}_\lambda$, by Lemma 4.4.17 there exists a doubly stochastic matrix S such that $x = S\lambda$. Therefore, we have

$$\|x\|_\infty \leq \|S\|_\infty \|\lambda\|_\infty = \|\lambda\|_\infty,$$

which means that $\mathcal{K}_\lambda$ is bounded.

(3) Suppose that $x_m \in \mathcal{K}_\lambda$ and $x \in \mathbf{R}^n$ satisfy that $\lim_{m\to\infty} x_m = x$. By Lemma 4.4.17 there exists a doubly stochastic matrix S_m such that $x_m = S_m\lambda$ for every $m = 1, 2, \ldots$ Note that $\|S_m\|_\infty = 1$ for $m = 1, 2, \ldots$ It follows that there exists a subsequence $\{S_{m_i}\}_{i=1}^\infty$ of $\{S_m\}_{m=1}^\infty$ and an $n \times n$ real matrix S such that $\lim_{m\to\infty} S_{m_i} = S$. It is easy to verify that S is a doubly stochastic matrix. Consequently, we have

$$x = \lim_{i\to\infty} x_{m_i} = \lim_{i\to\infty} S_{m_i}\lambda = S\lambda \in \mathcal{K}_\lambda,$$

and so $\mathcal{K}_\lambda$ is closed.

(4) Let $x, y \in \mathcal{K}_\lambda$. By Lemma 4.4.17 there exist doubly stochastic matrices S_1 and S_2 such that $S_1\lambda = x$ and $S_2\lambda = y$. Thus

$$tx + (1-t)y = (tS_1 + (1-t)S_2)\lambda, \quad 0 \leq t \leq 1.$$

Note that $tS_1 + (1-t)S_2$ is also a doubly stochastic matrix. It follows that $tx + (1-t)y \in \mathcal{K}_\lambda$, and hence $\mathcal{K}_\lambda$ is a convex set. This completes the proof of Lemma 4.4.18. □

With the foregoing preparations we can now establish a sufficient condition for the solvability of Problem **M**.

Theorem 4.4.19 *Suppose that $\alpha_{ii} = 1$ for $i = 1, \ldots, n$ and $\lambda = (\lambda_1, \ldots, \lambda_n)^T$ with $0 < \lambda_1 < \cdots < \lambda_n$ in Problem* **M**. *If*

$$d(\lambda) \geq \lambda_n \left(\lambda_{\max}(A^{(0)}) - \lambda_{\min}(A^{(0)})\right), \tag{4.4.40}$$

then there exists a vector $c = (c_1, \ldots, c_n) \in \mathbf{R}^n$ with $c_i > 0$ such that the matrix DA has eigenvalues $\lambda_1, \ldots, \lambda_n$, where $D = \operatorname{diag}(c_1, \ldots c_n)$, $d(\lambda)$ is defined as in (4.4.22), and $A^{(0)} = A - I$.

Proof Without loss of generality we may assume that $A^{(0)} \neq 0$. Set

$$\mathcal{D}_\lambda = \left\{x = (\xi_1, \ldots, \xi_n)^T \in \mathbf{R}^n \ : \ x \in \mathcal{K}_\lambda, \quad \xi_1 \leq \cdots \leq \xi_n\right\},$$

where $\mathcal{K}_\lambda$ is defined as in Lemma 4.4.18. By Lemma 4.4.18, it follows that $\mathcal{D}_\lambda$ is a nonempty convex compact set in $\mathbf{R}^n$.

Let $x \in \mathcal{D}_\lambda$. Then $x \in \mathcal{K}_\lambda$, i.e., $x \succ \lambda$. By the definition of majorization, we have

$$\sum_{i=1}^{k} \xi_i \geq \sum_{i=1}^{k} \lambda_i, \qquad k = 1, 2, ..., n-1,$$

and

$$\sum_{i=1}^{n} \xi_i = \sum_{i=1}^{n} \lambda_i,$$

and so we have

$$\xi_1 \geq \lambda_1 > 0$$

and

$$\xi_n = \sum_{i=1}^{n} \xi_i - \sum_{i=1}^{n-1} \xi_i = \sum_{i=1}^{n} \lambda_i - \sum_{i=1}^{n-1} \xi_i \leq \sum_{i=1}^{n} \lambda_i - \sum_{i=1}^{n-1} \lambda_i = \lambda_n.$$

Consequently, $x \in \mathcal{D}_\lambda$ implies

$$\lambda_1 \leq \xi_i \leq \lambda_n, \qquad 1 \leq i \leq n. \tag{4.4.41}$$

As has been defined in the proof of Theorem 4.4.14, let f be the continuous vector-valued function with

$$f : \mathcal{D}_\lambda \to \mathbf{R}^n \quad \text{and} \quad f(x) = \big(f_1(x), \dots, f_n(x)\big)^T,$$

where $f_i(x) = \lambda_i + \xi_i - \lambda_i(x)$, $\lambda_i(x)$ is the ith eigenvalue of the matrix $D^{\frac{1}{2}} A D^{\frac{1}{2}}$ in the sense that $\lambda_1(x) \leq \cdots \leq \lambda_n(x)$, $D = \operatorname{diag}(\xi_1, \dots, \xi_n)$. We now prove that $x \in \mathcal{D}_\lambda$ implies $f(x) \in \mathcal{D}_\lambda$.

For any given $x \in \mathcal{D}_\lambda$, applying Theorem 1.5.4 to $D^{\frac{1}{2}} A D^{\frac{1}{2}} = D^{\frac{1}{2}} A^{(0)} D^{\frac{1}{2}} + D$, we have

$$\xi_i + \lambda_{\min}(D^{\frac{1}{2}} A^{(0)} D^{\frac{1}{2}}) \leq \lambda_i(x) \leq \xi_i + \lambda_{\max}(D^{\frac{1}{2}} A^{(0)} D^{\frac{1}{2}}) \tag{4.4.42}$$

for $i = 1, 2, \dots, n$.

Since $A^{(0)} \neq 0$ but $\operatorname{tr}(A^{(0)}) = 0$, it follows that

$$\lambda_{\max}(A^{(0)}) > 0 \quad \text{and} \quad \lambda_{\min}(A^{(0)}) < 0.$$

Thus, by Theorem 1.5.3, we have

$$\begin{aligned}
\lambda_{\max}(D^{\frac{1}{2}} A^{(0)} D^{\frac{1}{2}}) &= \max_{y \neq 0} \frac{y^T D^{\frac{1}{2}} A^{(0)} D^{\frac{1}{2}} y}{y^T y} = \max_{y \neq 0} \frac{y^T A^{(0)} y}{y^T D^{-1} y} \\
&\leq \max_{y \neq 0} \frac{y^T y}{y^T D^{-1} y} \max_{y \neq 0} \frac{y^T A^{(0)} y}{y^T y} = \xi_n \lambda_{\max}(A^{(0)}) \\
&\leq \lambda_n \lambda_{\max}(A^{(0)})
\end{aligned} \tag{4.4.43}$$

and

$$\lambda_{\min}(D^{\frac{1}{2}}A^{(0)}D^{\frac{1}{2}}) = \min_{y\neq 0} \frac{y^T D^{\frac{1}{2}} A^{(0)} D^{\frac{1}{2}} y}{y^T y} = \min_{y\neq 0} \frac{y^T A^{(0)} y}{y^T D^{-1} y}$$
$$= \min_{y\neq 0} \frac{y^T y}{y^T D^{-1} y} \frac{y^T A^{(0)} y}{y^T y}. \tag{4.4.44}$$

Notice that

$$\min_{y\neq 0} \frac{y^T A^{(0)} y}{y^T y} = \lambda_{\min}(A^{(0)}) < 0 \quad \text{and} \quad 0 < \frac{y^T y}{y^T D^{-1} y} \leq \xi_n \leq \lambda_n, \qquad \text{for any } y \neq 0.$$

We have

$$\frac{y^T y}{y^T D^{-1} y} \frac{y^T A^{(0)} y}{y^T y} \geq \frac{y^T y}{y^T D^{-1} y} \min_{y\neq 0} \frac{y^T A^{(0)} y}{y^T y}$$
$$\geq \lambda_n \min_{y\neq 0} \frac{y^T A^{(0)} y}{y^T y} = \lambda_n \lambda_{\min}(A^{(0)}). \tag{4.4.45}$$

Combining (4.4.44) and (4.4.45) yields

$$\lambda_{\min}(D^{\frac{1}{2}}A^{(0)}D^{\frac{1}{2}}) \geq \lambda_n \lambda_{\min}(A^{(0)}). \tag{4.4.46}$$

Consequently, it follows from (4.4.42), (4.4.43) and (4.4.46) that $x \in \mathcal{D}_\lambda$ implies

$$\begin{aligned} f_{i+1}(x) - f_i(x) &= (\lambda_{i+1} - \lambda_i) + \big(\xi_{i+1} - \lambda_{i+1}(x) - \xi_i + \lambda_i(x)\big) \\ &\geq d(\lambda) + \Big(-\lambda_n \lambda_{\max}(A^{(0)}) + \lambda_n \lambda_{\min}(A^{(0)})\Big) \\ &> 0 \end{aligned} \tag{4.4.47}$$

for all $i = 1, 2, \ldots, n-1$. Note that $x = (\xi_1, \ldots, \xi_n)^T$ is the vector of diagonal elements of $D^{\frac{1}{2}}AD^{\frac{1}{2}}$. By Lemma 4.4.16 we know that $\lambda \succ \lambda(x)$; that is,

$$\sum_{i=1}^{k} \xi_i \geq \sum_{i=1}^{k} \lambda_i(x) \tag{4.4.48}$$

for all $k = 1, 2, \ldots, n$ with equality for $k = n$. Thus, it follows from (4.4.47) and (4.4.48) that $x \in \mathcal{D}_\lambda$ implies

$$\begin{aligned} &\min\left\{ \sum_{j=1}^{k} f_{i_j}(x) \;:\; 1 \leq i_1 < \cdots < i_k \leq n \right\} \\ &= \sum_{j=1}^{k} f_j(x) = \sum_{i=1}^{k} (\lambda_i + \xi_i - \lambda_i(x)) \geq \sum_{i=1}^{k} \lambda_i \\ &= \min\left\{ \sum_{j=1}^{k} \lambda_{i_j} \;:\; 1 \leq i_1 < \cdots < i_k \leq n \right\} \end{aligned} \tag{4.4.49}$$

for all $k = 1, 2, \ldots, n$ with equality for $k = n$. This, together with (4.4.47), proves that $x \in \mathcal{D}_\lambda$ implies that $f(x) \in \mathcal{D}_\lambda$. Consequently, applying Brouwer's fixed-point theorem, we thus have $x \in \mathcal{D}_\lambda$ such that $f(x) = x$; that is, $\lambda(x) = \lambda$. In other words, Problem M has a solution $x \in \mathcal{D}_\lambda$. This completes the proof of Theorem 4.4.19. □

Similarly, we can establish a sufficient condition for the solvability of Problem SG.

Theorem 4.4.20 *Let $A_k = [\alpha_{ij}^{(k)}] \in \mathbf{R}^{n\times n}$, $k = 0, 1, \ldots, n$, be symmetric with $\alpha_{ii}^{(k)} = \delta_{ik}$, $i, k = 1, 2, \ldots, n$, and let $\lambda = (\lambda_1, \ldots, \lambda_n)^T \in \mathbf{R}^n$ with $\lambda_1 < \cdots < \lambda_n$. If*

$$d(\lambda) \geq \widetilde{\lambda}_n - \widetilde{\lambda}_1 + \lambda_0 \sum_{k=1}^{n} \left(\lambda_{\max}(A_k^{(0)}) - \lambda_{\min}(A_k^{(0)}) \right), \tag{4.4.50}$$

then there exists a vector $c = (c_1, \ldots, c_n)^T \in \mathbf{R}^n$ such that the matrix $A_0 + \sum_{k=1}^n c_k A_k$ has eigenvalues $\lambda_1, \ldots, \lambda_n$, where $d(\lambda)$ is defined as in (4.4.22), and

$$\lambda_0 = \max\{|\lambda_1|,\ |\lambda_n|\}, \quad A_k^{(0)} = A_k - \operatorname{diag}(\alpha_{11}^{(k)}, \ldots, \alpha_{nn}^{(k)}),$$
$$\widetilde{\lambda}_1 = \lambda_{\min}\left(A^{(0)} - \sum_{k=1}^{n} \alpha_{kk}^{(0)} A_k^{(0)} \right), \quad \widetilde{\lambda}_n = \lambda_{\max}\left(A^{(0)} - \sum_{k=1}^{n} \alpha_{kk}^{(0)} A_k^{(0)} \right).$$

Applying Theorem 4.4.20 to the additive inverse eigenvalue problem, we immediately get the following result.

Corollary 4.4.21 *Let $\alpha_{ii} = 0$ for $i = 1, 2, \ldots, n$ and let $\lambda = (\lambda_1, \ldots, \lambda_n)^T$ with $\lambda_1 < \ldots < \lambda_n$ in Problem A. If*

$$d(\lambda) \geq \lambda_{\max}(A) - \lambda_{\min}(A), \tag{4.4.51}$$

then there exists a vector $c = (c_1, \ldots, c_n)^T \in \mathbf{R}^n$ such that the matrix $A + D$ has $\lambda_1, \ldots, \lambda_n$ as its eigenvalues, where $D = \operatorname{diag}(c_1, \ldots, c_n)$ and $d(\lambda)$ is defined as in (4.4.22).

Proof of Theorem 4.4.20 Let

$$\mathcal{D}_\lambda(a) = \left\{ x = (\xi_1, \ldots, \xi_n)^T \in \mathbf{R}^n \ :\ x + a \in \mathcal{K}_\lambda, \quad \xi_1 + \alpha_{11}^{(0)} \leq \cdots \leq \xi_n + \alpha_{nn}^{(0)} \right\},$$

where $a = \left(\alpha_{11}^{(0)}, \ldots, \alpha_{nn}^{(0)} \right)^T$. By Lemma 4.4.18, it follows immediately that $\mathcal{D}_\lambda(a)$ is a nonempty convex compact set in $\mathbf{R}^n$.

For any given $x \in \mathcal{D}_\lambda(a)$, let

$$A(x) = A_0 + \sum_{k=1}^{n} \xi_k A_k,$$

and let its eigenvalues be $\lambda_1(x) \le \cdots \le \lambda(x)$. As before, consider the continuous vector-valued function $f : \mathcal{D}_\lambda(a) \to \mathbf{R}^n$ defined by

$$f(x) = \big(f_1(x), \dots, f_n(x)\big)^T, \quad \text{with } f_i(x) = \lambda_i + \xi_i - \lambda_i(x), \quad i = 1, 2, \dots, n.$$

Notice that for any $x \in \mathcal{D}_\lambda(a)$ we have $x + a \in \mathcal{K}_\lambda$. Hence, $x \in \mathcal{D}_\lambda(a)$ implies that

$$\lambda_1 \le \xi_i + \alpha_{ii}^{(0)} \le \lambda_n$$

for all $i = 1, 2, \dots, n$, and so we have

$$\big|\xi_i + \alpha_{ii}^{(0)}\big| \le \max\big\{|\lambda_1|,\ |\lambda_n|\big\} = \lambda_0. \tag{4.4.52}$$

Moreover, applying Theorem 1.5.4, we have

$$\xi_i + \alpha_{ii}^{(0)} + \lambda_{\min}([A(x)]^{(0)}) \le \lambda_i(x) \le \xi_i + \alpha_{ii}^{(0)} + \lambda_{\max}([A(x)]^{(0)}). \tag{4.4.53}$$

for $i = 1, 2, \dots, n$, where $[A(x)]^{(0)} = A_0^{(0)} + \sum_{k=1}^{n} \xi_k A_k^{(0)}$ and $x \in \mathcal{D}_\lambda(a)$.

Now let $x \in \mathcal{D}_\lambda(a)$. We wish to prove that $f(x) \in \mathcal{D}_\lambda(a)$. For the given $x \in \mathcal{D}_\lambda(a)$, there are three possibilities:

(1) $\xi_i + \alpha_{ii}^{(0)} \ge 0$ for $i = 1, 2, \dots, n$;

(2) $\xi_i + \alpha_{ii}^{(0)} \le 0$ for $i = 1, 2, \dots, n$; or

(3) $\xi_1 + \alpha_{11}^{(0)} \le 0$ and $\xi_n + \alpha_{nn}^{(0)} \ge 0$.

Without loss of generality we may assume that x satisfies condition (3). Then there exists a $j_0 \in \{1, 2, \dots, n\}$ such that $\xi_{j_0} + \alpha_{j_0 j_0}^{(0)} \le 0$ but $\xi_{j_0+1} + \alpha_{j_0+1, j_0+1}^{(0)} \ge 0$.

We thus have

$$
\begin{aligned}
\lambda_{\max}([A(x)]^{(0)}) &= \max_{\|y\|_2=1} y^T [A(x)]^{(0)} y \\
&= \max_{\|y\|_2=1} y^T \left(A^{(0)} + \sum_{k=1}^{n} \xi_k A_k^{(0)} \right) y \\
&= \max_{\|y\|_2=1} \left(y^T \left(A^{(0)} - \sum_{k=1}^{n} \alpha_{kk}^{(0)} A_k^{(0)} \right) y + \sum_{k=1}^{n} (\alpha_{kk}^{(0)} + \xi_k) y^T A_k^{(0)} y \right) \\
&\le \max_{\|y\|_2=1} y^T \left(A^{(0)} - \sum_{k=1}^{n} \alpha_{kk}^{(0)} A_k^{(0)} \right) y + \sum_{k=1}^{n} \max_{\|y\|_2=1} (\alpha_{kk}^{(0)} + \xi_k) y^T A_k^{(0)} y \\
&= \widetilde{\lambda}_n - \sum_{k=1}^{j_0} |\alpha_{kk}^{(0)} + \xi_k| \min_{\|y\|_2=1} y^T A_k^{(0)} y + \sum_{k=j_0+1}^{n} |\alpha_{kk}^{(0)} + \xi_k| \max_{\|y\|_2=1} y^T A_k^{(0)} y \\
&= \widetilde{\lambda}_n - \sum_{k=1}^{j_0} |\alpha_{kk}^{(0)} + \xi_k| \lambda_{\min}(A_k^{(0)}) + \sum_{k=j_0+1}^{n} |\alpha_{kk}^{(0)} + \xi_k| \lambda_{\max}(A_k^{(0)}).
\end{aligned}
$$

Note that $\lambda_{\max}(A_k^{(0)}) \ge 0$ and $\lambda_{\min}(A_k^{(0)}) \le 0$. Using (4.4.52), we get

$$
\lambda_{\max}([A(x)]^{(0)}) \le \widetilde{\lambda}_n - \lambda_0 \sum_{k=1}^{j_0} \lambda_{\min}(A_k^{(0)}) + \lambda_0 \sum_{k=j_0+1}^{n} \lambda_{\max}(A_k^{(0)}). \tag{4.4.54}
$$

Similarly, we can prove that

$$
\lambda_{\min}([A(x)]^{(0)}) \ge \widetilde{\lambda}_1 - \lambda_0 \sum_{k=1}^{j_0} \lambda_{\max}(A_k^{(0)}) + \lambda_0 \sum_{k=j_0+1}^{n} \lambda_{\min}(A_k^{(0)}). \tag{4.4.55}
$$

Combining (4.4.53), (4.4.54), and (4.4.55) gives

$$
\begin{aligned}
\xi_i + \alpha_{ii}^{(0)} + \widetilde{\lambda}_1 - \lambda_0 \sum_{k=1}^{j_0} \lambda_{\max}(A_k^{(0)}) + \lambda_0 \sum_{k=j_0+1}^{n} \lambda_{\min}(A_k^{(0)}) \le \lambda_i(x) \\
\le \xi_i + \alpha_{ii}^{(0)} + \widetilde{\lambda}_n - \lambda_0 \sum_{k=1}^{j_0} \lambda_{\min}(A_k^{(0)}) + \lambda_0 \sum_{k=j_0+1}^{n} \lambda_{\max}(A_k^{(0)})
\end{aligned}
\tag{4.4.56}
$$

for all $i = 1, 2, \ldots, n$. Consequently, we have

$$
\begin{aligned}
&\left(f_{i+1,i+1}(x) + a_{i+1,i+1}^{(0)}\right) - \left(f_i(x) + \alpha_{ii}^{(0)}\right) \\
&\quad = \left(\lambda_{i+1} + \xi_{i+1} - \lambda_{i+1}(x) + \alpha_{i+1,i+1}^{(0)}\right) - \left(\lambda_i + \xi_i - \lambda_i(x) + \alpha_{ii}^{(0)}\right) \\
&\quad \geq d(\lambda) + \Bigg(-\widetilde{\lambda}_n + \lambda_0 \sum_{k=1}^{j_0} \lambda_{\min}(A_k^{(0)}) - \lambda_0 \sum_{k=j_0+1}^{n} \lambda_{\max}(A_k^{(0)}) \\
&\qquad\quad + \widetilde{\lambda}_1 - \lambda_0 \sum_{k=1}^{j_0} \lambda_{\max}(A_k^{(0)}) + \lambda_0 \sum_{k=j_0+1}^{n} \lambda_{\min}(A_k^{(0)})\Bigg) \\
&\quad = d(\lambda) - \left(\widetilde{\lambda}_n - \widetilde{\lambda}_1 + \lambda_0 \sum_{k=1}^{n} \left(\lambda_{\max}(A_k^{(0)}) - \lambda_{\min}(A_k^{(0)})\right)\right) \\
&\quad \geq 0
\end{aligned}
\tag{4.4.57}
$$

for $i = 1, 2, \ldots, n-1$. Note that $x + a$ is the vector of diagonal entries of $A(x)$. It follows from Lemma 4.4.16 that

$$
\sum_{i=1}^{k} (\xi_i + \alpha_{ii}^{(0)}) \geq \sum_{i=1}^{k} \lambda_i(x) \tag{4.4.58}
$$

for all $k = 1, \ldots, n$ with equality for $k = n$. Thus, from (4.4.57) and (4.4.58) we obtain

$$
\begin{aligned}
&\min\left\{\sum_{j=1}^{k} (f_{i_j}(x) + \alpha_{i_j,i_j}^{(0)}) \ : \ 1 \leq i_1 < \cdots < i_k \leq n\right\} \\
&\quad = \sum_{j=1}^{k} (f_j(x) + \alpha_{jj}^{(0)}) = \sum_{j=1}^{k} (\lambda_j + \xi_j - \lambda_j(x) + \alpha_{jj}^{(0)}) \geq \sum_{j=1}^{k} \lambda_j \\
&\quad \geq \min\left\{\sum_{j=1}^{k} \lambda_{i_j} \ : \ 1 \leq i_1 < \cdots < i_k \leq n\right\}
\end{aligned}
$$

for all $k = 1, \ldots, n$ with equality for $k = n$. This, together with (4.4.57), implies that $f(x) \in \mathcal{D}_\lambda(a)$. Consequently, applying Brouwer's fixed-point theorem, we thus have $x \in \mathcal{D}_\lambda(a)$ such that $f(x) = x$; that is, Problem SG has a solution $x \in \mathcal{D}_\lambda(a)$. This proves Theorem 4.4.20. □

Finally, it is worthwhile to point out that all the theorems in this section guarantee only existence, and not uniqueness, of solutions. In fact, generally speaking, if the solution exists, then the solutions are not unique and there are at most $n!$ solutions except for some very special cases.

Example 4.4.1 Let

$$A = \begin{bmatrix} 0 & 1 & 0 \\ 1 & 0 & 1 \\ 0 & 1 & 0 \end{bmatrix}, \quad \lambda_1 = -2,\ \lambda_2 = 0,\ \lambda_3 = 2.$$

Find a diagonal matrix $D = \mathrm{diag}\,(c_1, c_2, c_3)$ such that $A + D$ has the given values λ_1, λ_2, λ_3 as its eigenvalue. It is easy to verify that none of the sufficient conditions developed in this section is satisfied for these given data. However, this problem has the following six solutions:

$$\begin{cases} c_1 = \sqrt{2}, \\ c_2 = 0, \\ c_3 = -\sqrt{2}, \end{cases} \qquad \begin{cases} c_1 = -\sqrt{2}, \\ c_2 = 0, \\ c_3 = \sqrt{2}, \end{cases} \qquad \begin{cases} c_1 = -\frac{1}{2}(\sqrt{5}-1), \\ c_2 = -1, \\ c_3 = \frac{1}{2}(\sqrt{5}+1), \end{cases}$$

$$\begin{cases} c_1 = \frac{1}{2}(\sqrt{5}+1), \\ c_2 = -1, \\ c_3 = -\frac{1}{2}(\sqrt{5}-1), \end{cases} \qquad \begin{cases} c_1 = \frac{1}{2}(\sqrt{5}-1), \\ c_2 = 1, \\ c_3 = -\frac{1}{2}(\sqrt{5}+1), \end{cases} \qquad \begin{cases} c_1 = -\frac{1}{2}(\sqrt{5}+1), \\ c_2 = 1, \\ c_3 = \frac{1}{2}(\sqrt{5}-1). \end{cases}$$

4.5 Sensitivity Analysis

Let $A_k = [\alpha_{ij}^{(k)}] \in \mathbf{R}^{n\times n}$, $k = 0, 1, \ldots, n$, be symmetric with $\alpha_{ii}^{(k)} = \delta_{ik}$, $i, k = 1, 2, \ldots, n$, and let $\lambda = (\lambda_1, \ldots, \lambda_n)^T \in \mathbf{R}^n$ with $\lambda_1 < \lambda_2 < \cdots < \lambda_n$. Assume that $c = (c_1, \ldots, c_n)^T \in \mathbf{R}^n$ is a solution to Problem SG with the given data; that is, the matrix

$$A(c) = A_0 + \sum_{k=1}^{n} c_k A_k \tag{4.5.1}$$

has $\lambda_1, \ldots, \lambda_n$ as its eigenvalues. In this section we investigate the variation in the solution c with respect to changes in the given data $\{A_k\}_{k=0}^n$ and $\{\lambda_i\}_{i=1}^n$.

The assumptions imply that there exists an orthogonal matrix $Q(c) = [q_{ij}(c)] = [q_1(c), \ldots, q_n(c)] \in \mathbf{R}^{n\times n}$ such that

$$Q(c)^T A(c) Q(c) = \mathrm{diag}\,(\lambda_1, \ldots, \lambda_n). \tag{4.5.2}$$

Now define an $n \times n$ matrix $J(c) = [J_{ij}(c)]$ by

$$J_{ij}(c) = q_i(c)^T A_j q_i(c), \quad i, j = 1, 2, \ldots, n. \tag{4.5.3}$$

Clearly, $J(c)$ is uniquely determined by c. If $J(c)$ is nonsingular, then the solution c is called a **regular solution** to Problem SG.

Notice that in the case where $A_k = e_k e_k^T$ for $k = 1, 2, \ldots, n$ (i.e., the given problem is an additive inverse eigenvalue problem) we have

$$J(c) = \begin{bmatrix} q_{11}(c)^2 & q_{21}(c)^2 & \cdots & q_{n1}(c)^2 \\ q_{12}(c)^2 & q_{22}(c)^2 & \cdots & q_{n2}(c)^2 \\ \vdots & \vdots & & \vdots \\ q_{1n}(c)^2 & q_{2n}(c)^2 & \cdots & q_{nn}(c)^2 \end{bmatrix}. \tag{4.5.4}$$

In this section we only investigate the sensitivity of a regular solution to Problem SG. This is because the changes of a nonregular solution may be very complicated when the given data has some small changes and, moreover, perturbation bounds for this case do not appear to be known.

The question that arises immediately is: under what conditions is a solution to Problem SG regular. This is a very difficult problem. Here we use the theory developed in the previous section to give a condition to guarantee that Problem SG has a regular solution.

Lemma 4.5.1 *Let $B = [\beta_{ij}] \in \mathbf{R}^{n\times n}$ satisfy*

$$\delta = \min_{1\le i\le n}\left(|\beta_{ii}| - \sum_{\substack{j=1\\ j\neq i}}^{n} |\beta_{ij}|\right) > 0.$$

Then B is nonsingular and $\|B^{-1}\|_\infty \le \dfrac{1}{\delta}$.

Proof Let $x = (\xi_1, \ldots, \xi_n) \in \mathbf{R}^n$ with $|\xi_k| = \|x\|_\infty > 0$. Then we have

$$\begin{aligned} 0 < \delta|\xi_k| &\le \left(|\beta_{kk}| - \sum_{\substack{j=1\\ j\neq k}}^{n} |\beta_{kj}|\right)|\xi_k| \\ &\le \left|\sum_{j=1}^{n} \beta_{kj}\xi_j\right| \le \max_{1\le i\le n}\left|\sum_{j=1}^{n} \beta_{ij}\xi_j\right|, \end{aligned}$$

that is, $\|Bx\|_\infty \ge \delta\|x\|_\infty > 0$ for any $0 \neq x \in \mathbf{R}^n$, which implies that B is nonsingular and $\|B^{-1}\|_\infty \le \dfrac{1}{\beta}$. □

Theorem 4.5.2 *Let l_j be defined as in Theorem 4.4.9, and let*

$$d_j(\lambda) = \min_{i\neq j}|\lambda_j - \lambda_i|, \quad \delta = \max_{1\le j\le n} d_j(\lambda) \quad \eta = \min_{1\le j\le n}(d_j(\lambda) - 2l_j).$$

If the given data satisfies

$$\lambda_{i+1} - \lambda_i > 2\max\{l_i, l_{i+1}\}, \qquad i = 1, 2, \ldots, n-1, \tag{4.5.5}$$

and

$$\sum_{k=1}^{n} \|A_k^{(0)}\|_2 < \frac{\eta}{\delta}, \tag{4.5.6}$$

where $A_k^{(0)} = A_k - e_k e_k^T$, $k = 1, 2, \ldots, n$, *and if the solution* $c = (c_1, \ldots, c_n)^T \in \mathbf{R}^n$ *satisfies*

$$\lambda_1 \le \alpha_{11}^{(0)} + c_1 \le \cdots \le \alpha_{nn}^{(0)} + c_n \le \lambda_n, \tag{4.5.7}$$

then c *is regular and*

$$\|J(c)^{-1}\|_\infty \le \frac{\eta}{\eta - \delta \sum_{k=1}^{n} \|A_k^{(0)}\|_2}. \tag{4.5.8}$$

Proof From (4.5.2) we have

$$A(c) - \lambda_j I = Q(c)\mathrm{diag}\,(\lambda_1 - \lambda_j, \ldots, \lambda_n - \lambda_j) Q(c)^T,$$

and so we have

$$\|(A(c) - \lambda_j I)e_j\|_2 = \|\mathrm{diag}\,(\lambda_1 - \lambda_j, \ldots, \lambda_n - \lambda_j) Q(c)^T e_j\|_2,$$

that is,

$$(a_{jj}^{(0)} + c_j - \lambda_j)^2 + \sum_{\substack{i=1\\ i\neq j}}^{n} \left(a_{ij}^{(0)} + \sum_{k=1}^{n} c_k a_{ij}^{(k)} \right)^2 = \sum_{\substack{i=1\\ i\neq j}}^{n} (\lambda_i - \lambda_j)^2 q_{ji}^2(c) \tag{4.5.9}$$

for $j = 1, 2, \ldots, n$. By Lemma 4.4.5, it follows from (4.5.5) and (4.5.7) that

$$\left|\alpha_{jj}^{(0)} + c_j - \lambda_j\right| \le l_j, \qquad j = 1, \ldots, n. \tag{4.5.10}$$

On the other hand, we have

$$\begin{aligned} &\sum_{\substack{i=1\\ i\neq j}}^{n} \left(\alpha_{ij}^{(0)} + \sum_{k=1}^{n} c_k a_{ij}^{(k)} \right)^2 \\ &\quad \le \left(\sum_{\substack{i=1\\ i\neq j}}^{n} \left(\left| \alpha_{ij}^{(0)} - \sum_{k=1}^{n} \alpha_{kk}^{(0)} \alpha_{ij}^{(k)} \right| + \sum_{k=1}^{n} \left|\alpha_{kk}^{(0)} + c_k\right| \left|\alpha_{ij}^{(k)}\right| \right) \right)^2 \\ &\quad \le l_j^2 \end{aligned} \tag{4.5.11}$$

for $j = 1, 2, \ldots, n$. Combining (4.5.9), (4.5.10), and (4.5.11) gives

$$d_j^2(\lambda) \sum_{\substack{i=1 \\ i \neq j}}^{n} q_{ji}^2(c) \leq 2l_j^2, \quad j = 1, \ldots, n. \tag{4.5.12}$$

From the orthogonality of $Q(c)$, it follows that

$$\sum_{i=1}^{n} q_{ji}^2(c) = 1 = \sum_{j=1}^{n} q_{ji}^2(c), \quad i, j = 1, \ldots, n,$$

and so

$$q_{jj}^2(c) = 1 - \sum_{\substack{i=1 \\ i \neq j}}^{n} q_{ji}^2(c), \quad j = 1, 2, \ldots, n.$$

This, together with (4.5.12), gives

$$q_{jj}^2(c) - \sum_{\substack{i=1 \\ i \neq j}}^{n} q_{ji}^2(c) \geq 1 - \frac{4l_j^2}{d_j^2(\lambda)} \geq \frac{d_j(\lambda) - 2l_j}{d_j(\lambda)} \geq \frac{\eta}{\delta} \tag{4.5.13}$$

for all $j = 1, 2, \ldots, n$.

Notice that the matrix $J(c)$ can be split into

$$J(c) = G(c) + J_0(c) \tag{4.5.14}$$

where

$$G(c) = \left[g_{ij}\right], \quad \text{with } g_{ij} = q_{ji}^2(c), \ i, j = 1, \ldots, n, \tag{4.5.15}$$

and

$$J_0(c) = \left[q_i(c)^T A_j^{(0)} q_i(c)\right]. \tag{4.5.16}$$

By Lemma 4.5.1, it follows from (4.5.13) that $G(c)$ is nonsingular and

$$\|G(c)^{-1}\|_\infty \leq \frac{\delta}{\eta}. \tag{4.5.17}$$

Thus it follows from (4.5.6) that

$$\|G(c)^{-1} J_0(c)\|_\infty \leq \|G(c)^{-1}\|_\infty \|J_0(c)\|_\infty \leq \frac{\delta}{\eta} \sum_{k=1}^{n} \|A_k^{(0)}\|_2 < 1.$$

Consequently, $J(c) = G(c) + J_0(c)$ is nonsingular and

$$\|J(c)^{-1}\|_\infty \le \frac{\delta}{\eta} \Big/ \left(1 - \frac{\delta}{\eta}\sum_{k=1}^{n} \|A_k^{(0)}\|_2\right),$$

which is the desired result. □

Applying Theorem 4.5.2 to the additive inverse eigenvalues problem we immediately get the following corollary .

Corollary 4.5.3 *Let $A = [\alpha_{ij}] \in \mathbf{R}^n$ be symmetric with $\alpha_{ii} = 0$ for all i, let $\lambda_1 < \cdots < \lambda_n$, let $d_j(\lambda)$ and δ be defined as in Theorem 4.5.2, let $g_i(A)$ be defined by (4.4.4), and let*

$$\omega = \min_{1\le j\le n} \big(d_j(\lambda) - 2g_j(A)\big).$$

If

$$\lambda_{i+1} - \lambda_i > 2\max\big(g_i(A),\, g_{i+1}(A)\big), \quad i = 1, 2, \dots, n-1,$$

and if the solution $c = (c_1, \dots, c_n)^T$ to Problem A with the given data satisfies

$$\lambda_1 \le c_1 \le \cdots \le c_n \le \lambda_n,$$

then c is regular and

$$\|J(c)^{-1}\|_\infty \le \frac{\delta}{\omega},$$

where $J(c)$ is defined by (4.5.4).

We turn now to the establishment of perturbation theorems. Assume c is a regular solution of Problem SG with the given data $\{A_k\}_{k=0}^n$ and $\{\lambda_i\}_{i=1}^n$. Let $E_k \in \mathbf{R}^{n\times n}$, $k = 0, 1, \dots, 2$, be symmetric with all the diagonal elements of E_k being zero, $k = 1, 2, \dots, n$, and let $\varepsilon_i \in \mathbf{R}$, $i = 1, 2, \dots, n$. Define

$$A_k(t) = A_k + tE_k, \quad k = 0, 1, 2, \dots, n,$$

$$\lambda_i(t) = \lambda_i + t\varepsilon_i, \quad i = 1, 2, \dots, n, \quad t \in [-1,\ 1].$$

Assume further that the ε_i are sufficiently small so that $\lambda_1(t) < \lambda_2 < \cdots < \lambda_n(t)$ for every $t \in [-1,\ 1]$.

Now consider the matrix-valued function

$$A(t; x) = A_0(t) + \sum_{k=1}^{n} \xi_k A_k(t), \quad t \in [-1,\ 1], \quad x = (\xi_1, \dots, \xi_n)^T \in \mathbf{R}^n.$$

It is clear that $A(t;x)$ is analytic and $A(0;c) = A_0 + \sum_{k=1}^n c_k A_k$ has distinct eigenvalues $\lambda_1, \dots, \lambda_n$. Define $f : [-1, 1] \times \mathbf{R}^n \longrightarrow \mathbf{R}^n$ by

$$f(t;x) = \left(\lambda_1(t;x) - \lambda_1(t), \dots, \lambda_n(t;x) - \lambda_n(t)\right)^T,$$

where $\lambda_1(t;x) \le \lambda_2(t;x) \le \cdots \le \lambda_n(t;x)$ are the ordered eigenvalues of $A(t;x)$. Clearly, $f(0;c) = 0$. Moreover, by Theorem 1.9.2, it follows that there exists a positive number ε_1 and a neighborhood $\mathcal{B}(c)$ of c such that $f(t;x)$ is analytic in $(-\varepsilon_1, \varepsilon_1) \times \mathcal{B}(c)$ and, moreover,

$$f'_x(0;c) = J(c),$$

which is nonsingular since c is regular. Consequently, by Theorem 1.8.2, there exists a positive number $\varepsilon_0 < \varepsilon_1$ and a unique vector-valued function $c(t) = \left(c_1(t), \dots, c_n(t)\right)^T \in \mathbf{R}^n$ such that

(1) $c(t)$ is analytic in $(-\varepsilon_0, \varepsilon_0)$;

(2) $c(0) = c$;

(3) $f(t; c(t)) = 0$ for every $t \in (-\varepsilon_0, \varepsilon_0)$.

By the definition of f, (3) means that for every $t \in (-\varepsilon_0, \varepsilon)$, $A(t; c(t))$ has $\lambda_1(t), \dots, \lambda_n(t)$ as its eigenvalues; that is, $c(t)$ is a solution of Problem SG with data $\{A_k(t)\}_{k=0}^n$ and $\{\lambda_i(t)\}_{i=1}^n$. Thus there exists an orthogonal matrix $Q(t) = [q_1(t), \dots, q_n(t)] \in \mathbf{R}^{n\times n}$ such that

$$Q(t)^T A(t) Q(t) = \operatorname{diag}\left(\lambda_1(t), \dots, \lambda_n(t)\right), \tag{4.5.18}$$

where $A(t) = A(t; c(t))$. Since $A(t; c(t))$ is analytic and $\lambda_1(t) < \lambda_2(t) < \cdots < \lambda_n(t)$ for any $t \in (-\varepsilon_0, \varepsilon_0)$, it follows from Theorem 1.9.2 that the $q_i(t)$ can be chosen as analytic vector-valued functions with $q_i(0) = q_i(c)$.

Equating the diagonal elements of (4.5.18) gives

$$\lambda_i(t) = q_i(t)^T A(t) q_i(t), \quad i = 1, 2, \dots, n. \tag{4.5.19}$$

Differentiating (4.5.19) at $t = 0$ and using the fact that $q_i(t)^T q(t) = 1$ for any $t \in (-\varepsilon_0, \varepsilon_0)$, we have

$$\dot\lambda_i(0) = q_i(c)^T \dot A(0) q_i(c), \quad i = 1, 2, \dots, n. \tag{4.5.20}$$

Notice that

$$\dot A(0) = E_0 + \sum_{k=1}^n \dot c_k(0) A_k + \sum_{k=1}^n c_k E_k, \tag{4.5.21}$$

$$\dot\lambda_i(0) = \varepsilon_i, \quad i = 1, 2, \dots, n. \tag{4.5.22}$$

Substituting (4.5.21) and (4.5.22) into (4.5.20) and writing it in vector form, we have

$$e = \widetilde{b} + J(c)\dot{c}(0) + \widetilde{J}c, \tag{4.5.23}$$

where

$$e = (\varepsilon_1, \dots, \varepsilon_n)^T, \quad \widetilde{b} = \left(q_1(c)^T E_0 q_1(c), \dots, q_n(c)^T E_0 q_n(c)\right)^T,$$

$$\widetilde{J} = \begin{bmatrix} q_1(c)^T E_1 q_1(c) & q_1(c)^T E_2 q_1(c) & \cdots & q_1(c)^T E_n q_1(c) \\ q_2(c)^T E_1 q_2(c) & q_2(c)^T E_2 q_2(c) & \cdots & q_2(c)^T E_n q_2(c) \\ \vdots & \vdots & & \vdots \\ q_n(c)^T E_1 q_n(c) & q_n(c)^T E_2 q_n(c) & \cdots & q_n(c)^T E_n q_n(c) \end{bmatrix}.$$

Consequently, we have

$$\begin{aligned} \|c(t) - (c)\|_\infty &= \|\dot{c}(0)t + O(t^2)\|_\infty \\ &\le \|J(c)^{-1}\|_\infty \left(\|te\|_\infty + \|t\widetilde{b}\|_\infty + \|t\widetilde{J}\|_\infty \|c\|_\infty\right) + O(t^2) \\ &\le \|J(c)^{-1}\|_\infty \left(\|te\|_\infty + \|tE_0\|_2 + \|c\|_\infty \sum_{k=1}^{n} \|tE_k\|_2\right) + O(t^2). \end{aligned} \tag{4.5.24}$$

Letting $t = 0$ in (4.5.19) and writing it in vector form, we get

$$\lambda = J(c)c + b,$$

where

$$b = \left(q_1(c)^T A_0 q_1(c), \dots, q_n(c)^T A_0 q_n(c)\right)^T, \tag{4.5.25}$$

and so, we have

$$\|\lambda - b\|_\infty \le \|J(c)\|_\infty \|c\|_\infty .$$

This, together with (4.5.24), gives

$$\frac{\|c(t) - c\|_\infty}{\|c\|_\infty} \le \|J(c)^{-1}\|_\infty \|J(c)\|_\infty \left(\frac{\|te\|_\infty + \|tE_0\|_2}{\|\lambda - b\|_\infty} + \frac{\sum_{k=1}^{n} \|tE_k\|_2}{\|J(c)\|_\infty}\right) + O(t^2).$$

Thus we have proved the following theorem.

Theorem 4.5.4 *Let* $A_k = [\alpha_{ij}^{(k)}]$, $\widetilde{A}_k = [\widetilde{\alpha}_{ij}^{(k)}] \in \mathbf{R}^{n\times n}$, $k = 0, 1, 2, \dots, n$, *be symmetric with* $\alpha_{ij}^{(k)} = \widetilde{\alpha}_{ij}^{(k)} = \delta_{ik}$, $i, k = 1, 2, \dots, n$, *and let* $\lambda = (\lambda_1, \dots, \lambda_n)^T$, $\widetilde{\lambda} = (\widetilde{\lambda}_1, \dots, \widetilde{\lambda}_n)^T \in \mathbf{R}^n$ *with* $\lambda_1 < \lambda_2 < \cdots < \lambda_n$ *and* $\widetilde{\lambda}_1 < \widetilde{\lambda}_2 < \cdots < \widetilde{\lambda}_n$.

Assume further that c is a regular solution to Problem SG *with data* $\{A_k\}_{k=0}^n$ *and* $\{\lambda_i\}_{i=1}^n$. *If*

$$\delta = \|\lambda - \widetilde{\lambda}\|_\infty + \sum_{k=0}^{n} \|A_k - \widetilde{A}_k\|_2$$

is sufficiently small, then there exists a solution $\widetilde{c}$ to Problem SG *with data* $\{\widetilde{A}_k\}_{k=0}^n$ *and* $\{\lambda_i\}_{i=1}^n$ *such that*

$$\frac{\|\widetilde{c} - c\|_\infty}{\|c\|_\infty} \le \kappa(c)\left(\frac{\|\lambda - \widetilde{\lambda}\|_\infty + \|A_0 - \widetilde{A}_0\|_2}{\|\lambda - b\|_\infty} + \frac{\sum_{k=1}^n \|A_k - \widetilde{A}_k\|_2}{\|J(c)\|_\infty}\right) + O(\delta^2),$$

where $J(c)$ and b are defined by (4.5.3) *and* (4.5.25), *respectively, and*

$$\kappa(c) = \|J(c)^{-1}\|_\infty \|J(c)\|_\infty. \tag{4.5.26}$$

Remark 4.5.1 In view of Theorem 4.5.4, the scalar $\kappa(c)$ defined by (4.5.26) is referred to as the **condition number** of the solution c to Problem SG with the given data $\{A_k\}_{k=0}^n$ and $\{\lambda_i\}_{i=1}^n$. If $\kappa(c)$ is large, then we will call the corresponding solution c **ill-conditioned.**

Applying Theorem 4.5.4 to the additive inverse eigenvalue problem, we immediately obtain the following result.

Corollary 4.5.5 *Let $A, \widetilde{A} \in \mathbf{R}^{n\times n}$ be symmetric, and let $\lambda = (\lambda_1 \dots, \lambda_n)^T$, $\widetilde{\lambda} = (\widetilde{\lambda}_1 \dots, \widetilde{\lambda}_n) \in \mathbf{R}^n$ with*

$$\lambda_1 < \lambda_2 < \dots < \lambda_n, \quad \widetilde{\lambda}_1 < \widetilde{\lambda}_2 < \dots < \widetilde{\lambda}_n.$$

Assume further that $c = (c_1, \dots, c_n)^T \in \mathbf{R}^n$ is a regular solution to Problem A *with data A and* $\{\lambda_i\}_{i=1}^n$. *If*

$$\varepsilon = \|\lambda - \widetilde{\lambda}\|_\infty + \|A - \widetilde{A}\|_2$$

is sufficiently small, then there exists a solution $\widetilde{c}$ to Problem A *with data $\widetilde{A}$ and* $\{\widetilde{\lambda}_i\}_{i=1}^n$ *such that*

$$\frac{\|c - \widetilde{c}\|_\infty}{\|c\|_\infty} \le \kappa_A(c)\frac{\|\lambda - \widetilde{\lambda}\|_\infty + \|A - \widetilde{A}\|_2}{\|\lambda - b_A\|_\infty} + O(\varepsilon^2),$$

where

$$b_A = (q_1^T A q_1, \dots, q_n^T A q_n)^T,$$
$$\kappa_A(c) = \|J(c)^{-1}\|_\infty \|J(c)\|_\infty = \|J(c)^{-1}\|_\infty.$$

Similarly, we can develop a perturbation bound for a regular solution to Problem **M** as follows.

Theorem 4.5.6 *Let* $A = [\alpha_{ij}]$, $\widetilde{A} = [\widetilde{\alpha}_{ij}] \in \mathbf{R}^{n\times n}$ *be symmetric positive definite with* $\alpha_{ii} = \widetilde{\alpha}_{ii} = 1$, $i = 1, 2, \ldots, n$, *and let* $\lambda = \lambda_1, \ldots, \lambda_n)^T$, $\widetilde{\lambda} = \widetilde{\lambda}_1, \ldots, \widetilde{\lambda}_n) \in \mathbf{R}^n$ *with*

$$0 < \lambda_1 < \lambda_2 < \cdots < \lambda_n, \quad 0 < \widetilde{\lambda}_1 < \cdots < \widetilde{\lambda}_n.$$

Assume further that there exists a vector $c = (c_1, \ldots, c_n) \in \mathbf{R}^n$ *with* $c_i > 0$ *and an orthogonal matrix* $U = [u_{ij}] \in \mathbf{R}^{n\times n}$ *with*

$$W = \begin{bmatrix} u_{11}^2 & u_{21}^2 & \cdots & u_{n1}^2 \\ u_{12}^2 & u_{22}^2 & \cdots & u_{n2}^2 \\ \vdots & \vdots & & \vdots \\ u_{1n}^2 & u_{2n}^2 & \cdots & u_{nn}^2 \end{bmatrix}$$

nonsingular such that

$$D^{\frac{1}{2}} A D^{\frac{1}{2}} = U \Lambda U^T,$$

where $D = \operatorname{diag}(c_1, \ldots, c_n)$; *that is, c is a* **regular solution** *to Problem* M *with data A and* $\{\lambda_i\}_{i=1}^n$. *If*

$$\widetilde{\varepsilon} = \|\lambda - \widetilde{\lambda}\|_\infty + \|A - \widetilde{A}\|_2$$

is sufficiently small, then there exists a solution $\widetilde{c}$ *to Problem* M *with data* $\widetilde{A}$ *and* $\{\widetilde{\lambda}_i\}_{i=1}^n$, *such that*

$$\frac{\|c - \widetilde{c}\|_\infty}{\|c\|_\infty} \le \kappa_M(c)\left(\frac{\|\lambda - \widetilde{\lambda}\|_\infty}{\|\lambda\|_\infty} + \|A - \widetilde{A}\|_2\right) + O(\widetilde{\varepsilon}^2),$$

where $\kappa_M(c) = \dfrac{\lambda_n}{\lambda_1}\|W^{-1}\|_\infty$.

Proof Left to the reader as an exercise. □

4.6 Newton-like Methods

From now on we will devote ourselves to the development of the numerical methods for solving the inverse eigenvalue problems described in Section 4.1. In this section we will develop several methods based on Newton's method.

4.6.1 Methods for Solving Problem SG

Let $n+1$ symmetric matrix $A_0, \dots, A_k \in \mathbf{R}^{n\times n}$ and n ordered real numbers $\lambda_1 < \cdots < \lambda_n$ be given. Assume that Problem SG with the given data $\{A_k\}_{k=0}^n$ and $\{\lambda_i\}_{i=1}^n$ has a regular solution $c = (c_1, \dots, c_n)^T \in \mathbf{R}^n$.

Now define $f : \mathbf{R}^n \longrightarrow \mathbf{R}^n$ by

$$f(x) = \big(\lambda_1(x) - \lambda_1, \dots, \lambda_n(x) - \lambda_n\big)^T, \quad x = (\xi_1, \dots, \xi_n)^T \in \mathbf{R}^n,$$

where $\lambda_1(x) \le \lambda_2(x) \le \cdots \le \lambda_n(x)$ are the ordered eigenvalues of $A(x) = A_0 + \sum_{k=1}^n \xi_k A_k$. Clearly, x is a solution of Problem SG if and only if $f(x) = 0$. Therefore, we can formulate Problem SG as a system of nonlinear equations

$$f(x) = 0. \tag{4.6.1}$$

The first method we will describe consists of applying Newton's method to (4.6.1). By Theorem 1.9.2, it follows that under the above assumptions the function $f(x)$ is analytic in some neighborhood of c and

$$J(x) = f'(x) = [q_i(x)^T A_j q_i(x)], \tag{4.6.2}$$

where $Q(x) = [q_1(x), \dots, q_n(x)]$ is the orthogonal matrix of eigenvectors of $A(x)$. Thus, one step of **Newton's method** is defined by

$$J(x_m)\big(x_{m+1} - x_m\big) = -f(x_m). \tag{4.6.3}$$

Notice that

$$q_i(x)^T A_0 q_i(x) + \sum_{k=1}^n \xi_k q_i(x)^T A_k q_i(x) = \lambda_i(x), \quad i = 1, 2, \dots, n,$$

or equivalently

$$J(x)x = \lambda(x) - b(x), \tag{4.6.4}$$

where

$$\lambda(x) = \big(\lambda_1(x), \dots, \lambda_n(x)\big)^T,$$
$$b(x) = \big(q_1(x)^T A_0 q_1(x), \dots, q_n(x)^T A_0 q_n(x)\big)^T.$$

Utilizing (4.6.4), we can write (4.6.3) as

$$J(x_m)x_{m+1} = \lambda - b(x_m), \tag{4.6.5}$$

where $\lambda = (\lambda_1, \dots, \lambda_n)^T$.

In summarizing, we obtain the following algorithm.

Algorithm 4.6.1 Given $n+1$ real symmetric $n \times n$ matrices A_k, $k = 0, 1, \ldots, n$ and given n real numbers $\lambda_1 \leq \lambda_2 \leq \cdots \leq \lambda_n$, the following algorithm is to use Newton's method to compute a vector $c = (c_1, \ldots, c_n)^T$ such that $\lambda_1, \ldots, \lambda_n$ are the eigenvalues of $A_0 + \sum_{k=1}^n c_k A_k$.

Choose a starting vector x_0, an iteration maximum $k_{\max}$, and a precision
    requirement $\varepsilon < 1$;
$r = 1$; $k = 0$; $c = x_0$;
while $k < k_{\max}$ and $r > \varepsilon$
    $A = A_0$;
    for $j = 1 : n$
        $A = A + c(j)A_j$;
    end
    Compute the factorization

$$A = Q\text{diag}\,(\mu_1, \ldots, \mu_n)Q^T,$$

    where $Q = [q_1, \ldots, q_n]$ is orthogonal and $\mu_1 < \cdots < \mu_n$;
    for $i = 1 : n$
        $b(i) = \lambda_i - q_i^T A_0 q_i$;
        for $j = 1 : n$
            $J(i,j) = q_i^T A_j q_i$;
        end
    end
    Solve $Jx = b$ for x;
    $r = \dfrac{\|x - c\|_\infty}{\|c\|_\infty}$; $c = x$; $k = k + 1$;
end

The following theorem shows that this method is locally quadratically convergent.

Theorem 4.6.1 *Assume that c is a regular solution to Problem* SG *and the given eigenvalues are distinct. Then there exists* $\delta > 0$ *such that, if* $\|x_0 - c\|_2 < \delta$, *the iterates* $\{x_m\}$ *generated by* (4.6.3) *converge to c and*

$$\|x_{m+1} - c\|_2 \leq C\|x_m - c\|_2^2$$

for $m = 0, 1, 2, \ldots$, *where* C *is a constant independent of* m.

Proof By Theorem 1.9.2, the assumptions of Theorem 4.6.1 imply that $f(x)$ and its Jacobian matrix $J(x)$ are analytic in some neighborhood of c. Consequently, there exists $\delta_0 > 0$ such that $\|x - c\|_2 \leq \delta_0$ implies

$$\|f(x) - J(c)(x - c)\|_2 \leq C_1\|x - c\|_2^2,$$
$$\|J(x) - J(c)\|_2 \leq C_1\|x - c\|_2, \quad \text{and} \quad \|J(x)^{-1}\|_2 \leq C_1.$$

Where C_1 is a constant independent of x. From (4.6.3) we easily get

$$J(x_m)(x_{m+1} - c) = \big(J(x_m) - J(c)\big)(x_m - c) + \big(J(c)(x_m - c) - f(x_m)\big),$$

and so

$$\|x_{m+1} - c\|_2 \le 2C_1^2 \|x_m - c\|_2^2$$

as long as $\|x_m - c\|_2 \le \delta_0$. Thus let $C = 2C_1^2$ and $\delta = \min\{\delta_0, \frac{1}{C+1}\}$. Using mathematical induction, it is easy to verify that $\|x_0 - c\|_2 \le \delta$ implies

$$\|x_{m+1} - c\|_2 \le C\|x_m - c\|_2^2 \le C\delta\|x_m - c\|_2 \le \|x_m - c\|_2 \le \delta \le \delta_0$$

for all $m = 0, 1, \ldots,$ from which the theorem follows immediately. □

Here, for the sake of simplicity, we assume that the given eigenvalues are distinct. In fact, this assumption is not necessary. Of course, when multiple eigenvalues are present, the eigenvalues are not, in general, differentiable at c. Furthermore, the eigenvectors are not unique, and they cannot generally be defined to be continuous functions. However, as long as the eigenvalues of $A(x_m)$ are distinct for all iterates x_m, the above algorithm remains well defined. Moreover, in this case the convergence is generally quadratic in practice.

Example 4.6.1 Consider Problem A with the given data:

$$A = \begin{bmatrix} 0 & -2 & 2 \\ -2 & 0 & 4 \\ 2 & 4 & 0 \end{bmatrix}, \qquad \lambda = (-6, 3, 3)^T.$$

It is easy to verify that $c = (2, -1, -1)^T$ is an exact solution of this problem. Applying Algorithm 4.6.1 to this problem with the starting vector $x_0 = (1.8, -0.7, -1.2)^T$ we find

k	$\|x_k - c\|_2$	$\|\lambda - \lambda^{(k)}\|_2$
1	6.946403×10^{-2}	2.642512×10^{-2}
2	1.790298×10^{-3}	9.023822×10^{-4}
3	1.995217×10^{-6}	7.642480×10^{-7}
4	1.459136×10^{-12}	7.241390×10^{-13}

Here $\lambda^{(k)}$ denotes the vector of the ordered eigenvalues of $A(x_k)$ according to increasing size. Observe the quadratic-like convergence.

In Algorithm 4.6.1 all the eigenvectors of $A(x)$ have to be computed per step, which is very time consuming. Therefore, instead of computing these eigenvectors at each step we should consider approximating them. One possibility is to use inverse iteration. Suppose that we have determined an estimate x_m of c and an approximation $Q_m = [q_1^{(m)}, \ldots, q_n^{(m)}]$ to $Q(x_m)$, the orthogonal matrix of eigenvectors of $A(x_m)$. Here we assume that $Q_0 = Q(x_0)$. Now define

$$J_m = [(q_i^{(m)})^T A_j q_i^{(m)}] \in \mathbf{R}^{n\times n}, \tag{4.6.6}$$

$$b_m = \left((q_1^{(m)})^T A_0 q_1^{(m)}, \ldots, (q_n^{(m)})^T A_0 q_n^{(m)}\right)^T. \tag{4.6.7}$$

Then we compute a new estimate x_{m+1} by solving

$$J_m x_{m+1} = \lambda - b_m. \tag{4.6.8}$$

To update our approximations to the eigenvectors, we apply one step of **inverse iteration**; that is, we compute v_i, $i = 1, \ldots, n$ by solving

$$\left(A(x_{m+1}) - \lambda_i I\right) v_i = q_i^{(m)}, \quad i = 1, 2, \ldots, n. \tag{4.6.9}$$

We then define $Q_{m+1} = [q_i^{(m+1)}, \ldots, q_n^{(m+1)}]$ by

$$q_i^{(m+1)} = \frac{v_i}{||v_i||_2}, \quad i = 1, 2, \ldots, n.$$

Thus we are performing a **Newton-like iteration** where instead of computing the exact eigenvectors of $A(x)$ at each step we update an approximation to them by performing one step of inverse iteration. Note that the cost of determining Q_{m+1} is $O(n^4)$ flops. To reduce the cost, we may first reduce $A(x_{m+1})$ into tridiagonal form T, then apply one step inverse iteration to $T - \lambda_i I$ for each $i = 1, 2, \ldots, n$. In this way the cost can be reduced to $O(n^3)$ flops.

Summarizing the above discussion, we obtain the following algorithm.

Algorithm 4.6.2 Given $n+1$ real symmetric $n \times n$ matrices A_k, $k = 0, 1, \ldots, n$ and given n real numbers $\lambda_1 \le \lambda_2 \le \cdots \le \lambda_n$, the following algorithm is to use Newton-like method to compute a vector $c = (c_1, \ldots, c_n)^T$ such that $\lambda_1, \ldots, \lambda_n$ are the eigenvalues of $A_0 + \sum_{k=1}^n c_k A_k$.

Choose a starting vector x_0, an iteration maximum $k_{\max}$, and a precision requirement $\varepsilon < 1$;

$c = x_0$; $A = A_0 + \sum_{k=1}^{n} c_k A_k$;

Compute the factorization

$$A = Q\mathrm{diag}(\mu_1, \ldots, \mu_n)Q^T,$$

where $Q = [q_1, \ldots, q_n]$ is orthogonal and $\mu_1 \le \cdots \le \mu_n$;

$r = \left(\sum_{i=1}^{n} (\lambda_i - \mu_i)^2\right)^{\frac{1}{2}}$; $k = 0$;

while $k < k_{\max}$ and $r > \varepsilon$
  if $k \geq 1$
    $A = A_0$;
    for $j = 1 : n$
      $A = A + c(j)A_j$;
    end
    Compute the factorization

$$A = UTU^T,$$

    where U is orthogonal and T is tridiagonal;
    for $i = 1 : n$
      Solve $(T - \lambda_i I)y = U^T q_i$ for y;

$$q_i = \frac{Uy}{\|y\|_2};$$

    end
  end
  for $i = 1 : n$
    $b(i) = \lambda_i - q_i^T A_0 q_i$;
    for $j = 1 : n$
      $J(i, j) = q_i^T A_j q_i$;
    end
  end
  Solve $Jx = b$ for x;

$$r = \frac{\|x - c\|_\infty}{\|c\|_\infty}; \quad c = x; \quad k = k + 1;$$

end

The local behavior of this algorithm is very similar to Algorithm 4.6.1. In fact, we can prove the following convergence result for the Newton-like method.

Theorem 4.6.2 *Assume that c is a regular solution to Problem* SG *and the given eigenvalues are distinct. Then there exists $\delta > 0$ such that, if $\|x_0 - c\|_2 < \delta$, the iterates $\{x_m\}$ generated by* (4.6.6)–(4.6.9) *converge to c and, moreover,*

$$\|x_{m+1} - c\|_2 \leq C\|x_m - c\|_2^2 \tag{4.6.10}$$

for $m = 0, 1, 2, \ldots$, where C is a constant independent of m.

Proof Let $Q(x) = [q_1(x), \ldots, q_n(x)] \in \mathbf{R}^{n\times n}$ be the orthogonal matrix of eigenvectors of $A(x)$, i.e.,

$$Q(x)^T Q(x) = I, \quad A(x)Q(x) = Q(x)\Lambda(x),$$

where $\Lambda(x) = \operatorname{diag}\big(\lambda_1(x), \ldots, \lambda_n(x)\big)$ with $\lambda_1(x) \leq \cdots \leq \lambda_n(x)$. From the assumptions we know that $\lambda_i(c) = \lambda_i$, $i = 1, 2, \ldots, n$, and $J(c)$ is nonsingular.

Hence, by the continuity of the matrix inverse and the analyticity of a simple eigenvalue and the corresponding eigenvectors, there exist positive numbers δ_0, C_0, N_0, and γ such that :

(1) $\max_{1\le i\le n} \|q_i^{(m)} - q_i(c)\|_2 \le \delta_0$ implies that J_m is nonsigular and

$$\|J_m^{-1}\|_F \le C_0; \tag{4.6.11}$$

(2) $\|x - c\|_2 \le \delta_0$, $x \in \mathbf{R}^{n\times n}$, implies

$$|\lambda_i(x) - \lambda_i| \le N_0\|x - c\|_2, \quad i = 1, 2, \dots, n, \tag{4.6.12}$$

$$\|q_i(x) - q_i(c)\|_2 \le N_0\|x - c\|_2, \quad i = 1, 2, \dots, n, \tag{4.6.13}$$

$$|\lambda_i(x) - \lambda_j| > \gamma > 0, \quad i \ne j. \tag{4.6.14}$$

In addition, there exists a positive number N_1 such that $|t| \le \frac{1}{2}$ implies

$$\left|(1+t)^{-\frac{1}{2}} - 1\right| \le N_1|t|.$$

Notice that the iterate x_{m+1} generated by (4.6.8) is independent of the signs of the $q_i^{(m)}$. Therefore, without loss of generality we may assume that

$$\left(q_i^{(m)}\right)^T q_i(x_m) \ge 0, \qquad m = 0, 1, 2, \dots, \quad i = 1, 2, \dots, n.$$

Let now

$$C_1 = \frac{2\sqrt{2N_1} + \gamma}{\gamma} N_0, \qquad C = 2nC_0C_1^2\|\lambda\|_\infty,$$

$$\delta = \min\left\{\delta_0, \frac{\delta_0}{C_1}, \frac{1}{C+1}, \frac{1}{N_0 + C_1}, \frac{\gamma}{4N_0}\right\}.$$

We will use the mathematical induction to prove that $\|x_0 - c\|_2 \le \delta$ implies

$$\|q_i^{(m)} - q_i(c)\|_2 \le C_1\|x_m - c\|_2, \quad i = 1, 2, \dots, n, \tag{4.6.15}$$

$$\|x_{m+1} - c\|_2 \le C\|x_m - c\|_2^2, \tag{4.6.16}$$

for all $m = 0, 1, 2, \dots$. It is clear that (4.6.16) implies Theorem 4.6.2.

Let

$$Q(c)^T Q_m = I + E_m.$$

Then we have

$$\begin{aligned} Q_m^T A(c) Q_m &= Q_m^T Q(c) \Lambda Q(c)^T Q_m \\ &= (I + E_m)^T \Lambda (I + E_m) \\ &= \Lambda + E_m^T \Lambda + \Lambda E_m + E_m^T \Lambda E_m, \end{aligned} \tag{4.6.17}$$

where $\Lambda = \operatorname{diag}(\lambda_1, \ldots, \lambda_n)$. The diagonal equations of (4.6.17) give

$$J_m c = \lambda - b_m + \widetilde{e}_m,$$

where $\widetilde{e}_m = (\varepsilon_1, \ldots, \varepsilon_n)^n$ is the vector of diagonal elements of $E_m^T \Lambda + \Lambda E_m + E_m^T \Lambda E_m$, i.e.,

$$\varepsilon_i = 2\lambda_i \varepsilon_{ii} + \sum_{j=1}^{n} \lambda_j \varepsilon_{ij}^2, \qquad i = 1, 2, \ldots, n.$$

Here ε_{ij} denotes the (i, j) element of E_m. The new iterate x_{m+1} is defined by

$$J_m x_{m+1} = \lambda - b_m,$$

and thus

$$J_m(c - x_{m+1}) = \widetilde{e}_m. \tag{4.6.18}$$

By the definition of E_m and the fact that $Q(c)$ is orthogonal, we have

$$I + E_m^T + E_m + E_m^T E_m = (I + E_m)^T (I + E_m) = Q_m^T Q_m.$$

Since $q_i^{(m)}$ are unit vectors, we see that the main diagonal elements of $Q_m^T Q_m$ are 1. Hence the main diagonal elements of $E_m^T + E_m + E_m^T E_m$ are zeros. Therefore, we get

$$2\varepsilon_{ii} = -\sum_{j=1}^{n} \varepsilon_{ij}^2, \qquad i = 1, 2, \ldots, n.$$

Thus, we have

$$\begin{aligned}
\|\widetilde{e}_m\|_2^2 &= \sum_{i=1}^{n} \varepsilon_i^2 \le 2 \sum_{i=1}^{n} \left(4\lambda_i^2 \varepsilon_{ii}^2 + \left(\sum_{j=1}^{n} \lambda_j \varepsilon_{ij}^2 \right)^2 \right) \\
&\le 4 \max_{1 \le i \le n} |\lambda_i|^2 \sum_{i=1}^{n} \left(\sum_{j=1}^{n} \varepsilon_{ij}^2 \right)^2 \\
&\le 4\|\lambda\|_\infty^2 \left(\sum_{i=1}^{n} \sum_{j=1}^{n} \varepsilon_{ij}^2 \right)^2 = 4\|\lambda\|_\infty^2 \|E_m\|_F^4 \\
&= 4\|\lambda\|_\infty^2 \|Q(c)^T Q_m - I\|_F^4 = 4\|\lambda\|_\infty^2 \|Q_m - Q(c)\|_F^4 \\
&= 4\|\lambda\|_\infty^2 \left(\sum_{j=1}^{n} \|q_j^{(m)} - q_j(c)\|_2^2 \right)^2 \\
&\le 4\|\lambda\|_\infty^2 n^2 \max_{1 \le j \le n} \|q_j^{(m)} - q_j(c)\|_2^4,
\end{aligned}$$

and so

$$\|\widetilde{e}_m\|_2 \leq 2n\|\lambda\|_\infty \max_{1\leq j\leq n} \|q_j^{(m)} - q_j(c)\|_2^2. \tag{4.6.19}$$

Noting that $\|x_0 - c\|_2 \leq \delta \leq \delta_0$ and $q_i^{(0)} = q_i(x_0)$, the eigenvectors of $A(x_0)$, it follows from (4.6.13) that

$$\|q_i^{(0)} - q_i(c)\|_2 \leq N_0\|x_0 - c\|_2 \leq C_1\|x_0 - c\|_2.$$

This, together with (4.6.11), (4.6.18), and (4.6.19), implies

$$\|x_1 - c\|_2 \leq 2n\|\lambda\|_\infty C_0 C_1^2 \|x_0 - c\|_2^2 = C\|x_0 - c\|_2^2.$$

Hence (4.6.15) and (4.6.16) are true for $m = 0$. Now we show that these inequalities hold for the integer $m+1$, assuming they are true for all nonnegative integers less than or equal to m.

We first prove (4.6.15) is true for $m+1$. Since $q_1(x_{m+1}), \ldots, q_n(x_{m+1})$ form an orthogonal basis for $\mathbf{R}^n$, we can write $q_i^{(m)}$ as

$$q_i^{(m)} = \sum_{j=1}^{n} \alpha_{ij} q_j(x_{m+1}), \quad i = 1, 2, \ldots, n.$$

Note that $\|x_{m+1} - c\|_2 \leq C\|x_m - c\|_2^2 \leq \|x_m - c\|_2 \leq \delta \leq \delta_0$, and so it follows from the induction hypothesis and (4.6.13) that

$$\begin{aligned}
\|q_i^{(m)} - q_i(x_{m+1})\|_2 &\leq \|q_i^{(m)} - q_i(c)\|_2 + \|q_i(c) - q_i(x_{m+1})\|_2 \\
&\leq C_1\|x_m - c\|_2 + N_0\|x_{m+1} - c\|_2 \\
&= (N_0 + C_1)\|x_m - c\|_2;
\end{aligned}$$

that is,

$$\sum_{\substack{j=1 \\ j\neq i}}^{n} \alpha_{ij}^2 + (1 - \alpha_{ii})^2 = 2(1 - \alpha_{ii}) \leq (N_0 + C_1)^2\|x_m - c\|_2^2,$$

which implies

$$\alpha_{ii} \geq 1 - \frac{1}{2}(N_0 + C_1)^2\|x_m - c\|_2^2 \geq \frac{1}{2}. \tag{4.6.20}$$

Here we use the fact that $\sum_{j=1}^{n} \alpha_{ij}^2 = \|q_i^{(m)}\|_2^2 = 1$. From (4.6.9) we have

$$v_i = \left(A(x_{m+1}) - \lambda_i I\right)^{-1} q_i^{(m)} = \sum_{j=1}^{n} \frac{\alpha_{ij}}{\lambda_j(x_{m+1}) - \lambda_i} q_j(x_{m+1}),$$

and hence

$$q_i^{(m+1)} = \frac{v_i}{\|v_i\|_2} = \sum_{j=1}^{n} \beta_{ij} q_j(x_{m+1}),$$

where

$$\beta_{ij} = \frac{\alpha_{ij}}{\lambda_j(x_{m+1}) - \lambda_i} \left| \frac{\lambda_i(x_{m+1}) - \lambda_i}{\alpha_{ii}} \right| \left(1 + \sum_{\substack{j=1 \\ j \neq i}}^{n} \frac{\alpha_{ij}^2 \left(\lambda_i(x_{m+1}) - \lambda_i \right)^2}{\alpha_{ii}^2 \left(\lambda_j(x_{m+1}) - \lambda_i \right)^2} \right)^{-\frac{1}{2}}.$$

This, together with (4.6.12), (4.6.14), and (4.6.20), implies

$$\begin{aligned} |1 - \beta_{ii}| &\leq N_1 \sum_{\substack{j=1 \\ j \neq i}}^{n} \frac{\alpha_{ij}^2 \left(\lambda_i(x_{m+1}) - \lambda_i \right)^2}{\alpha_{ii}^2 \left(\lambda_j(x_{m+1}) - \lambda_i \right)^2} \\ &\leq N_1 \frac{4N_0^2}{\gamma^2} \|x_{m+1} - c\|_2^2 \sum_{\substack{j=1 \\ j \neq i}}^{n} \alpha_{ij}^2 \\ &\leq \frac{4N_1 N_0^2}{\gamma^2} \|x_{m+1} - c\|_2^2. \end{aligned}$$

Here we use the fact that $\|x_{m+1} - c\|_2 \leq C\|x_m - c\|_2^2 \leq \|x_m - c\|_2 \leq \delta \leq \delta_0$ and $\left(q_i^{(m+1)}\right)^T q_i(x_{m+1}) \geq 0$ implies

$$\beta_{ii} = \left(1 + \sum_{\substack{j=1 \\ j \neq i}}^{n} \frac{\alpha_{ij}^2 \left(\lambda_i(x_{m+1}) - \lambda_i \right)^2}{\alpha_{ii}^2 \left(\lambda_j(x_{m+1}) - \lambda_i \right)^2} \right)^{-\frac{1}{2}}.$$

Consequently, we have

$$\begin{aligned} \|q_i^{(m+1)} - q_i(c)\|_2 &\leq \|q_i^{(m+1)} - q_i(x_{m+1})\|_2 + \|q_i(x_{m+1}) - q_i(c)\|_2 \\ &\leq (2 - 2\beta_{ii})^{\frac{1}{2}} + N_0 \|x_{m+1} - c\|_2 \\ &\leq \frac{2\sqrt{2N_1} N_0}{\gamma} \|x_{m+1} - c\|_2 + N_0 \|x_{m+1} - c\|_2 \\ &= C_1 \|x_{m+1} - c\|_2. \end{aligned}$$

This, together with (4.6.11), (4.6.18), and (4.6.19), implies

$$\|x_{m+2} - c\|_2 \leq 2nC_0 C_1^2 \|\lambda\|_\infty \, \|x_{m+1} - c\|_2^2 = C\|x_{m+1} - c\|_2^2.$$

Thus we have proved that (4.6.15) and (4.6.16) are also true for the integer $m+1$. By induction principle, (4.6.15) and (4.6.16) hold for all nonnegative integers, and consequently the theorem is established. □

Example 4.6.2 Applying Algorithm 4.6.2 to the problem as in Example 4.6.1 with the same starting vector $x_0 = (1.8, -0.7, -1.2)^T$, we find

k	$\|\|x_k - c\|\|_2$	$\|\|\lambda - \lambda^{(k)}\|\|_2$
1	6.946403×10^{-2}	2.642512×10^{-2}
2	2.138676×10^{-2}	7.811738×10^{-3}
3	4.577806×10^{-4}	1.867384×10^{-4}
4	1.252708×10^{-7}	5.833784×10^{-8}
5	1.692935×10^{-14}	1.456721×10^{-14}

As before, here $\lambda^{(k)}$ also denotes the vector of the ordered eigenvalues of $A(x_k)$ according to increasing size. Observe that the speed of convergence is slightly slower than Algorithm 4.6.1.

4.6.2 Methods for Solving Problem G

Following the same idea as in the derivation of Algorithm 4.6.2 we can develop a Newton-like method for solving Problem G. As before, assume that given eigenvalues are distinct and there exists a solution c to Problem G. Since in the present case the matrix $A(x)$ may have complex eigenvalues, we consider the following formulation of Problem G: solve the nonlinear system

$$g(x) = \begin{bmatrix} \sigma_{\min}(A(x) - \lambda_1 I) \\ \sigma_{\min}(A(x) - \lambda_2 I) \\ \cdots\cdots\cdots\cdots \\ \sigma_{\min}(A(x) - \lambda_n I) \end{bmatrix} = 0, \tag{4.6.21}$$

where $\sigma_{\min}(B)$ denotes the smallest singular value of a matrix B.

Now we apply Newton's Method to (4.6.23). By Theorem 1.9.3, we know that if $\sigma_{\min}(A(x) - \lambda_i I) \neq 0$, then $\sigma_{\min}(A(x) - \lambda_i I)$ is analytic in some neighborhood of x and

$$\frac{\partial \sigma_{\min}(A(x) - \lambda_i I)}{\partial \xi_j} = u_i(x)^T A_j v_i(x) \tag{4.6.22}$$

where $v_i(x)$, $u_i(x) \in \mathbf{R}^n$ are the right and left unit singular vectors associated with the smallest singular value $\sigma_{\min}(A(x) - \lambda_i I)$ of $A(x) - \lambda_i I$, respectively. Thus the Jacobian matrix of g is

$$J(x) = \left[u_i(x)^T A_j v_i(x)\right],$$

and one step of Newton's method is defined by

$$J(x_m)(x_{m+1} - x_m) = -g(x_m). \tag{4.6.23}$$

To apply this method, n singular values and $2n$ singular vectors have to be computed per step, which needs approximately $22n^4$ flops. One way to reduce the cost is to use inverse iteration with a zero shift to approximate these singular values and vectors. Let us first recall the inverse iteration for computing the smallest singular value of a matrix and the corresponding singular vectors. Let $u_0 \in \mathbf{R}^n$ be a given unit vector and assume $B \in \mathbf{R}^{n\times n}$ is nonsingular. The following is referred to as **inverse iteration**:

for $k = 1, 2, \ldots$
  Solve $Bw = u_{k-1}$ for w;
  $v_k = \dfrac{w}{\|w\|_2}$;
  Solve $B^T y = v_k$ for y;
  $\sigma_k = \dfrac{1}{\|y\|_2}$;
  $u_k = \sigma_k y$;
end

Generally speaking, only one iteration is typically required to produce adequate approximations to a smallest singular value and the corresponding singular vectors.

Thus, suppose that x_m is our current estimate of c, $\sigma_i^{(m)}$ is an approximation to $\sigma_{\min}(A(x_m) - \lambda_i I)$ and that $u_i^{(m)}$ and $v_i^{(m)}$ are approximations to $u_j(x_m)$ and $v_j(x_m)$, the left and right unit singular vectors of $A(x_m) - \lambda_i I$, $i = 1, 2, \ldots, n$, respectively. To compute an new estimate x_{m+1} we form

$$J_m = \left[(u_i^{(m)})^T A_j v_i^{(m)}\right],$$

and solve

$$J_m(x_{m+1} - x_m) = -(\sigma_1^{(m)}, \ldots, \sigma_n^{(m)})^T.$$

To update our approximations to singular values and singular vectors we apply one step of inverse iteration:

for $i = 1 : n$
  Solve $\left(A(x_{m+1}) - \lambda_i I\right) w = u_i^{(m)}$ for w;
  $v_i^{(m+1)} = \dfrac{w}{\|w\|_2}$;
  Solve $\left(A(x_{m+1}) - \lambda_i I\right)^T y = v_i^{(m+1)}$ for y;
  $\sigma_i^{(m+1)} = \dfrac{1}{\|y\|_2}$;
  $u_i^{(m+1)} = \sigma_i^{(m+1)} y$;
end

In addition, we can considerably reduce the cost of above iteration if we had reduced $A(x_{m+1})$ to Hessenberg form in advance.

Overall we obtain the following algorithm.

Algorithm 4.6.3 Given $n+1$ real $n \times n$ matrices A_k, $k = 0, 1, \ldots, n$, and given n real numbers $\lambda_1 \leq \lambda_2 \leq \cdots \leq \lambda_n$, the following algorithm is to compute a vector $c = (c_1, \ldots, c_n)^T \in \mathbf{R}^n$ such that $\lambda_1, \ldots, \lambda_n$ are the eigenvalues of $A_0 + \sum_{k=1}^{n} c_k A_k$.

Choose a starting vector x_0, a starting orthogonal matrix $U = [u_1, \ldots, u_n]$, an iteration maximum $k_{\max}$, and a precision requirement $\varepsilon < 1$;
$k = 0$; $c = x_0$; $r = 1$;
while $k < k_{\max}$ and $r > \varepsilon$
  $A = A_0$:
  for $j = 1 : n$
    $A = A + c(j)A_j$;
  end
  Compute the factorization

$$A = QHQ^T,$$

  where Q is orthogonal and H is upper Hessenberg;
  for $i = 1 : n$
    Solve $(H - \lambda_i I)w = Q^T u_i$ for w;
    $v_i = \dfrac{w}{\|w\|_2}$;
    Solve $(H - \lambda_i I)^T y = v_i$ for y;
    $g(i) = \dfrac{1}{\|y\|_2}$; $u_i = g(i)Qy$; $v_i = Qv_i$;
  end
  for $i = 1 : n$
    for $j = 1 : n$
      $J(i,j) = u_i^T A_j v_i$;
    end
  end
  Solve $Jx = -g$ for x;
  $r = \dfrac{\|x\|_\infty}{\|c\|_\infty}$; $c = c + x$; $k = k + 1$;
end

To the best of our knowledge, the convergence analysis of this algorithm has not been given yet. But numerical evidence indicates that the algorithm behaves like Algorithm 4.6.2.

Example 4.6.3 Let

$$A_0 = 0, \quad A_k = b_k e_k^T, \quad k = 1, 2, 3, \quad \lambda = (1, 2, 3)^T,$$

where

$$B = [b_1, b_2, b_3] = Q\text{diag}\,(1, 2, 3)Q^T,$$

$$Q = \begin{bmatrix} -0.2811022 & -0.7613087 & 0.5842864 \\ 0.6636202 & -0.5940103 & -0.4547087 \\ 0.6932458 & 0.2599247 & 0.6721975 \end{bmatrix},$$

which is an orthogonal matrix generated by MATLAB functions **randn** and **qr**. Clearly, $c = (1,1,1)^T$ is an exact solution to Problem G with these given data. Applying Algorithm 4.6.3 to this problem, starting with $x_0 = (0.8, 0.9, 1.1)^T$, we find

k	$\|x_k - c\|_2$
1	4.889354×10^{-2}
2	4.980550×10^{-3}
3	6.495111×10^{-5}
4	3.034955×10^{-8}
5	2.047150×10^{-15}

Observe that the convergent behavior is very similar to Algorithm 4.6.2.

4.7 Homotopy Method

As has been shown in the preceding section, the Newton-like methods are rapidly convergent even for the case when multiple eigenvalues are presented. However, these methods inherit the usual shortcomings of local convergence. In many cases, it often leads to the failure of the algorithms since it is hard to select a valid starting point. Therefore, how to select a valid starting point becomes a very important problem. One way to solve this problem is to apply homotopy method to the inverse eigenvalue problems. The basic idea of the homotopy method is to perform a continuous deformation of a difficult problem to a simple problem. In this section we will present a homotopy method to solve the additive inverse eigenvalue problem.

4.7.1 Homotopy Theory

Let $A = [\alpha_{ij}] \in \mathbf{R}^{n\times n}$ be symmetric with $\alpha_{ii} = 0$ for all $i = 1, 2, \ldots, n$, and let $\lambda = (\lambda_1, \ldots, \lambda_n)^T$, $\mu = (\mu_1, \ldots, \mu_n)^T \in \mathbf{R}^n$ with $\lambda_1 < \lambda_2 < \cdots < \lambda_n$, $\mu_1 < \mu_2 < \cdots < \mu_n$. For any $t \in [0,1]$ and any $x = (\xi_1, \ldots, \xi_n)^T \in \mathbf{R}^n$, we define

$$\begin{aligned} A(t;x) &= tA + \operatorname{diag}(\xi_1, \ldots, \xi_n), \\ \lambda(t) &= \left(\lambda_1(t), \ldots, \lambda_n(t)\right)^T = (1-t)\mu + t\lambda, \qquad (4.7.1) \\ h(t;x) &= \left(\lambda_1(t;x) - \lambda_1(t), \ldots, \lambda_n(t;x) - \lambda_n(t)\right)^T, \end{aligned}$$

where $\lambda_1(t;x) \le \lambda_2(t;x) \le \cdots \le \lambda_n(t;x)$ are the ordered eigenvalues of $A(t;x)$. Clearly, h is a continuous vector-valued function from $[0,1] \times \mathbf{R}^n$ to $\mathbf{R}^n$ and $h(0;x) = 0$ if and only if there exists a permutation π of $\{1, 2, \ldots, n\}$ such that $\xi_i = \mu_{\pi(i)}$ for $i = 1, 2, \ldots, n$. Moreover, we have

Theorem 4.7.1 *If the given data satisfies*

$$\min_{1\le i\le n-1} (\lambda_{i+1} - \lambda_i) > 2 \max_{1\le i\le n} g_i(A), \tag{4.7.2}$$

where $g_i(A)$ is defined by (4.4.4), then for each permutation π of $\{1, 2, \ldots, n\}$ there exists an analytic vector-valued function $c : [0,1] \longrightarrow \mathbf{R}^n$ such that

(1) $c(0) = (\mu_{\pi(1)}, \ldots, \mu_{\pi(n)})^T$;

(2) $h\big(t, c(t)\big) = 0$ *for any* $t \in [0,1]$;

(3) $\lambda_1(t) \le c_{\pi(1)}(t) < c_{\pi(2)}(t) < \cdots < c_{\pi(n)}(t) \le \lambda_n(t)$ *for any* $t \in [0,1]$.

Proof For the sake of simplicity, without loss of generality we may assume that $\pi(i) = i$ for $i = 1, 2, \ldots, n$. Notice that $A(0;\mu) = \operatorname{diag}(\mu_1, \ldots, \mu_n)$ has distinct eigenvalues. Hence, by Theorem 1.9.2, $h(t;x)$ is analytic in some neighborhood of $(0;\mu)$ and $h'_x(0;\mu) = I$. Thus, applying Theorem 1.8.2, we know that there exists a real number δ with $0 < \delta \le 1$ and a unique analytic vector-valued function $c : [0,\delta] \longrightarrow \mathbf{R}^n$ such that

(a) $c(0) = \mu$;

(b) $\lambda_1(t) \le c_1(t) < c_2(t) < \cdots < c_n(t) \le \lambda_n(t)$ for any $t \in [0,\delta]$;

(c) $h\big(t; c(t)\big) = 0$ but $h'_x\big(t; c(t)\big)$ is nonsingular for any $t \in [0,\delta]$.

Notice that (b) implies that for any $t \in [0,\delta]$,

$$||c(t)||_\infty \le \max\{|\lambda_1(t)|, |\lambda_n(t)|\} \le \max\{|\lambda_1|, |\mu_1|, |\lambda_n|, |\mu_n|\}.$$

Hence, it follows that there exists a sequence $\{t_m\} \subset [0,\delta]$ and a vector $c(\delta) \in \mathbf{R}^n$ such that

$$\lim_{m\to\infty} t_m = \delta \quad \text{and} \quad \lim_{m\to\infty} c(t_m) = c(\delta). \tag{4.7.3}$$

By the continuity of h, we have $h\big(\delta; c(\delta)\big) = 0$; that is, $\lambda_1(\delta) < \cdots < \lambda_n(\delta)$ are the ordered eigenvalues of $A\big(\delta; c(\delta)\big)$. Moreover, from (b) we have

$$\lambda_1(\delta) \le c_1(\delta) \le \cdots \le c_n(\delta) \le \lambda_n(\delta).$$

On the other hand, for any $t \in (0, 1]$ and any $x \in \mathbf{R}^n$ we have

$$\begin{aligned} \lambda_{i+1}(t) - \lambda_i(t) &= (1-t)(\mu_{i+1} - \mu_i) + t(\lambda_{i+1} - \lambda_i) \\ &\ge t(\lambda_{i+1} - \lambda_i) > 2t \max_{1\le i\le n} g_i(A) \ge 2\max\{tg_i(A), tg_{i+1}(A)\} \\ &= 2\max\{g_i\big(A(t;x)\big), g_{i+1}\big(A(t;x)\big)\}. \end{aligned} \tag{4.7.4}$$

Consequently, by Lemma 4.4.5, it follows that

$$|\lambda_i(\delta) - c_i(\delta)| \le g_i\big(A(\delta); c(\delta)\big) = \delta g_i(A)$$

for each $i = 1, 2, \ldots, n$. This, together with (4.7.4), implies.

$$c_i(\delta) \ne c_{i+1}(\delta), \quad i = 1, 2, \ldots, n-1,$$

and so

$$\lambda_1(\delta) \le c_1(\delta) < c_2(\delta) < \cdots < c_n(\delta) \le \lambda_n(\delta). \tag{4.7.5}$$

By Theorem 1.9.2, we know that $h(t; x)$ is analytic in some neighborhood of $\big(\delta, c(\delta)\big)$ and

$$h'_x\big(\delta; c(\delta)\big) = J\big(c(\delta)\big),$$

which is defined by (4.5.4). By Corollary 4.5.2, (4.7.4) implies that $h'_x\big(\delta; c(\delta)\big)$ is nonsingular. Hence, there exist positive numbers δ_0 and η such that for any $(t, x) \in [0, 1] \times \mathbf{R}^n$, $\sqrt{\|x - c(\delta)\|_2^2 + |t - \delta|^2} < \delta_0$ implies

$$\|h'_x(t; x)^{-1}\|_2 \|h'_t(t; x)\|_2 \le \eta. \tag{4.7.6}$$

On the other hand, for any given ε with $0 < \frac{3}{2}\varepsilon < \delta_0$, it follows from (4.7.3) that there exists a positive integer m_0 such that

$$0 < \delta - t_{m_0} < \frac{\varepsilon}{2(\eta + 1)} \tag{4.7.7}$$

and

$$\|c(t_{m_0}) - c(\delta)\|_2 < \frac{\varepsilon}{2}. \tag{4.7.8}$$

Now, we can claim that

$$\|c(t) - c(\delta)\|_2 < \varepsilon \tag{4.7.9}$$

for any $t \in (t_{m_0}, \delta)$. In fact, if (4.7.9) were false, then there would be some $\widetilde{t} \in (t_{m_0}, \delta)$ such that

$$\|c(\widetilde{t}) - c(\delta)\|_2 \ge \varepsilon. \tag{4.7.10}$$

Let

$$\alpha = \inf\{\widetilde{t} \in (t_{m_0}, \delta) \;:\; \widetilde{t} \text{ satisfies } (4.7.10)\}.$$

Then we have

$$t_{m_0} < \alpha < \delta \quad \text{and} \quad \|c(\alpha) - c(\delta)\|_2 = \varepsilon. \tag{4.7.11}$$

Differentiating $h(t; c(t)) = 0$ gives

$$h'_x(t; c(t))\dot{c}(t) + h'_t(t; c(t)) = 0$$

for any $t \in (0, \delta)$, and so

$$\dot{c}(t) = -h'_x(t; c(t))^{-1} h'_t(t; (t)), \quad t \in (0, \delta). \tag{4.7.12}$$

Applying the Mean-value Formula (see Theorem 1.8.1) to $c(t)$ on $[t_{m_0}, \alpha]$ and using (4.4.12), (4.7.7), and (4.7.6), we get

$$\|c(t_{m_0}) - c(\alpha)\|_2 < \frac{\varepsilon}{2}.$$

Hence, (4.7.8) and (4.7.11) provide the contradiction that

$$\|c(t_{m_0}) - c(\alpha)\|_2 \geq \frac{\varepsilon}{2}.$$

Thus we have proved that

$$\lim_{t \to \delta - 0} c(t) = c(\delta).$$

This, together with (4.7.5) and $h'_x(\delta; c(\delta))$ nonsingular, allows us to extend $c(t)$ to $[0, 1]$ so that the requirements of Theorem 4.7.1 are satisfied by using Theorem 1.8.2. □

Remark 4.7.1 Theorem 4.7.1 implies that if the given data satisfies (4.7.2), then Problem A has at least $n!$ solutions. Friedland [72] has proved that Problem A has at most $n!$ solutions. Consequently, under the assumption of Theorem 4.7.1, Problem A has exactly $n!$ solutions. Moreover, Theorem 4.7.1 says that if a solution c_π to Problem A satisfies $c_{\pi(1)} < c_{\pi(2)} < \cdots < c_{\pi(n)}$, then there is a unique smooth homotopy curve emanating from $\mu_\pi = (\mu_{\pi(1)}, \cdots, \mu_{\pi(n)})$ such that it terminates at c_π.

4.7.2 Homotopy Algorithm

If we want to find a solution $c_\pi = (c_1, \ldots, c_n)^T$ to Problem A with $c_{\pi(1)} \leq \cdots \leq c_{\pi(n)}$ for some permutation π of $\{1, 2, \ldots, n\}$, then Theorem 4.7.1 ensures that we can reach the solution c_π by following the homotopy curve $c(t)$ emanating from $\mu_\pi = (\mu_{\pi(1)}, \cdots, \mu_{\pi(n)})$. We here present a algorithm for following the homotopy curve $c(t)$. The main steps of the algorithm are (1) prediction, (2) correction, (3) checking, and (4) stepsize selection. We now give a detailed description of these steps.

(1) Prediction Assume that after i steps an approximation $\widetilde{c}(t_i)$ to $c(t_i)$ on the homotopy curve $c(t)$ at t_i is known and the next stepsize h is determined. We want to find an approximation $\widetilde{c}(t_{i+1})$ to $c(t_{i+1})$ on the homotopy curve $c(t)$ at $t_{i+1} = t_i + h$. Let

$$A(t;x) = Q(t;x)\mathrm{diag}\,\big(\lambda_1(t;x),\dots,\lambda_n(t;x)\big)Q(t;x)^T, \tag{4.7.13}$$

where

$$Q(t;x) = [q_1(t;x),\dots,q_n(t;x)] = [q_{ij}(t;x)] \in \mathbf{R}^{n\times n}$$

is orthogonal. The diagonal equations of (4.7.13) give

$$x = G(t;x)^T \lambda(t;x), \tag{4.7.14}$$

where

$$\lambda(t;x) = \big(\lambda_1(t;x),\lambda_2(t;x),\dots,\lambda_n(t;x)\big)^T.$$

$$G(t;x) = \begin{bmatrix} q_{11}(t;x)^2 & q_{21}(t;x)^2 & \cdots & q_{n1}(t;x)^2 \\ q_{12}(t;x)^2 & q_{22}(t;x)^2 & \cdots & q_{n2}(t;x)^2 \\ \vdots & \vdots & & \vdots \\ q_{1n}(t;x)^2 & q_{2n}(t;x)^2 & \cdots & q_{nn}(t;x)^2 \end{bmatrix}, \tag{4.7.15}$$

Since $h\big(t;c(t)\big) = 0$, we have

$$\lambda_i\big(t;c(t)\big) = \lambda_i(t), \quad i = 1,2,\dots,n,$$

and so letting $x = c(t)$ in (4.7.14), we get

$$c(t) = G\big(t;c(t)\big)^T \lambda(t).$$

Consequently, we can use

$$x_0 = G\big(t_i;\widetilde{c}(t_i)\big)^T \lambda(t_{i+1}) \tag{4.7.16}$$

to predict the $c(t_{i+1})$.

(2) Correction As a corrector, we use Newton method as described in Section 4.6, starting with x_0. To be more precise, we solve

$$h'_x(t_{i+1};x_k)(x_{k+1} - x_k) = -h(t_{i+1};x_k),$$

or equivalently,

$$G(t_{i+1};x_k)x_{k+1} = \lambda(t_{i+1}) - b(t_{i+1};x_k), \tag{4.7.17}$$

where $G(t;x)$ is defined by (4.7.15) and

$$b(t;x) = \left(q_1(t;x)^T tAq_1(t;x), \dots, q_n(t;x)^T tAq_n(t;x)\right).$$

We repeat the above process to within the precision requirement. The stop point x_{k+1} will be taken as an approximation $\widetilde{c}(t_{i+1})$ to $c(t_{i+1})$. The quadratic convergence rate of Newton method makes the corrector highly efficient.

(3) Checking To check to see if x_0 is a valid starting point, when x_2 is known, we check to see if

$$\|x_2 - x_1\|_2 \le \|x_1 - x_0\|_2;$$

moreover, to ensure that we follow the homotopy curve $c(t)$ from beginning to end, when $\widetilde{c}(t_{i+1})$ is known, we check to see if

$$\widetilde{c}_{\pi(j)}(t_{i+1}) \le \widetilde{c}_{\pi(j+1)}(t_{i+1}), \quad j = 1, 2, \dots, n-1.$$

If the checking or Newton iteration fails, we reduce the stepsize to $\frac{1}{2}h$ and repeat the whole process once again, beginning with the prediction in (1).

(4) Stepsize selection In the first attempt, we always choose stepsize $h = 1 - t_i$ at $t_i < 1$. After the prediction and correction steps, if the checking step fails, we reduce the stepsize to $h/2$ as mentioned in (3). Very small stepsize can also cause inefficiency of the algorithm. Therefore, we impose a minimum δ on stepsize h. If $h < \delta$, we give up following the homotopy curve, modify the initial vector μ or the permutation π, and then repeat the whole process again, beginning at $t = 0$.

Notes and References

In 1956, Downing and Householder [58] first proposed the additive and multiplicative inverse eigenvalue problems for Hermitian matrices. Afterwards some generalizations of these problems appeared, see [17], [15], and [217]. Some examples of applications can be found in [76].

For the solvability of the additive and multiplicative inverse eigenvalue problem for complex matrices, Friedland ([70]–[72]) obtained the following excellent results:

(1) For any $n \times n$ complex matrix A, there exists a complex diagonal matrix D such that the spectrum of $A + D$ is a given set $\{\lambda_1, \dots, \lambda_n\} \subset \mathbf{C}$; Moreover, the number of different matrices D is at most $n!$.

(2) If all of the principal minors of a given $n \times n$ complex matrix A are distinct from zero, then there exists a complex diagonal matrix D such that the eigenvalues of DA are n given complex numbers $\lambda_1, \dots, \lambda_n$; moreover, the number of different matrices D does not exceed $n!$.

For the solvability of Problem G for complex matrices, Xu [225] proved that for almost all of $n \times n$ complex matrices $A_0, A_1, \ldots, A_n$, there exists a complex vector $c = (c_1, \ldots, c_n)^T$ such that the eigenvalues of $A_0 + \sum_{k=1}^{n} c_k A_k$ are n given complex numbers $\lambda_1, \ldots, \lambda_n$; moreover, the number of different vectors c is at most $n!$.

Section 4.3 is mainly based on [226] with some significant changes; see also [50], [53], [72], and [163].

The degree theory we introduced are detailed in Chapter 6 of [168]. Subsection 4.4.1 is mainly based on [203] and [41], but here some significant improvement has been made. Corollary 4.4.8 originally appeared in [14], but there $\pi(i) = i$, $i = 1, 2, \ldots, n$, and the coefficient in (4.4.23) is not 2 but 4. Theorem 4.4.9 is due to Sun [203]; Corollary 4.4.10 and Theorem 4.4.11 are due to Hadeder [98], but there $\pi(i) = i$, $i = 1, 2, \ldots, n$, and all the inequalities of (4.4.26) and (4.4.27) are strict. Theorem 4.4.14 is taken from [14], Corollary 4.4.15 is due to Hadeler [95], and Theorems 4.4.19 and 4.4.20 originally appeared in [227]. Other related references include [50], [97], [129], [134], [135], [163], and [185].

As we have noted in Sections 4.4 and 4.5, the presently known sufficient conditions all require that the previously given eigenvalues should be distinct, but the necessary conditions do not imply such a requirement. Therefore, it is very natural to ask whether the inverse eigenvalues problems as described in Section 4.1 are solvable when previously given eigenvalues have multiple eigenvalues. Theoretically, it has been proved that in this case it is unsolvable almost everywhere; see [188], [199], [210], and [239]. However, this case frequently occurs in engineering applications. Consequently, it is very important how to modify Problem G such that it can be applied to more extensive practical problems. This problem has been studied in [76] and [229].

Section 4.5 is mainly based on [200] and [228] with some significant changes; see also [230].

Algorithms 4.6.1 and 4.6.2 are taken from [76]. The proof of Theorem 4.6.2 is mainly based on [30]. Although Theorem 3.3 of [76] also claimed that Algorithm 4.6.2 is quadratically convergent, the proof is incorrect. This is because the authors of [76] mistook the matrix Q_m generated by inverse iteration for an orthogonal matrix. Algorithm 4.6.3 is mainly based on [237]. Section 4.7 is mainly based on [231]; see also [34] and [232]. Those readers wishing a review of homotopy method are referred to [2] or [218]. Other papers concerned with numerical methods include [15], [17], [136], [137], [164], [167], [217], and [238].

There are also some papers related to some special cases of Problem G. Toeplitz matrix inverse eigenvalue problem is discussed in [75], [130], and [133]. The additive inverse eigenvalue problem for some special matrices is studied in [107], [122], and [211]. Inverse singular value problems are touched in [35] and [51].

Chapter 5

Nonnegative Matrix Inverse Eigenvalue Problems

5.1 Introduction

In this chapter we consider the following inverse eigenvalue problem for nonnegative matrices.

Problem *Determine necessary and sufficient conditions for a set of n complex numbers, $\{\lambda_1, \ldots, \lambda_n\} \subset \mathbf{C}$, to be the set of eigenvalues of some nonnegative $n \times n$ matrix.*

Theoretically, this problem is interesting and useful. However, the problem of obtaining sufficient conditions for a set of complex numbers to be the spectrum of a nonnegative matrix seems to be very difficult. To the best of our knowledge, up to now the only direction in which some progress has been made is the problem of the existence of a nonnegative matrix with a prescribed real spectrum. Consequently, this chapter mainly considers the problem of finding sufficient conditions for a set of real numbers to be the spectrum of a nonnegative matrix.

It is well-known that the Perron-Frobenius theory is the most important aspect of the spectral theory of nonnegative matrices. It will play an important role in this chapter. However, the detailed description of this theory is beyond the scope of this book, and therefore only the Perron-Frobenius theorem is stated without proof in Section 5.2.

In Section 5.3 we introduce the smallest realizable spectral radius and discuss its basic properties. In our opinion, perhaps this is one of the most important developments in the field of the inverse eigenvalue problems for nonnegative matrices in the recent years. As will be shown, finding necessary and sufficient conditions for a set of complex numbers to be the spectrum of a nonnegative matrix is equivalent to finding the corresponding smallest realizable spectral radius. Moreover, the continuity of the smallest realizable spectral radius allows us to approximate it.

Thus, maybe some numerical methods for solving the foregoing inverse eigenvalue problem can be developed .

Since a real symmetric matrix has many elegant properties, some more general sufficient conditions for a set of n real numbers to be the spectrum of a symmetric nonnegative matrix have been developed. Some interesting and useful results in this area will be described in Section 5.4.

5.2 Perron-Frobenius Theorem

In this section we state without proof a well-known result of Perron-Frobenius on the spectral properties of a nonnegative matrix, which is the base of this chapter. See [12], [118], or [151] for further details.

We begin with some basic definitions. Let $A = [\alpha_{ij}] \in \mathbf{R}^{n\times m}$. A is said to be **nonnegative** if all $\alpha_{ij} \geq 0$, and **positive** if all $\alpha_{ij} > 0$. Especially, if $x = (\xi_1, \dots, \xi_n)^T \in \mathbf{R}^n$ satisfies all $\xi_i \geq 0$ ($\xi_i > 0$), then x is said to be a **nonnegative (positive) vector**. A nonnegative $n \times n$ matrix A, $n \geq 2$, is called **reducible** if there exists a permutation matrix P such that

$$P^T AP = \begin{bmatrix} B & C \\ 0 & D \end{bmatrix}$$

where B and D are square submatrices. Otherwise, A is **irreducible.**

Now, the Perron-Frobenius Theorem can be stated as follows.

Theorem 5.2.1 (Perron-Frobenius Theorem) *Let $A \in \mathbf{R}^{n\times n}$ be irreducible and nonnegative. Then*

(1) $\rho(A) > 0$;

(2) *$\rho(A)$ is a simple eigenvalue of A;*

(3) *There is a positive vector x such that $Ax = \rho(A)x$;*

(4) *If an eigenvector of A is nonnegative, then the corresponding eigenvalue must be $\rho(A)$.*

Note that A^T is irreducible if and only if A is irreducible, so an irreducible nonnegative matrix also has a positive left eigenvector.

Applying the theorem, by the definition of reducibility, we immediately obtain the following result.

Corollary 5.2.2 *Let $A \in \mathbf{R}^{n\times n}$ be nonnegative. Then*

(1) *$\rho(A)$ is an eigenvalue of A;*

(2) *there exists a nonnegative vector x with $x \neq 0$ such that $Ax = \rho(A)x$.*

5.3 The Smallest Realizable Spectral Radius

We now consider the inverse eigenvalue problem of finding necessary and sufficient conditions for a set $\{\lambda_1, \ldots, \lambda_n\}$ to be the spectrum of a nonnegative matrix. The following theorem gives four necessary conditions.

Theorem 5.3.1 *If $\mathcal{L} = \{\lambda_1, \ldots, \lambda_n\}$ is the spectrum of an $n \times n$ nonnegative matrix A, then we have*

(1) *$\mathcal{L}$ is closed under complex conjugation, i,e., $\lambda \in \mathcal{L}$ implies $\overline{\lambda} \in \mathcal{L}$;*

(2) $\max\{|\lambda| : \lambda \in \mathcal{L}\} \in \mathcal{L}$;

(3) $s_k \equiv \sum_{i=1}^n \lambda_i^k \geq 0$ *for every positive integer k;*

(4) $(s_k)^m \leq n^{m-1} s_{km}$ *for any positive integers k and m.*

Proof (1) follows from the well-known property of eigenvalues of real matrices. (2) is implied by the Perron-Frobenius Theorem. (3) is true since the trace of any positive integer power of a nonnegative matrix must be nonnegative:

$$s_k = \operatorname{tr} A^k \geq 0, \quad k = 1, 2, \ldots .$$

(4) Let $B = [\beta_{ij}] = A^k$. Applying Hölder's inequality, we have

$$\begin{aligned} s_k = \sum_{i=1}^n \lambda_i^k = \sum_{i=1}^n \beta_{ii} &\leq n^{\frac{m-1}{m}} \left(\sum_{i=1}^n \beta_{ii}^m \right)^{\frac{1}{m}} \\ &\leq n^{\frac{m-1}{m}} (\operatorname{tr} B^m)^{\frac{1}{m}} = n^{\frac{m-1}{m}} \left(\sum_{i=1}^n \lambda_i^{km} \right)^{\frac{1}{m}}, \end{aligned}$$

and so

$$s_k^m \leq n^{m-1} s_{km},$$

which is the desired result. □

Note that for $k = 1$, condition (4) becomes

$$s_m \geq (s_1)^m / n^{m-1}, \quad m = 1, 2, \ldots .$$

Thus (4), together with the condition $s_1 \geq 0$, implies condition (3). However, the following example shows that the condition (4) is independent of conditions (1), (2), and (3).

Example 5.3.1 Let $\mathcal{L} = \{\sqrt{2}, i, -i\}$. It is easy to verify that the conditions (1), (2), and (3) are satisfied. However, if we take $k = 1$ and $m = 2$, then

$$s_1^2 = 2 > 3\, s_2 = 0,$$

and condition (4) does not hold.

For $n \leq 3$, conditions (1) and (2) are also sufficient. In general, however, these two conditions, even with conditions (3) and (4), are not sufficient.

Example 5.3.2 The set $\{\sqrt{2}, \sqrt{2}, i, -i\}$ satisfies all the conditions but it cannot be the spectrum of any 4×4 nonnegative matrix,. In fact, if it were the spectrum of a nonnegative matrix A, then by the Perron-Frobenius theorem, the matrix A would have to be reducible, and thus $\{\sqrt{2}, i, -i\}$ would be the spectrum of a 3×3 nonnegative matrix, which is clearly impossible since $\{\sqrt{2}, i, -i\}$ does not satisfy condition (4) as shown in Example 5.3.1.

For the sake of statement simplicity, we now introduce the following notation. Let

$$\mathcal{E}_{n-1} = \{\mathcal{L} = \{\lambda_2, \dots, \lambda_n\} \subset \mathbf{C} \ : \ \lambda \in \mathcal{L} \ \text{implies} \ \overline{\lambda} \in \mathcal{L}\}. \tag{5.3.1}$$

For $\mathcal{L} = \{\lambda_2, \dots, \lambda_n\} \in \mathcal{E}_{n-1}$ and $\lambda_1 \in \mathbf{R}$, if there exists a nonnegative matrix A such that $\rho(1) = \lambda_1$ and $\lambda(1) = \{\lambda_1; \mathcal{L}\}$, we say that the set $\{\lambda_1; \mathcal{L}\}$ is **realized** by A. Denote by $\mathcal{M}_n$ the collection of all sets $\{\lambda_1; \mathcal{L}\}$ which are realized by a nonnegative matrix. The following lemma shows that if $\{\lambda_1; \mathcal{L}\} \in \mathcal{M}_n$, then we can find a special nonnegative matrix which realizes the set $\{\lambda_1; \mathcal{L}\}$.

Lemma 5.3.2 *Let $\{\lambda_1; \mathcal{L}\} \in \mathcal{M}_n$. Then there exists an $n \times n$ nonnegative matrix B with $Be = \lambda_1 e$ such that $\lambda(B) = \{\lambda_1; \mathcal{L}\}$, where $e = (1, 1, \dots, 1)^T \in \mathbf{R}^n$.*

Proof Suppose that $A = [\alpha_{ij}] \in \mathbf{R}^{n \times n}$ is nonnegative and realizes $\{\lambda_1; \mathcal{L}\}$, and let $A(\varepsilon) = [\alpha_{ij} + \varepsilon]$ for any given $\varepsilon > 0$. Then $A(\varepsilon)$ is positive, and so, by the Perron-Frobenius theorem, there exists a positive vector

$$x(\varepsilon) = \big(\xi_1(\varepsilon), \dots, \xi_n(\varepsilon)\big)^T$$

such that

$$A(\varepsilon)x(\varepsilon) = \rho(A(\varepsilon))x(\varepsilon).$$

Define

$$B(\varepsilon) = X(\varepsilon)^{-1}A(x)X(\varepsilon),$$

where

$$X(\varepsilon) = \operatorname{diag}\big(\xi_1(\varepsilon), \dots, \xi_n(\varepsilon)\big).$$

Clearly, $B(\varepsilon)$ is still nonnegative and satisfies

$$\lambda\big(B(\varepsilon)\big) = \lambda\big(A(\varepsilon)\big) \quad \text{and} \quad B(\varepsilon)e = \rho(A(\varepsilon))e.$$

This implies

$$\|B(\varepsilon)\|_\infty = \rho(A(\varepsilon)) \longrightarrow \rho(A) = \lambda_1, \quad \varepsilon \longrightarrow 0,$$

and hence there exists a sequence $\{\varepsilon_m\}$ with $\lim_{m\to\infty} \varepsilon_m = 0$ such that $\{B(\varepsilon_m)\}$ is convergent. Let $B = \lim_{m\to\infty} B(\varepsilon_m)$. Then B is the desired nonnegative matrix, and so the lemma is established. □

In what follows we use e to denote the vector $(1, 1, \dots, 1)^T \in \mathbf{R}^n$.

Lemma 5.3.3 *Let $\mathcal{L} = \{\lambda_2, \dots, \lambda_n\} \in \mathcal{E}_{n-1}$. Then there exists a $\lambda_1 \in \mathbf{R}$ such that $\{\lambda_1; \mathcal{L}\} \in \mathcal{M}_n$.*

Proof Let $\delta = \max\{|\lambda| \ : \ \lambda \in \mathcal{L}\}$. Assume that $\lambda_{2k} = \overline{\lambda}_{2k+1} = \mu_k + i\nu_k$, $k = 1, 2, \dots, s$, are complex and $\lambda_{2s+2}, \dots, \lambda_n$ are real. Define

$$D = \operatorname{diag}(0, D_1, \dots, D_s, D_{s+1}) \in \mathbf{R}^{n\times n},$$

where

$$D_k = \begin{bmatrix} \mu_k & \nu_k \\ -\nu_k & \mu_k \end{bmatrix}, \quad k = 1, 2, \dots, s, \quad D_{s+1} = \operatorname{diag}(\lambda_{2s+2}, \dots, \lambda_n).$$

Then it is easy to verify that the matrix

$$2\delta ee^T + D - (De)e_1^T = P\big(2\delta(n-1)e_1e_1^T + 2\delta e_1e^T + D\big)P^{-1}$$

is nonnegative and realizes $\{2\delta n\,; \mathcal{L}\}$, where $P = [e, e_2, \dots, e_{n-1}]$. □

By Lemma 5.3.3, for any given $\mathcal{L} \in \mathcal{E}_{n-1}$ the scalar

$$R(\mathcal{L}) = \inf\{\lambda_1 \in \mathbf{R} \ : \ \{\lambda_1; \mathcal{L}\} \in \mathcal{M}_n\} \tag{5.3.2}$$

is well-defined and is referred as to the **smallest realizable spectral radius** of $\mathcal{L}$. Using this concept we have

Theorem 5.3.4 *For any $\mathcal{L} \in \mathcal{E}_{n-1}$, $\{\lambda_1\,; \mathcal{L}\} \in \mathcal{M}_n$ if and only if $\lambda_1 \geq R(\mathcal{L})$.*

Proof By the definition of $R(\mathcal{L})$, it is clear that $\{\lambda_1\,; \mathcal{L}\} \in \mathcal{M}_n$ implies $\lambda_1 \geq R(\mathcal{L})$. Now assume that $\lambda_1 \geq R(\mathcal{L})$. Then for any $\varepsilon > 0$ there exists a $\widetilde{\lambda}_1 \in \mathbf{R}$ with $\lambda_1 + \varepsilon > \widetilde{\lambda}_1 > R(\mathcal{L})$ such that $\{\lambda_1\,; \mathcal{L}\} \in \mathcal{M}_n$. By Lemma 5.3.2, we may assume that $B \in \mathbf{R}^{n\times n}$ with $Be = \widetilde{\lambda}_1 e$ is nonnegative and realizes $\{\widetilde{\lambda}_1\,; \mathcal{L}\}$. Let $P = [e, e_2, \dots, e_{n-1}]$. Then

$$P^{-1}BP = \begin{bmatrix} \widetilde{\lambda}_1 & * \\ 0 & B_1 \end{bmatrix},$$

which implies $\lambda(B_1) = \mathcal{L}$. Thus, the matrix

$$B(\varepsilon) = B + \frac{1}{n}(\lambda_1 + \varepsilon - \widetilde{\lambda}_1)ee^T = P \begin{bmatrix} \lambda_1 + \varepsilon & * \\ 0 & B_1 \end{bmatrix} P^{-1}$$

is positive and realizes $\{\lambda_1 + \varepsilon\,;\, \mathcal{L}\}$. Since $\{B(\varepsilon) : 0 < \varepsilon < \varepsilon_0\}$ is bounded for some $\varepsilon_0 > 0$, there must exist a sequence $\{\varepsilon_m\}$ with $\lim_{m\to\infty} \varepsilon_m = 0$ such that the matrix sequence $\{B(\varepsilon_m)\}$ is convergent. Let $\widetilde{B} = \lim_{m\to\infty} B(\varepsilon_m)$. Then $\widetilde{B}$ is nonnegative and realizes $\{\lambda_1\,;\, \mathcal{L}\}$. This completes the proof. □

In fact, in the above proof we have proved the following result.

Corollary 5.3.5 *For any $\mathcal{L} \in \mathcal{E}_{n-1}$, if $\lambda_1 > R(\mathcal{L})$, then there exists a positive matrix A with $Ae = \lambda_1 e$ such that $\{\lambda_1\,;\, \mathcal{L}\}$ is realized by A.*

The following theorem shows that $R(\mathcal{L})$ is continuous.

Theorem 5.3.6 *Let $\{\lambda_2^{(m)}, \ldots, \lambda_n^{(m)}\} \in \mathcal{E}_{n-1}$, $m = 0, 1, \ldots$. If $\lim_{m\to\infty} \lambda_k^{(m)} = \lambda_k^{(0)}$, $k = 2, \ldots, n$, then*

$$\lim_{m\to\infty} R(\{\lambda_2^{(m)}, \ldots, \lambda_n^{(m)}\}) = R(\{\lambda_2^{(0)}, \ldots, \lambda_n^{(0)}\}).$$

Proof We first prove that for any $\varepsilon > 0$ there exists a positive integer m_0 such that if $m > m_0$ then

$$R(\{\lambda_2^{(m)}, \ldots, \lambda_n^{(m)}\}) < R(\{\lambda_2^{(0)}, \ldots, \lambda_n^{(0)}\}) + \varepsilon. \tag{5.3.3}$$

Let $\lambda_1^{(0)} = R(\{\lambda_2^{(0)}, \ldots, \lambda_n^{(0)}\})$. For any given $\varepsilon > 0$, by Corollary 5.3.5, there exists a positive matrix $A(\varepsilon)$ such that $\{\lambda_1^{(0)} + \varepsilon; \lambda_2^{(0)}, \ldots, \lambda_n^{(0)}\}$ is realized by $A(\varepsilon)$. Let the Schur decomposition of $A(\varepsilon)$ be

$$A(\varepsilon) = Q(D + T)Q^T,$$

where $D = \operatorname{diag}(\lambda_1^{(0)} + \varepsilon, \lambda_2^{(0)}, \ldots, \lambda_n^{(0)})$, T is a strictly upper triangular matrix, and Q is an orthogonal matrix. Define

$$A_m(\varepsilon) = Q(D_m + T)Q^T,$$

where $D_m = \operatorname{diag}(\lambda_1^{(0)} + \varepsilon, \lambda_2^{(m)}, \ldots, \lambda_n^{(m)})$. Clearly,

$$\lambda(A_m(\varepsilon)) = \{\lambda_1^{(0)} + \varepsilon; \lambda_2^{(m)}, \ldots, \lambda_n^{(m)}\} \quad \text{and} \quad \rho(A_m(\varepsilon)) = \lambda_1^{(0)} + \varepsilon.$$

Moreover, we have

$$\|A_m(\varepsilon) - A(\varepsilon)\|_F = \left((\lambda_2^{(m)} - \lambda_2^{(0)})^2 + \cdots + (\lambda_n^{(m)} - \lambda_n^{(0)})^2\right)^{\frac{1}{2}} \to 0, \quad m \to \infty,$$

and hence there exists a positive integer m_0 such that if $m > m_0$, then $A_m(\varepsilon)$ is positive. This implies that if $m > m_0$, then

$$R\big(\{\lambda_2^{(m)}, \ldots, \lambda_n^{(m)}\}\big) \le \lambda_1^{(0)} + \varepsilon,$$

which is the desired result.

Next, we prove that for any $\varepsilon > 0$ there exists a positive integer $\widetilde{m}_0$ such that if $m > \widetilde{m}_0$, then

$$R\big(\{\lambda_2^{(m)}, \ldots, \lambda_n^{(m)}\}\big) > R\big(\{\lambda_2^{(0)}, \ldots, \lambda_n^{(0)}\}\big) - \varepsilon. \tag{5.3.4}$$

Let $\lambda_1^{(m)} = R\big(\{\lambda_2^{(m)}, \ldots, \lambda_n^{(m)}\}\big)$, $m = 0, 1, \ldots$. If (5.3.4) were not true, then there would exist a positive number ε_0 and a subsequence $\{m_i\}$ of $\{m\}$ such that

$$\lambda_1^{(m_i)} < \lambda_1^{(0)} - \varepsilon_0, \quad i = 1, 2, \ldots. \tag{5.3.5}$$

Let $A_{m_i} \in \mathbf{R}^{n\times n}$ with $A_{m_i} e = \lambda_1^{(m_i)} e$ be nonnegative and realize

$$\{\lambda_1^{(m_i)}; \lambda_2^{(m_i)}, \ldots, \lambda_n^{(m_i)}\}.$$

Moreover, without loss of generality we may assume that $\{A_{m_i}\}$ is convergent. Let $A = \lim_{i\to\infty} A_{m_i}$. Then A is nonnegative and realizes $\{\rho(A); \lambda_2^{(0)}, \ldots, \lambda_n^{(0)}\}$, and so

$$\lambda_1^0 \le \rho(A) = \lim_{i\to\infty} \rho(A_{m_i}) = \lim_{i\to\infty} \lambda_1^{(m_i)} \le \lambda_1^{(0)} - \varepsilon_0 < \lambda_1^{(0)}.$$

This contradiction demonstrates that (5.3.4) is true.

Combining (5.3.3) and (5.3.4) gives

$$\lim_{m\to\infty} R\big(\{\lambda_2^{(m)}, \ldots, \lambda_n^{(m)}\}\big) = R\big(\{\lambda_2^{(0)}, \ldots, \lambda_n^{(0)}\}\big).$$

This establishes the theorem. □

The following theorem is useful.

Theorem 5.3.7 *Let $\{\lambda_2, \ldots, \lambda_n\} \in \mathcal{E}_{n-1}$ with $\lambda_2 \in \mathbf{R}$. Then for any $\delta \in \mathbf{R}$ we have*

$$R\big(\{\lambda_2 + \delta, \ldots, \lambda_n\}\big) \le R\big(\{\lambda_2, \ldots, \lambda_n\}\big) + |\delta|. \tag{5.3.6}$$

Proof Let $\lambda_1 = R(\{\lambda_2, \dots, \lambda_n\})$. By Corollary 5.3.5, for any given $\varepsilon > 0$ there exists a positive matrix $A(\varepsilon) \in \mathbf{R}^{n\times n}$ with $A(\varepsilon)e = (\lambda_1+\varepsilon)e$ such that $A(\varepsilon)$ realizes $\{\lambda_1+\varepsilon; \lambda_2, \dots, \lambda_n\}$. Let $x = (\xi_1, \dots, \xi_n)^T \in \mathbf{R}^n$ be an eigenvector of $A(\varepsilon)$ corresponding to λ_2. Then, by the Perron-Frobenius theorem, we know that

$$\max_{1\le i\le n}\{\xi_i\} > 0 \quad \text{and} \quad \min_{1\le i\le n}\{\xi_i\} < 0.$$

For simplicity, without loss of generality we may assume that $\delta < 0$ and

$$\xi_1 = \max_{1\le i\le n}\{\xi_i\} \quad \text{and} \quad \xi_2 = \min_{1\le i\le n}\{\xi_i\}.$$

Let $P = [e, x, \widetilde{P}] \in \mathbf{R}^{n\times n}$ be nonsingular. Then $P^{-1}A(\varepsilon)P$ has the form

$$P^{-1}A(\varepsilon)P = \left[\begin{array}{cc|c} \lambda_1+\varepsilon & 0 & \multirow{2}{*}{$*$} \\ 0 & \lambda_2 & \\ \hline \multicolumn{2}{c|}{0} & B(\varepsilon) \end{array}\right]$$

where $\lambda\big(B(\varepsilon)\big) = \{\lambda_3, \dots, \lambda_n\}$. Now define

$$y_1^T = \left(\frac{\xi_1}{\xi_1-\xi_2}, -\frac{\xi_2}{\xi_1-\xi_2}, 0, \cdots, 0\right)P,$$
$$y_2^T = \left(\frac{1}{\xi_1-\xi_2}, -\frac{1}{\xi_1-\xi_2}, 0, \cdots, 0\right)P.$$

Clearly, $y_1^T = (1, \eta_2, \dots, \eta_n)$, $y_2^T = (0, 1, \zeta_3, \dots, \zeta_n)$. Thus, the following matrix

$$A = P\left(\left[\begin{array}{cc|c} \lambda_1+\varepsilon & 0 & \multirow{2}{*}{$*$} \\ 0 & \lambda_2 & \\ \hline \multicolumn{2}{c|}{0} & B(\varepsilon) \end{array}\right] + |\delta| e_1 y_1^T + \delta e_2 y_2^T\right)P^{-1}$$

has $\lambda_1 + |\delta| + \varepsilon$, $\lambda_2 + \delta$, $\lambda_3, \dots, \lambda_n$ as its eigenvalues. Moreover, it is clear that

$$A = A(\varepsilon) + \frac{|\delta|}{\xi_1 - \xi_2}[\xi_1 e - x, x - \xi_2 e, 0, \dots, 0]$$

is nonnegative. Consequently,

$$R(\{\lambda_2+\delta, \lambda_3, \dots, \lambda_n\}) \le \lambda_1 + |\delta| + \varepsilon$$

for any $\varepsilon > 0$, and so (5.3.6) follows immediately. □

By repeated application of Theorem 5.3.7, we obtain

Corollary 5.3.8 *If* $\{\lambda_2, \ldots, \lambda_n\} \subset \mathbf{R}$, *then*

$$R(\{\lambda_2, \ldots, \lambda_n\}) \leq |\lambda_2| + \cdots + |\lambda_n|.$$

As a simple consequence, we have the following result, which is the first, and perhaps the most important, sufficient condition.

Corollary 5.3.9 *If* $\{\lambda_2, \ldots, \lambda_n\} \subset \mathbf{R}$ *with* $\lambda_i \leq 0$, *then* $\{\lambda_1; \lambda_2, \ldots, \lambda_n\} \in \mathcal{M}_n$ *if and only if*

$$\lambda_1 + \lambda_2 + \cdots + \lambda_n \geq 0. \tag{5.3.7}$$

Proof If $\{\lambda_1; \lambda_2, \ldots, \lambda_n\} \in \mathcal{M}_n$, by Theorem 5.3.1, it must have that $\lambda_1 + \lambda_2 + \cdots + \lambda_n \geq 0$. Conversely, by Corollary 5.3.8, condition (5.3.7) implies that

$$\lambda_1 \geq -(\lambda_2 + \cdots + \lambda_n) \geq R(\{\lambda_2, \ldots, \lambda_n\}).$$

and therefore, by Theorem 5.3.4, $\{\lambda_1; \lambda_2, \ldots, \lambda_n\} \in \mathcal{M}_n$. □

We conclude this section with the sufficient and necessary conditions for a set $\{\lambda_1; \lambda_2, \lambda_3\}$ to be the spectrum of a 3×3 nonnegative matrix.

Theorem 5.3.10 *Let* $\{\lambda_2, \lambda_3\} \in \mathcal{E}_2$. *Then*

$$R(\{\lambda_2, \lambda_3\}) = \begin{cases} -(\lambda_2 + \lambda_3), & \text{if } \lambda_2, \lambda_3 \in \mathbf{R} \text{ with } \lambda_2, \lambda_3 < 0, \\ \max\{|\lambda_2|, |\lambda_3|\}, & \text{if } \lambda_2, \lambda_3 \in \mathbf{R} \text{ with } \lambda_2 \geq 0 \text{ or } \lambda_3 \geq 0, \\ \max\{-2\mu, \mu + \sqrt{3}\nu\}, & \text{if } \lambda_2 = \overline{\lambda_3} = \mu + i\nu \text{ with } \nu > 0. \end{cases}$$

That is, when $\lambda_2, \lambda_3 \in \mathbf{R}$ *with* $\lambda_2, \lambda_3 < 0$, $\{\lambda_1; \lambda_2, \lambda_3\} \in \mathcal{M}_3$ *if and only if* $\lambda_1 + \lambda_2 + \lambda_3 \geq 0$; *when* $\lambda_3, \lambda_2 \in \mathbf{R}$ *with* $\lambda_2 \geq 0$ *or* $\lambda_3 \geq 0$, $\{\lambda_1; \lambda_2, \lambda_3\} \in \mathcal{M}_3$ *if and only if* $\lambda_1 \geq \max\{|\lambda_2|, |\lambda_3|\}$; *when* $\lambda_2 = \overline{\lambda_3} = \mu + i\nu$ *with* $\nu > 0$, $\{\lambda_1; \lambda_2, \lambda_3\} \in \mathcal{M}_3$ *if and only if* $\lambda_1 \geq \max\{-2\mu, \mu + \sqrt{3}\nu\}$.

Proof The theorem is clear if $\lambda_2, \lambda_3 \in \mathbf{R}$. To prove the theorem for the case when $\lambda_2 = \overline{\lambda_3} = \mu + i\nu$ with $\nu > 0$, let

$$Q = \begin{bmatrix} \frac{1}{\sqrt{3}} & \frac{1}{\sqrt{2}} & \frac{1}{\sqrt{6}} \\ \frac{1}{\sqrt{3}} & -\frac{1}{\sqrt{2}} & \frac{1}{\sqrt{6}} \\ \frac{1}{\sqrt{3}} & 0 & -\frac{2}{\sqrt{6}} \end{bmatrix},$$

which is a 3×3 orthogonal matrix. A simple calculation gives

$$A = Q \begin{bmatrix} \lambda_1 & 0 & 0 \\ 0 & \mu & -\nu \\ 0 & \nu & \mu \end{bmatrix} Q^T$$

$$= \frac{1}{3} \begin{bmatrix} \lambda_1 + 2\mu & \lambda_1 - \mu - \sqrt{3}\nu & \lambda_1 - \mu + \sqrt{3}\nu \\ \lambda_1 - \mu + \sqrt{3}\nu & \lambda_1 + 2\mu & \lambda_1 - \mu - \sqrt{3}\nu \\ \lambda_1 - \mu - \sqrt{3}\nu & \lambda_1 - \mu + \sqrt{3}\nu & \lambda_1 + 2\mu \end{bmatrix},$$

and hence, if $\lambda_1 \geq \max\{-2\mu, \mu + \sqrt{3}\nu\}$, then A is nonnegative and realizes $\{\lambda_1; \lambda_2, \lambda_3\}$. This shows that

$$R(\{\lambda_2, \lambda_3\}) \leq \max\{-2\mu, \mu + \sqrt{3}\nu\}. \tag{5.3.8}$$

On the other hand, let $\lambda_1 = R(\{\lambda_2, \lambda_3\})$, by Theorems 5.3.4 and 5.3.1, it must have

$$\lambda_1 + \lambda_2 + \lambda_3 \geq 0 \quad \text{and} \quad (\lambda_1 + \lambda_2 + \lambda_3)^2 \leq 3(\lambda_1^2 + \lambda_2^2 + \lambda_3^2),$$

which implies $\lambda_1 + 2\mu \geq 0$ and $\lambda_1 - \mu - \sqrt{3}\nu \geq 0$, that is

$$\lambda_1 = R(\{\lambda_2, \lambda_3\}) \geq \max\{-2\mu, \mu + \sqrt{3}\nu\}.$$

This, together with (5.3.8), gives

$$R(\{\lambda_2, \lambda_3\}) = \max\{-2\mu, \mu + \sqrt{3}\nu\}.$$

This completes the proof of the theorem. □

5.4 Symmetric Problem

In this section we consider the existence of a symmetric nonnegative matrix with prescribed eigenvalues. For simplicity, we first introduce the following notation.

Let $\{\lambda_1; \lambda_2, \ldots, \lambda_n\} \subset \mathbf{R}$. If a symmetric nonnegative matrix $A \in \mathbf{R}^{n \times n}$ satisfies that $\lambda(A) = \{\lambda_1, \lambda_2, \ldots, \lambda_n\}$ and $\rho(A) = \lambda_1$, we say that $\{\lambda_1; \lambda_2, \ldots, \lambda_n\}$ is **realized** by A. Let $\mathcal{S}_n$ denote the collection of all sets $\{\lambda_1; \lambda_2, \ldots, \lambda_n\}$ which are realized by a symmetric nonnegative matrix. For any given $\mathcal{L} = \{\lambda_2, \lambda_3, \ldots, \lambda_n\} \subset \mathbf{R}$, define

$$S(\mathcal{L}) = \inf\{\lambda_1, : \{\lambda_1; \mathcal{L}\} \in \mathcal{S}_n\}. \tag{5.4.1}$$

It is easy to verify that $S(\mathcal{L})$ is well-defined. In fact, let $Q = [q_1, q_2, \dots, q_n] \in \mathbf{R}^{n\times n}$ be orthogonal, with q_1 positive. Then for any given $\mathcal{L} = \{\lambda_2, \lambda_3, \dots, \lambda_n\} \subset \mathbf{R}$ there must exist a positive number λ_1 with $\lambda_1 \geq |\lambda_i|$, $i = 2, 3, \dots, n$, such that the matrix

$$A = \lambda_1 q_1 q_1^T + \lambda_2 q_2 q_2^T + \cdots + \lambda_n q_n q_n^T$$

is positive since $q_1 q_1^T$ is positive. That is, for any $\mathcal{L} = \{\lambda_2, \lambda_3, \dots, \lambda_n\} \subset \mathbf{R}$ there exists a positive number λ_1 such that $\{\lambda_1; \mathcal{L}\} \in \mathcal{S}_n$, which implies that $S(\mathcal{L})$ is well-defined.

Theorem 5.4.1 *Let $\mathcal{L} = \{\lambda_2, \lambda_3, \dots, \lambda_n\} \subset \mathbf{R}$. Then $\{\lambda_1; \mathcal{L}\} \in \mathcal{S}_n$ if and only if $\lambda_1 \geq S(\mathcal{L})$.*

Proof If $\{\lambda; \mathcal{L}\} \in \mathcal{S}_n$, it is clear that $\lambda_1 \geq S(\mathcal{L})$, by (5.4.1).

Conversely, assume that $\lambda_1 \geq S(\mathcal{L})$. Then for any $\varepsilon > 0$, there exists a positive number $\widetilde{\lambda}_1$ such that $\lambda_1 + \varepsilon > \widetilde{\lambda}_1 > S(\mathcal{L})$ and $\{\widetilde{\lambda}_1; \mathcal{L}\} \in \mathcal{S}_n$. Let $A \in \mathbf{R}^{n\times n}$ be a symmetric nonnegative matrix and realize $\{\widetilde{\lambda}_1; \mathcal{L}\} \in \mathcal{S}_n$, and let q, $\|q\|_2 = 1$, be a nonnegative eigenvector corresponding to $\widetilde{\lambda}_1$. Then it is easy to see that the matrix

$$A(\varepsilon) = A + (\lambda_1 + \varepsilon - \widetilde{\lambda}_1) q q^T$$

is a symmetric nonnegative matrix and realizes $\{\lambda_1 + \varepsilon; \mathcal{L}\}$. Noting that

$$\|A(\varepsilon)\|_F^2 = (\lambda_1 + \varepsilon)^2 + \lambda_2^2 + \cdots + \lambda_n^2,$$

it follows that there exists a sequence $\{\varepsilon_m\}$ with all $\varepsilon_m > 0$ and $\lim_{m\to\infty} \varepsilon_m = 0$ such that $\{A(\varepsilon_m)\}$ is convergent. Let $B = \lim_{m\to\infty} A(\varepsilon_m)$. Then B is a symmetric nonnegative matrix and realizes $\{\lambda_1; \mathcal{L}\}$, and so $\{\lambda_1; \mathcal{L}\} \in \mathcal{S}_n$. □

In order to give some sufficient conditions for a set of real numbers to be the spectrum of a symmetric nonnegative matrix, we need the following basic result.

Theorem 5.4.2 *Let $\{\lambda_1; \lambda_2, \dots, \lambda_m\}$, $\{\mu_1; \mu_2, \dots, \mu_n\} \subset \mathbf{R}$. Assume that $\lambda_1 \geq S(\{\lambda_2, \dots, \lambda_n\})$, $\mu_1 \geq S(\{\mu_2, \dots, \mu_n\})$, and $\lambda_1 \geq \mu_1$. Then for any $\delta \geq 0$ we have*

$$S(\{\mu_1 - \delta, \mu_2, \dots, \mu_n, \lambda_2, \dots, \lambda_m\}) \leq \lambda_1 + \delta. \tag{5.4.2}$$

Moreover, if both $\{\lambda_1; \lambda_2, \dots, \lambda_n\}$ and $\{\mu_1; \mu_2, \dots, \mu_n\}$ can be realize by some positive symmetric matrices, then for any $\delta > 0$, $\{\mu_1 - \delta, \mu_2, \dots, \mu_n; \lambda_2, \dots, \lambda_m\}$ is also able to be realized by a positive symmetric matrix.

Proof Let $A \in \mathbf{R}^{n\times n}$ be a symmetric nonnegative matrix which realizes $\{\mu_1; \mu_2, \ldots, \mu_n\}$, and let the spectral decomposition of A be

$$U^T AU = \operatorname{diag}(\mu_1, \ldots, \mu_n),$$

where $U = [u_1, U_2] \in \mathbf{R}^{n\times n}$ is orthogonal, with $u_1 \in \mathbf{R}^n$ nonnegative, which is a nonnegative eigenvector of A corresponding to $\mu_1 = \rho(A)$. Similarly, let $B \in \mathbf{R}^{n\times n}$ be a nonnegative symmetric matrix which realizes $\{\lambda_1; \lambda_2, \ldots, \lambda_n\}$, and let the spectral decomposition of B be

$$V^T BV = \operatorname{diag}(\lambda_1, \lambda_2, \ldots, \lambda_n),$$

where $V = [v_1, V_2] \in \mathbf{R}^{m\times m}$ is orthogonal, with $v_1 \in \mathbf{R}^m$ nonnegative, which satisfies $Bv_1 = \lambda_1 v_1$. Let $\eta = \big(\delta(\lambda_1 - \mu_1 + \delta)\big)^{\frac{1}{2}}$. Then it is easy to verify that the matrix

$$C = \begin{bmatrix} \mu_1 & \eta \\ \eta & \lambda_1 \end{bmatrix}$$

has $\lambda_1 + \delta$ and $\mu_1 - \delta$ as its eigenvalues. Let the spectral decomposition of C be

$$C = \begin{bmatrix} c & s \\ -s & c \end{bmatrix} \begin{bmatrix} \lambda_1 + \delta & 0 \\ 0 & \mu_1 - \delta \end{bmatrix} \begin{bmatrix} c & -s \\ s & c \end{bmatrix},$$

where $c^2 + s^2 = 1$. Now define

$$\widetilde{A} = \begin{bmatrix} A & \eta u_1 v_1^T \\ \eta v_1 u_1^T & B \end{bmatrix}, \qquad Q = \begin{bmatrix} cu_1 & su_1 & U_2 & 0 \\ -sv_1 & cv_1 & 0 & V_2 \end{bmatrix}.$$

Clearly, $\widetilde{A}$ is nonnegative. Moreover, it is straightforward to verify that Q is orthogonal and

$$Q^T \widetilde{A} Q = \operatorname{diag}(\lambda_1 + \delta, \mu_1 - \delta, \mu_2, \ldots, \mu_n, \lambda_2, \ldots, \lambda_m).$$

This implies that

$$S(\mu_1 - \delta, \mu_2, \ldots, \mu_n, \lambda_2, \ldots, \lambda_m) \le \lambda_1 + \delta.$$

In addition, it is clear that if both A and B are positive, then $\widetilde{A}$ is also positive, since in this case u_1 and v_1 are positive. □

Applying this theorem we can prove.

Theorem 5.4.3 *Let $\mathcal{L} = \{\lambda_2, \ldots, \lambda_n\} \subset \mathbf{R}$. If $\lambda_1 > S(\mathcal{L})$, then $\{\lambda_1; \mathcal{L}\}$ can be realized by a positive symmetric matrix.*

Proof The proof is by induction on n. For $n = 2$ the assumption that $\lambda_1 > S(\mathcal{L})$ becomes $\lambda_1 > |\lambda_2|$. Thus the matrix

$$\frac{1}{2}\begin{bmatrix} \lambda_1+\lambda_2 & \lambda_1-\lambda_2 \\ \lambda_1-\lambda_2 & \lambda_1+\lambda_2 \end{bmatrix}$$

is positive and realizes $\{\lambda_1; \lambda_2\}$.

To prove the theorem for $n > 2$, assume the result is true for all smaller sets. Let $\widetilde{\lambda}_1 = R(\mathcal{L})$, and let $A \in \mathbf{R}^{n\times n}$ be a symmetric nonnegative matrix which realizes $\{\widetilde{\lambda}_1; \mathcal{L}\}$. If A is irreducible, let $u \in \mathbf{R}^n$ with $\|u\|_2 = 1$ be a positive eigenvector of A, corresponding to $\rho(A) = \widetilde{\lambda}_1$. Then $\widetilde{A} = A + (\lambda_1 - \widetilde{\lambda}_1)uu^T$ is positive and realizes $\{\lambda_1; \mathcal{L}\}$.

If A is reducible, then we can assume that $\{\widetilde{\lambda}_1; \lambda_2, \ldots, \lambda_n\}$ are ordered in such a way that

$$\{\widetilde{\lambda}_1; \lambda_2, \ldots, \lambda_m\} \in \mathcal{S}_m \quad \text{and} \quad \{\lambda_{m+1}; \lambda_{m+2}, \ldots, \lambda_n\} \in \mathcal{S}_{n-m},$$

where $1 \le m < n$. Let $\varepsilon = \frac{1}{2}(\lambda_1 - \widetilde{\lambda}_1)$. Clearly, $\widetilde{\lambda}_1 + \varepsilon \ge \lambda_{m+1} + \varepsilon$. By the induction hypothesis, both $\{\widetilde{\lambda}_1+\varepsilon; \lambda_2, \ldots, \lambda_m\}$ and $\{\lambda_{m+1}+\varepsilon; \lambda_{m+2}, \ldots, \lambda_n\}$ can be realized by some positive symmetric matrices, and hence, applying Theorem 5.4.2 with $\delta = \varepsilon$, we know that $\{\lambda_1; \lambda_2, \ldots, \lambda_m, \lambda_{m+1}, \ldots, \lambda_n\}$ can be realized by a symmetric positive matrix. □

We now develop several sufficient conditions for a set of real numbers to be the set of eigenvalues of a nonnegative matrix. We first prove that under the hypothesis of Corollary 5.3.9 the set can be realized by a symmetric nonnegative matrix.

Theorem 5.4.4 *Let* $\{\lambda_1; \lambda_2, \ldots, \lambda_n\} \subset \mathbf{R}$ *satisfy*

$$\lambda_1 > 0 > \lambda_2 > \cdots > \lambda_n \tag{5.4.3}$$

and

$$\lambda_1 + \lambda_2 + \cdots + \lambda_n > 0. \tag{5.4.4}$$

Then $\{\lambda_1; \lambda_2, \ldots, \lambda_n\} \in \mathcal{S}_n$.

Proof To prove the theorem, by Theorem 5.4.1, it is sufficient to prove that

$$S(\{\lambda_2, \ldots, \lambda_n\}) \le -(\lambda_2 + \cdots + \lambda_n). \tag{5.4.5}$$

We now prove (5.4.5) by induction on n. For $n = 2$ it is clear that $S(\{\lambda_2\}) = -\lambda_2$. Now assume that the result is true for the integer $n-1$. Then, by the induction hypothesis,

$$S(\{\lambda_2, \dots, \lambda_{n-1}\}) \leq -(\lambda_2 + \cdots + \lambda_{n-1}).$$

Clearly,

$$S(\{0\}) = 0 \leq -(\lambda_2 + \cdots + \lambda_{n-1}).$$

Applying theorem 5.4.2 with $\delta = -\lambda_n$, we obtain that

$$S(\{\lambda_2, \dots, \lambda_{n-1}, \lambda_n\}) \leq -(\lambda_2 + \cdots + \lambda_{n-1} + \lambda_n),$$

which is the desired result. □

Remark 5.4.1 This theorem, together with Corollary 5.3.9, shows that if

$$\{\lambda_2, \dots, \lambda_n\} \subset \mathbf{R}, \quad \text{with } \lambda_i \leq 0, \quad i = 2, \dots, n,$$

then

$$S(\{\lambda_2, \dots, \lambda_n\}) = R(\{\lambda_2, \dots, \lambda_n\}).$$

Thus, it is natural to ask whether for any $\mathcal{L} = \{\lambda_2, \dots, \lambda_n\} \subset \mathbf{R}$, $S(\mathcal{L}) = R(\mathcal{L})$; that is, whether $\{\lambda_1; \mathcal{L}\}$ is the spectrum of a symmetric nonnegative matrix if it is the spectrum of some nonnegative matrix. This question is open and seemingly very difficult.

To give more general sufficient condition, we need the following lemma, which extends Theorem 5.4.4.

Lemma 5.4.5 *Let* $\{\lambda_2, \dots, \lambda_n\} \subset \mathbf{R}$ *satisfy that*

$$\lambda_2 \geq \lambda_3 \geq \cdots \geq \lambda_m > 0 > \lambda_{m+1} \geq \cdots \geq \lambda_n \tag{5.4.6}$$

with

$$1 \leq m \leq \frac{n+1}{2} \quad \text{and} \quad \lambda_j + \lambda_{n+2-j} < 0, \quad j = 2, \dots, m. \tag{5.4.7}$$

Then

$$S(\{\lambda_2, \dots, \lambda_n\}) \leq \max\{S_1, S_2\} \tag{5.4.8}$$

where

$$S_1 = \max\left\{-\sum_{j=2}^{k} \lambda_j - \sum_{j=1}^{k} \lambda_{n+1-j} \ : \ k = 1, 2, \dots, m-1\right\}, \tag{5.4.9}$$

$$S_2 = -\sum_{j=2}^{n} \lambda_j. \tag{5.4.10}$$

Remark 5.4.2 Theorem 5.4.4 follows from Lemma 5.4.5 when $m = 1$.

The proof of Lemma 5.4.5 The proof is by induction on n. If $n = 2$, then $m = 1$, and hence, it follows from Theorem 5.4.4 that $S(\{\lambda_2\}) = -\lambda_2 = S_2$, i.e., (5.4.8) is true. If $n = 3$, then $m = 1$ or 2. For $m = 1$ it follows from Theorem 5.4.4 that $S(\{\lambda_2, \lambda_3\}) = -\lambda_2 - \lambda_3$; and for $m = 2$ it is clear that $\{-\lambda_3; \lambda_2, \lambda_3\} \in \mathcal{S}_3$, which implies that $S(\{\lambda_2, \lambda_3\}) = -\lambda_3$. This shows that the lemma is true for $n = 3$.

To prove the lemma for $n > 3$, assume that the lemma is true for all smaller sets. Now let

$$\lambda_1 = \max\{S_1, S_2\}, \quad \varepsilon_1 = \lambda_1 + \lambda_n, \quad \varepsilon_2 = -(\lambda_2 + \lambda_n),$$

where S_1 and S_2 defined by (5.4.9) and (5.4.10). Clearly, $\varepsilon_1 \geq 0$, by the definition of λ_1, and $\varepsilon_2 > 0$, by (5.4.7). Define $\varepsilon = \min\{\varepsilon_1, \varepsilon_2\}$. Since $\lambda_1 - \lambda_2 = \varepsilon_1 + \varepsilon_2 \geq 2\varepsilon$, it follows that

$$\lambda_1 - \varepsilon \geq \lambda_2 + \varepsilon. \tag{5.4.11}$$

If $\varepsilon = \varepsilon_1$, clearly, the set $\{\lambda_3, \lambda_4, \dots, \lambda_{n-1}\}$ satisfies the assumptions of the lemma, and hence, by the induction hypothesis,

$$S(\{\lambda_3, \lambda_4, \dots, \lambda_{n-1}\}) \leq \max\{\widetilde{S}_1, \widetilde{S}_2\},$$

where

$$\widetilde{S}_1 = \max\left\{-\sum_{j=2}^{k} \lambda_{j+1} - \sum_{j=1}^{k} \lambda_{n-j} \ : \ k = 1, \dots, m-2\right\}, \quad \widetilde{S}_2 = -\sum_{j=3}^{n-1} \lambda_j.$$

Note that

$$\begin{aligned}
\widetilde{S}_1 - \varepsilon &= -\sum_{j=2}^{k_0} \lambda_{j+1} - \sum_{j=1}^{k_0} \lambda_{n-j} - (\lambda_1 + \lambda_n) \\
&= -\sum_{j=2}^{k_0+1} \lambda_j - \sum_{j=1}^{k_0+1} \lambda_{n+1-j} + (\lambda_2 - \lambda_1) \leq (S_1 - \lambda_1) + \lambda_2 \leq \lambda_2, \\
\widetilde{S}_2 - \varepsilon &= -\sum_{j=3}^{n-1} \lambda_j - (\lambda_1 + \lambda_n) \\
&= -\sum_{j=2}^{n} \lambda_j + (\lambda_2 - \lambda_1) = (S_2 - \lambda_1) + \lambda_2 \leq \lambda_2.
\end{aligned}$$

We obtain

$$S(\{\lambda_3, \dots, \lambda_{n-1}\}) \leq \lambda_2 + \varepsilon.$$

On the other hand, it is clear that

$$S(\{\lambda_n\}) = -\lambda_n = \lambda_1 - \varepsilon.$$

Noting (5.4.11) and applying Theorem 5.4.2 with $\delta = \varepsilon$ we have

$$S(\{\lambda_2, \lambda_3, \ldots, \lambda_{n-1}, \lambda_n\}) \leq \lambda_1 = \max\{S_1, S_2\}.$$

Similarly, if $\varepsilon = \varepsilon_2$, then we can prove that

$$S(\{\lambda_3, \ldots, \lambda_{n-1}\}) \leq \lambda_1 - \varepsilon, \quad S(\{\lambda_n\}) = -\lambda_n = \lambda_2 + \varepsilon,$$

and hence, by Theorem 5.4.2, here too,

$$S(\{\lambda_2, \lambda_3, \ldots, \lambda_{n-1}, \lambda_n\}) \leq \lambda_1 = \max\{S_1, S_2\}.$$

Thus, by induction principle, the theorem is true for all positive integers n. □

We also need the following lemma.

Lemma 5.4.6 *Let* $\{\lambda_1, \ldots, \lambda_n\}$ *satisfy that*

$$\lambda_1 \geq \lambda_2 \geq \cdots \geq \lambda_m \geq 0 > \lambda_{m+1} \geq \cdots \geq \lambda_n$$

with

$$n \geq m \geq \frac{n}{2} \quad and \quad \lambda_j + \lambda_{n+1-j} \geq 0, \quad j = m+1, \ldots, n.$$

Then

$$\{\lambda_1; \lambda_2, \ldots, \lambda_n\} \in \mathcal{S}_n$$

Proof Clearly,

$$\{\lambda_{n+1-j}; \lambda_j\} \in \mathcal{S}_2, \ j = m+1, \ldots, n, \ \{\lambda_{n-m+1}; \lambda_{n-m+2}, \ldots, \lambda_m\} \in \mathcal{S}_{n-2(n-m)},$$

and hence $\{\lambda_1; \lambda_2, \ldots, \lambda_n\} \in \mathcal{S}_n$. □

Combining Lemma 5.4.5 and 5.4.6, we immediately obtain the main result of this section.

Theorem 5.4.7 *Let* $\{\lambda_1; \lambda_2, \ldots, \lambda_n\} \subset \mathbf{R}$ *be ordered in such a way that*

$$\lambda_1 \geq \lambda_2 \geq \cdots \geq \lambda_p > 0 > \lambda_{p+1} \geq \cdots \geq \lambda_m$$

and

$$\lambda_{m+1} \geq \cdots \geq \lambda_{m+q} \geq 0 > \lambda_{m+q+1} \geq \cdots \geq \lambda_n,$$

with

$$1 \leq p \leq \frac{m+1}{2}, \quad \lambda_j + \lambda_{m+2-j} < 0, \; j = 2, \ldots, p,$$

$$\sum_{j=1}^{k} (\lambda_j + \lambda_{m+1-j}) \geq 0, \quad k = 1, \ldots, p-1, \quad \sum_{j=1}^{m} \lambda_j \geq 0,$$

$$n - m \geq q \geq \frac{n-m}{2}, \quad \lambda_1 \geq \lambda_{m+1},$$

$$\lambda_{m+j} + \lambda_{n+1-j} \geq 0, \quad j = q+1, \ldots, n-m.$$

Then we have $\{\lambda_1; \lambda_2, \ldots, \lambda_n\} \in \mathcal{S}_n$.

By Theorems 5.4.7 and 5.4.3 we now have the following result.

Corollary 5.4.8 *Let* $\{\lambda_1; \lambda_2, \ldots, \lambda_n\}$ *satisfy the assumptions of Theorem* 5.4.7, *and assume further that* (5.4.12) *and* (5.4.13) *hold with strict inequalities. Then* $\{\lambda_1; \lambda_2, \ldots, \lambda_n\}$ *can be realized by a symmetric positive* $n \times n$ *matrix.*

Notes and References

Most of Section 5.3 is based on [94], but here some significant changes and improvement have been made, and moreover, we named the scalar $R(\mathcal{L})$ defined by (5.3.2) the smallest reliable spectral radius. Examples 5.3.1 and 5.3.2 are taken from [151]; Corollary 5.3.9 is due to Suleimanova [196]; Theorem 5.3.10 originally appeared in [139].

Section 5.4 is mainly based on [12] with some significant changes. Theorem 5.4.7 is due to Fiedler [65].

Additional references on inverse eigenvalue problems for nonnegative matrices include [73], [125], [149], [150], [161], [176], and [184].

Bibliography

[1] J.C. Alexander, " The Additive Inverse Eigenvalue Problems and Topological Degree," *Pro. Amer. Math. Soc.*, **70**(1978), 5–7.

[2] E. Allgower and K. Georg, "Simplicial and Continuation Methods for Approximating Fixed Points and Solutions to Systems of Equations," *SIAM Review*, **22**(1980), 28–85.

[3] P. Arbenz, W. Gander, and G.H. Golub, "Restricted Rank Modification of the Symmetric Eigenvalue Problem: Theoretical Constructions," *Lin. Alg. and Its Applic.*, **104**(1988), 75–95.

[4] M. Arnold and B.N. Dalta, "An Algorithm for Multi-input Eigenvalue Assignment Problem," *IEEE Trans. Automat. Contr.*, **AC-35**(1990), 1149–1152.

[5] G.E. Backus and J.F. Gilbert, "Numerical Applications of a Formalism for Geophysical Inverse Problems," *Geophysical Journal of the Royal Astronomical Society*, **13**(1967), 247-276.

[6] V. Barcilon, "Iterative Solution of the Inverse Sturm-Liouville Equation," *J. Math. Phys.*, **15**(1974), 429–436.

[7] V. Barcilon, "On the Uniqueness of Inverse Eigenvalue Problems," *Geophysical Journal of the Royal Astronomical Society*, **38**(1974), 287–298.

[8] V. Barcilon, "On the Solution of Inverse Eigenvalue Problems of High Orders," *Geophysical Journal of the Royal Astronomical Society*, **39**(1974), 153–154.

[9] V. Barcilon, "Inverse Problem for a Vibrating Beam," *J. Appl. Math. Phys.*, 27(1976), 346–358.

[10] V. Barcilon, " Discrete Analog of an Iterative Method for Inverse Eigenvalue Problems for Jacobi Matrices," *J. Appl. Math. Phys.*, 29(1978), 295–300.

[11] R. Bellman, *Introduction to Matrix Analysis*, 2nd ed., McGraw-Hill, New York, 1970.

[12] A. Berman and R.J. Plemmons, *Nonnegative Matrices in the Mathematical Sciences*, SIAM, Philadelphia, 1994.

[13] F.W. Biegler-König, "Construction of Band Matrices from Spectral Data," *Lin. Alg. and Its Applic.*, **40**(1981), 79–87.

[14] F.W. Biegler-König, "Sufficient Conditions for the Solubility of Inverse Eigenvalue Problems," *Lin. Alg. and Its Applic.*, **40**(1981), 89–100.

[15] F.W. Biegler-könig, "A Newton Iteration Process for Inverse Eigenvalue Problems," *Numer. Math.*, **37**(1981), 349–354.

[16] S. Bochner and W.T. Martin, *Several Complex Variables*, Princeton University Press, Princeton, NJ, 1948.

[17] Z. Bohte, "Numerical Solution of the Inverse Algebraic Eigenvalue Problem," *Comput. J.*, **10**(1968), 385–388.

[18] D. Boley and G.H. Golub, "Inverse Eigenvalue Problems for Band Matrices," in Proc. Dundee Conf. on Numerical Analysis, Spring-Verlag, Berlin, New York, 1977,

[19] D. Boley and G.H. Golub, "The Matrix Inverse Eigenvalue Problem for Periodic Jacobi Matrices," in *Pro. Fourth Symposium on Basic Problems of Numerical Mathematics*, Pilsen, Czechoslovakia, 1978, 63–78.

[20] D. Boley and G.H. Golub, "A Modified Method for Reconstructing Periodic Jacobi Matrices," *Math. Comput.*, **42**(1984), 143–150.

[21] D. Boley and G.H. Golub, "A Survey of Matrix Inverse Eigenvalue Problems," *Inverse Problems*, **3**(1987), 595–622.

[22] C.F. Borges, R. Frezza, and W.B. Gragg, "Some Inverse Eigenproblems for Jacobi and Arrow Matrices," *Numer. Lin. Alg. Applic.*, **2**(1995), 195-203.

[23] R.W. Brockett and C.I. Byrnes, "Multivariable Nyquist Criteria, Root Loci, and Pole Placement: a Geometric Viewpoint," *IEEE Trans. Automat. Contr.*, **AC-26**(1981), 271–284.

[24] R. Bru, J. Mas, and A.M. Urbano, "An Algorithm for the Single-input Pole Assignment Problem," *SIAM J. Matrix Anal. Appl.*, **15**(1994), 393–407.

[25] J.R. Bunch, C.P. Nielsen, and D.C. Sorensen, "Rank One Modification of the Symmetric Eigenproblem," *Numer. Math.*, **31**(1978), 31–48.

[26] R. Byers and S.G. Nash, "Approaches to Robust Pole Assignment," *Int. J. Control*, **49**(1989), 97–117.

[27] C.I. Byrnes and X.C. Wang, "The Additive Inverse Eigenvalue Problem for Lie Perturbations," *SIAM J. Matrix Anal. Appl.*, **14**(1993), 113–117.

[28] D.Y. Cai and D.L. Chu, "A New Algorithm for Pole Placement Using State Feedback," *Math. Numer. Sinica*, **12**(1990), 86–90.

[29] G.Q. Cai and Q.J. Zhou, "A New Algorithm for the Robust Complex Pole Assignment of Multivariable Systems," *Control Theory and Applications*, **6**(1989), 57–64.

[30] R.H. Chan, X.S. Xu, and H.M. Zhou, "On the Convergence Rate of a Quasi-Newton Method for Inverse Eigenvalue Problems," to appear in *SIAM J. Numer. Anal.*

[31] X.W. Chang, C.C. Paige, and G.W. Stewart, "Perturbation Analysis for the QR Factorization," submitted to *SIAM J. Matrix Anal. Appl.*

[32] B.Chari and H. Hochstadt, "A Direct and Inverse Problem for a Hill's Matrix with Double Eigenvalues," *Lin. Alg. and Its Applic.*, **52/53**(1983), 177–197.

[33] C.H. Chen, "On a Numerical Method for Robust Pole Assignment by Linear Output Feedback," *J. Comput. Math.*, **6**(1988), 59–67.

[34] M.T. Chu, "Solving Additive Inverse Eigenvalue Problems for Symmetric Matrices by Homotopy Method," *IMA J. Numer. Anal.*, **9**(1990), 331–342.

[35] M.T. Chu, "Numerical Methods for Inverse Singular Value Problems," *SIAM J. Numer. Anal.*, **29**(1992), 885–903.

[36] C.L. Cox and W.F. Moss, "Backward Error Analysis for a Pole Assignment Algorithm," *SIAM J. Matrix Anal. Appl.*, **10**(1989), 446–456.

[37] C.L. Cox and W.F. Moss, "Backward Error Analysis for a Pole Assignment Algorithm II: The Complex Case," *SIAM J. Matrix Anal. Appl.*, **13**(1992), 1159–1171.

[38] J. Cullum and R.A. Willoughby, *Lanczos Algorithms for Larg Symmetric Eigenvalue Computations, Vol. I Theory*, Birkhaüser, Boston, 1985.

[39] J. Cullum and R.A. Willoughby, *Lanczos Algorithms for Larg Symmetric Eigenvalue Computations, Vol. II Programs*, Birkhaüser, Boston, 1986.

[40] H. Dai, "On the Construction of a Jacobi Matrix from Its Spectrum and a Submatrix," Report NHJB-89-5760, Nanjing Aeronautical Institute, Nanjing, China, 1989.

[41] H. Dai, "Sufficient Conditions for the Solubility of an Algebraic Inverse Eigenvalue Problem," *Math. Numer. Sinica*, **11**(1989), 333–336.

[42] H. Dai, "The Numerically Stable Construction of Jacobi Matrices from Spectral Data," *J. Num. Method and Comp. Appl.*, **11**(1990), 27–34, in Chinese; translation in *Chinese J. Num. Math. and Appl.*, **12**(1990), 58–66.

[43] H. Dai, "Numerically Stable Construction of Real Symmetric Matrices from Spectral Data," *Math. Numer. Sinica*, **12**(1990), 157–166.

[44] B.N. Datta, "An algorithm to Assign Eigenvalues in a Hessenberg Matrix: Single Input Case," *IEEE Trans. Automat. Contr.*, **AC-5**(1987), 414–417.

[45] B.N. Datta, "Parallel and Large-scale Matrix Computations in Control: Some Ideas," *Lin. Alg. and Its Applic.*, **121**(1989), 243–264.

[46] B.N. Datta and K. Data, "Efficient Parallel Algorithms for Controllability and Eigenvalue Assignment Problems," *Pro. 25th IEEE Conf. Decision and Control*, Athens, Greece, 1986, 1611–1616.

[47] E.J. Davison, "On Pole Assignment in Multi-variable Linear Systems," *IEEE Trans. Automat. Contr.*, **AC-13**(1968), 747–748.

[48] E.J. Davison, "On Pole Placement in Linear Systems with Incomplete State Feedback," *IEEE Trans. Automat. Contr.*, **AC-15**(1970), 348–351.

[49] C. de Boor and G.H. Golub, "The Numerical Stable Reconstruction of a Jacobi Matrix from Spectral Data," *Lin. Alg. and Its Applic.*, **21**(1978). 245–260.

[50] G.N. de Oliveira, "Note on an Inverse Characteristic Value Problem," *Numer. Math.*, **15**(1970), 345–347.

[51] G.N. de Oliveira, "Matrices with Prescribed Principal Elements and Singular Values," *Canad. Math. Bull.*, **14**(1971), 247–249.

[52] G.N. de Oliveira, "On the Multiplicative Inverse Eigenvalue Problem," *Canad. Math. Bull.*, **15**(1972), 189–193.

[53] G.N. de Oliveira, "Matrix Inequalities and the Additive Inverse Eigenvalue Problem," *Computing*, **9**(1972), 95–100.

[54] G.N. de Oliveira, "Matrices with Prescribed Entries and Eigenvalues I," *Proc. Amer. Math. Soc.*, **37**(1973), 380–386.

[55] G.N. de Oliveira, "Matrices with Prescribed Entries and Eigenvalues II," *SIAM J. Appl. Math.*, **24**(1973), 414–417.

[56] P. Deift and T. Nanda, "On the Determination of a Tridiagonal Matrix from Its Spectrum and a Submatrix," *Lin. Alg. and Its Applic.*, **60**(1984), 43–55.

[57] J.W. Demmel, "On Condition Numbers and the Distance to the Nearest Ill-posed Problem," *Numer. Math.*, **51**(1987), 251–389.

[58] A.C. Downing and A.S. Householder, "Some Inverse Characteristic Value Problems," *J. Assoc. Comp. Mach.*, **3**(1956), 203–207.

[59] G.R. Duan, "A Simple Algorithm for Robust Pole Assignment in Linear Output Feedback," *IEEE Proc.*, Part D, **139**(1992), 465–469.

[60] G.R. Duan, "Robust Eigenstructure Assignment via Dynamical Compensators," *Automatica*, **29**(1993), 469–474.

[61] P. Eberlein, "On Measures of Non Normality of Matrices," *Amer. Math. Monthly*, **72**(1965), 995–996.

[62] M.M. Fahmy and H.S. Tantawy, "Eigenstructure Assignment via Linear State-feedback Control," *INT. J. Control*, **40**(1984), 161–178.

[63] M.M. Fahmy and J. O'Reilly, "On Eigenstructure Assignment in Linear Multivariable Systems," *IEEE Trans. Automat. Contr.*, **AC-27**(1982), 690–693.

[64] W.E.Jr. Ferguson, "The Construction of Jacobi and Periodic Jacobi Matrices with Prescribed Spectral," *Math. Comput.*, **35**(1980), 1203–1220.

[65] M. Fiedler, "Eigenvalues of Nonnegative Symmetric Matrices," *Lin. Alg. and Its Applic.*, **9**(1974), 119–142.

[66] L.R. Fletcher, "An Intermediate Algorithm for Pole Placement by Output Feedback in Linear Multivariable Control Systems," *INT. J. Control*, **31**(1980), 1121–1136.

[67] L.R. Fletcher, "An Inverse Eigenvalue Problem from Control Theory," in *Numerical Treatment of Inverse Problems for Differential and Integral Equations*, eds. P. Deuflhard and E.Hairer, Birkhauser, Boston, 1983, 161–170.

[68] R. Fletcher and D.J.D. Powell, "A Rapidly Convergent Descent Method for Minimization," *Computer J.*, **6**(1963), 163–168.

[69] G.E. Forsythe and C. Moler, *Computer Solution of Linear Algebraic Systems*, Prentice-Hall, Englewood, Cliffs, NJ, 1967.

[70] S. Friedland, "Matrices with Prescribed Off-diagonal Elements," *Isr. J. Math.*, **11**(1972), 184–189.

[71] S. Friedland, "On Inverse Multiplicative Eigenvalue Problems for Matrices," *Lin. Alg. and Its Applic.*, **12**(1975), 127–137.

[72] S. Friedland, "Inverse Eigenvalue Problems," *Lin. Alg. and Its Applic.*, **17**(1977), 15–51.

[73] S. Friedland, "On an Inverse Problem for Nonnegative and Eventually Nonnegative Matrices," *Israel J. Math.*, **29**(1978), 43–60.

[74] S. Friedland, "The Reconstruction of a Symmetric Matrix from the Spectral Data," *J. Math. Anal. Appl.*, **71**(1979), 412–422.

[75] S. Friendland, "Inverse Eigenvalue Problems for Symmetric Toeplitz Matrices," *SIAM J. Matrix Anal. Appl.*, **13**(1992), 1142–1153.

[76] S. Friedland, J. Nocedal and M. L. Overton, "The Formulation and Analysis of Numerical Methods for Inverse Eigenvalue Problems," *SIAM J. Num. Anal.*, **24**(1987), 634-667.

[77] F.R. Gantmacher, *The Theory of Matrices*, Vols. 1 and 2, Chelsea, New York, 1959.

[78] W. Gautschi, "On Inverses of Vandermonde and Confluent Vandermonde Matrices," *Numer. Math.*, **4**(1962), 117–123.

[79] W. Gautschi, "Norm Estimates for Inverse of Vandermonde Matrices," *Numer. Math.*, **23**(1975), 337–347.

[80] I.M. Gelfand and B.M. Levitan, "On the Determination of a Differential Equation from its Spectral Function," *Amer. Math. Soc. Trans.*, Ser. z, **1**(1955), 253–304.

[81] A.J. Geurts, "A Contribution to Theory of Condition," *Numer. Math.*, **39**(1982), 85–96.

[82] G.M.L. Gladwell, "The Inverse Problem for the Vibrating Beam," *Proc. R. Soc.*, **A 393**(1984), 277–295.

[83] G.M.L. Gladwell, "The Inverse Mode Problem for Lumped-mass Systems," *Q. J. Mech. Appl. Math.*, **39**(1986), 297–307.

[84] G.M.L. Gladwell, *Inverse Problems in Vibration*, Martinus Nijhoff, Dordrecht, The Netherlands, Boston, MA, 1986.

[85] G.M.L. Gladwell and J. A. Gbadeyan, "On the Inverse Problem of the Vibrating String or Rod," *Q. J. Mech. Appl. Math.*, **38**(1985), 169–174.

[86] G.H. Golub, "Some Modified Matrix Eigenvalue Problems," *SIAM Rev.*, **15**(1973), 318–334.

[87] G.H. Golub, "Some Uese of the Lanczos Algorithm in Numerical Linear Algebra," in *Topics in Numerical Analysis*, ed. J.J.H. Miller, Academic Press, New York, 1974.

[88] G.H. Golub and D. Boley, "Inverse Eigenvalue Problems for Band Matrices," in *Numerical Analysis*, ed. G.A. Watson, Springer-Verlag, New York, 1977, 23–31.

[89] G.H. Golub and R.R. Underwood, "Block Lancozs Method for Computing Eigenvalues," in *Mathematical Software III*, ed. J.R. Rice, Academic Press, New York, 1977, 364–377.

[90] G.H. Golub and C.F. Van Loan, *Matrix Computations*, 2nd ed., Johns Hopkins University Press, Baltimore and London, 1989.

[91] G.H. Golub and J.H. Welsch, "Calculation of Gauss Quadrature Rules," *Math Comput.*, **23**(1969), 221–230.

[92] W.B. Gragg and W.J. Harrod, "The Numerically Stable Reconstruction of Jacobi Matrices from Spectral Data," *Numer. Math.*, **44**(1984), 317-335.

[93] L.J. Gray and D.G. Wilson, "Construction of a Jacobi Matrix from Spectral Data," *Lin. Alg. and Its Applic.*, **14**(1976), 131–134.

[94] W.W. Guo, "Inverse Eigenvalue Problems for Nonnegative Matrices," M.Sc. Thesis, Department of Mathematics, Peking University, Beijing, China, 1991.

[95] K.P. Hadeler, "Ein Inverse Eigenwertproblem," *Lin. Alg. and Its Applic.*, **1**(1968), 83–101.

[96] K.P. Hadeler, "Newton-Verfahren für inverse Eigenwertaufgaben," *Numer. Math.*, **12**(1968), 35–39.

[97] K.P. Hadeler, "Multiplikative Inverse Eigenwertprobleme," *Lin. Alg. and Its Applic.*, **2**(1969), 65–86.

[98] K.P. Hadeler, "Existenz- und Eindeutigkeitssätze für inverse Eigenwertaufgaben mit Hilfe des topologischen Abbildungsgrades," *Arch. Rat. Mech. and Anal.*, **42**(1971), 317–322.

[99] O.H. Hald, "On Discrete and Numerical Inverse Sturm-Liouville Problems," Ph.D. Dissertation, Department of Mathematics, New York University, New York, 1972.

[100] O.H. Hald, "Inverse Eigenvalue Problems for Jacobi Matrices," *Lin. Alg. and Its Applic.*, **14**(1976), 63—85.

[101] O.H. Hald, "Discrete Inverse Sturm-Liouville Problems," *Numer. Math.*, **27**(1977), 249–256.

[102] O.H. Hald, "Inverse Eigenvalue Problems for Layered Media," *Comm. Pure Appl. Math.*, **30**(1977), 69–94.

[103] O.H. Hald, "The Inverse Sturm-Liouville Problem and the Rayleigh-Ritz Method," *Math. Comp.*, **32**(1978), 687–705.

[104] O.H. Hald, "The Inverse Sturm-Liouville Problem with Symmetric Potentials," *Acta Math.*, **141**(1978), 263–291.

[105] O.H. Hald, "Discontinuous Inverse Eigenvalue Problems," *Comm. Pure Appl. Math.*, **37**(1984), 539–577.

[106] C. He, A.J. Laub, and V. Mehrmann, "Placing Plenty of Poles Is Pretty Preposterous," Report Spc **95-17**, Fakultät für Mathematik, TU Chemnitz-Zwickau, Chemnitz, Germany, 1995.

[107] M. Hegland and J.T. Marti, "Algorithms for the Reconstruction of Special Jacobi Matrices from Their Eigenvalues," *SIAM J. Matrix Anal. Appl.*, **10**(1989), 219–228.

[108] P. Henrici, "Bounds for Iterates, Inverse, Spectral Variation and Fields of Values of Nonnormal Matrices," *Numer. Math.*, **4**(1962), 24–40.

[109] M. Heymann, "On Pole Assignment in Multi-input Controllable Linear Systems," *IEEE Trans. Automat. Contr.*, **AC-13**(1968), 748–749.

[110] H. Hochstadt, "On Some Inverse Problems in Matrix Theory," *Arch. Math.*, **18**(1967), 201–207.

[111] H. Hochstadt, "The Inverse Sturm-Liouville Problem," *Comm. Pure Applied Math.*, **26**(1973), 715–729.

[112] H. Hochstadt, "On the Construction of a Jacobi Matrix from Spectral Data," *Lin. Alg. and Its Applic.*, **8**(1974), 435–446.

[113] H. Hochstadt, "Well Posed Inverse Spectral Problems," *Proc. Nat. Acad. Sci.*, **72**(1975), 2496–2497.

[114] H. Hochstadt, "On the Construction of a Jacobi Matrix from Mixed Given Data," *Lin. Alg. and Its Applic.*, **28**(1979), 113–115.

[115] H. Hochstadt and W. Goldberg, "An Inverse Problem for a Differential Operator with Mixed Spectrum," *J. Math. Anal. Appl.*, **105**(1985), 206–221.

[116] H. Hochstadt and M.W. Kim, "On a Singular Inverse Eigenvalue Problem," *Arch. Rat. Mech. Anal.*, **37**(1970), 243–254.

[117] H. Hochstadt and B. Lieberman, "An Inverse Sturm-Liouville Problem with Mixied Given Data," *SIAM J. Appl. Math.*, **34**(1978), 676–680.

[118] R.A. Horn and C.R. Johnson, *Matrix Analysis*, Cambridge University Press, Cambridge, 1985.

[119] R.A. Horn and C.R. Johnson, *Topics in Matrix Analysis*, Cambridge University Press, Cambridge, 1991.

[120] A.S. Householder, *The Theory of Matrices in Numerical Analysis*, Dover Publications, New York, 1974.

[121] E.X. Jiang, "On Spectral Variation of a Nonnormal Matrix," *Lin. Alg. and Its Applic.*, **42**(1982), 223–241.

[122] E.X. Jiang, "Inverse Eigenvalue Problems for Special Matrices," *Numer. Math., J. Chinese Univ.*, **9**(1987), 198–208.

[123] J. Kautsky, N.K. Nichols, P. Van Dooren, and L. Fletcher, "Numerical Methods for Robust Eigenstructure Assignment in Control System Design," in *Numerical Treatment of Inverse Problems for Differential and Integral Equations*, eds. P. Deuflhard and E.Hairer, Birkhauser Boston, 1983, 171–178.

[124] J. Kautsky, N.K. Nichols, and P. Van Dooren,"Robust Pole Assignment in Linear State Feedback,"*Int. J. Control*, **41**(1985), 1129–1155.

[125] R.B. Kellog, "Matrices Similar to a Positive or Essentially Positive Matrix," *Lin. Alg. and Its Applic.*, **4**(1971), 191–204.

[126] M.M. Konstantinov, N.D. Christov, and P. Hr. Petkov, "Perturbation Analysis of Linear Control Problems," IFAC 10th Congress, Munich, **9**(1987), 16–21.

[127] M.M. Konstantinov and P. Hr. Petkov, "Conditioning of Linear State Feedback," Report 93-61, Department of Engineering, Leicester University, 1993.

[128] B. Kouvaritakis and R. Cameron, "Pole Placement with Minimized Norm Controllers," *IEEE Proc.*, Part D, **127**(1980), 32–36.

[129] F. Laborde, "Sur un Problèm Inverse d'un Problème de valeurs Propres," *C.R. Acad. Sci. Paris*, **268**(1969), A-153–156.

[130] H.J. Landau, "The Inverse Eigenvalue Problem for Real Symmetric Toeplitz Matrices," *J. Amer. Math. Soc.*, **7**(1994), 749–767.

[131] A.J. Laub, "Numerical Linear Algebra Aspects of Control Design Computations," *IEEE Trans. Automat. Contr.*, **AC-30**(1983), 97–108.

[132] A.J. Laub and M. Wette, " Algorithms and Software for Pole Assignment and Observers," Report UCRL-15646, Department of Electrical and Computer Engineering, University of California, Santa Barbara, 1985.

[133] D.P. Laurie, "A Numerical Approach to the Inverse Toeplitz Eigenproblem," *SIAM J. Sci. Stat. Comput.*, **9**(1988), 401–405.

[134] L.L. Li, "Some Sufficient Conditions for the Solubility of Inverse Eigenvalue Problems," *Lin. Alg. and Its Applic.*, **148**(1991), 225–236.

[135] L.L. Li, "On Solubility of an Inverse Eigenvalue Problem," *J. Comput. Math.*, Supplementary Issue, 1992, 22–26.

[136] R.C. Li, "QR Decomposition and Nonlinear Eigenvalue Problems," *Math. Numer. Sinica*, **11**(1989), 374–385.

[137] R.C. Li, "Algorithms for Inverse Eigenvalue Problems," *J. Comput. Math.*, **10**(1992), 97–111.

[138] X.G. Liu, Perturbation Analysis for the Inverse Eigenvalue Problems for Jacobi Matrices, *J. Comput. Math.*, Supplementary Issue, 1992, 200-204.

[139] R. Loewy and D. London, " A Note on an Inverse Problem for Nonnegative Matrices," *Lin. Alg. and Its Applic.*, **6**(1978), 83–90.

[140] G. Loizou, "Nonnormality and Jordan Condition Numbers of Matrices," *J. Assoc. Comp. Mach.*, **16**(1969), 580–584.

[141] L.H. Loomis and S. Sternberc, *Advanced Calculus*, Addison-Wesley, Menlo Park, California, 1968.

[142] D.G. Luenberger, *Introduction to Linear and Nonlinear Programming*, Addison-Wesley, New York, 1973.

[143] D.G. Luenberger, *Introduction to Dynamic Systems*, John Wiley and Sons, New York, 1979.

[144] J.F. Magni and C. Champetier, "A Geometric Framework for Pole Assignment Algorithms,' *IEEE Trans. Automat. Contr.*, **AC-36**(1991), 1105–1111.

[145] M. Marcus, *Finite Dimensional Multilinear Algebra*, Part 1, Dekker, New York, 1973.

[146] M. Marcus, *Matrices and MATLAB*^{TM}*: a Tutorial*, Prentice-Hall, Englewood Cliffs, NJ, 1993.

[147] M.P. Mattis and H. Hochstadt, "On the Construction of Band Matrices from Spectral Data," *Lin. Alg. and Its Applic.*, **38**(1981), 109–119.

[148] V. Mehrmann and H. Xu, "An Analysis of the Pole Placement Problem: I. The Single-Input Case," Report Spc **95-32**, Fakultät für Mathematik, TU Chemnitz-Zwickau, Chemnitz, Germany, 1995.

[149] H. Minc, "Inverse Elementary Divisor Problem for Nonnegative Mtrices," *Prec. Amer. Math. Soc.*, **83**(1981), 665–670.

[150] H. Minc, "Inverse Elementary Divisor Problem for Doubly Stochastic Matrices," *Linear and Multilinear Alg.*, **11**(1982), 121–131.

[151] H. Minc, *Nonnegative Matrices*, John Wiley and Sons, New York, 1989.

[152] G.S. Miminis, "Numerical Algorithms for Controllability and Eigenvalue Location," M.Sc. Thesis, School of Computer Science, McGill University, Montreal, Quebee, Canada, 1981.

[153] G.S. Miminis, "Numerical Algorithms for the Pole Placement Problem," Ph.D. dissertation, School of Computer Science, McGill University, Montreal, Quebee, Canada, 1985.

[154] G.S. Miminis, "Deflation in Eigenvalue Assignment of Descriptor Systems Using State Feedback," *IEEE Trans. Automat. Contr.*, **AC-38**(1993), 1322–1336.

[155] G.S. Miminis and C.C. Paige, "An Algorithm for Pole Assignment of Time Invariant Linear Systems," *Int. J. Control*, **35**(1982), 341–354.

[156] G.S. Miminis and C.C. Paige, "An Algorithm for Pole Assignment of Time-invariant Multi-input Linear Systems," in *Proc. 21st IEEE Conf. on Decision and Control*, Orlando, Florida, **1**(1982), 62–67.

[157] G.S. Miminis and C.C. Paige, "A Direct Algorithm for Pole Assignment of Time-invariant Multi-input Linear Systems Using State Feedback," *Automatica*, **24**(1988), 343–356.

[158] G.S. Miminis and C.C. Paige, "A QR-like Approach for the Eigenvalue Assignment Problem," in *Proceedings of the 2nd Hellenic Conference on Mathematics and Informatics*, Athens, Greece, Sept., 1994.

[159] G.S. Miminis and M. Reid, "A FORTRAN Subroutine to Solve the Eigenvalue Assignment Problem for Single-input Systems," *ACM Trans. Math. Soft.*, **19**(1993), 224-232.

[160] G.S. Miminis and H. Roth, "A FORTRAN Subroutine to Solve the Eigenvalue Assignment Problem for Multi-input Systems Using State Feedback," to appear in *ACM Trans. Math. Soft.*

[161] J. Mirsky, "Inequalities and Existence Theorems in the Theory of Matrices," *J. Math. Anal. Appl.*, **9**(1964), 99–118.

[162] C.B. Moler, J.N. Little, and S. Bangert, *PC-Matlab User's Guide*, The Math Works Inc., Mass, 1987.

[163] P. Morel, "Sur le Problèm Inverse des Valeurs Propres," *Numer. Math.*, **23**(1974), 83–94.

[164] P. Morel, "Des Algorithms pour le Problèm Inverse des Valeurs Propres," *Lin. Alg. and Its Applic.*, **13**(1976), 251–273.

[165] F.D. Murnaghan and A. Wintner, "A Canonical Form for Real Matrices Under Orthogonal Transformations," *Proc. Nat. Acad. Sci.*, **17** (1931), 417–420.

[166] N.K. Nichols, "On Computational Algorithms for Pole Assignment," *IEEE Trans. Automat. Contr.*, **AC-31**(1986), 643–645.

[167] J. Nocedal and N.L. Overton, "Numerical Methods for Solving Inverse Eigenvalue Problems," in *Numerical Methods, Lecture Notes in Math.*, No. 1005, eds. V. Pereyra and A. Reinoza, Springer-Verlag, New York, 1983, 212–226.

[168] J.M. Ortega and W.C. Rheinboldt, *Iterative Solution of Nonlinear Equations in Several Variables*, Academic Press, New York, 1970.

[169] M.R. Osborne, "On the Inverse Eigenvalue Problem for Matrices and Related Problems for Difference and Differential Equations," *Conference on Application of Numerical Analysis, Lecture Notes in Math.*, No. 228, Springer-Verlag, Berlin, New York, 1971, 155–168.

[170] C.C. Paige, "The Computation of Eigenvalues and Eigenvectors of Very Large Sparse Matrices," Ph.D. dissertation, University of London, London, England, 1971.

[171] C.C. Paige, "Properties of Numerical Algorithms Related to Computing Controllability," *IEEE Trans. Automat. Contr.*, **AC-26**(1981), 130–138.

[172] J. Paine, "A Numerical Method for the Inverse Sturm-Liouville Problem," *SIAM J. Sci. Stat. Comput.*, **5**(1984), 149–156.

[173] R.L. Parker and K.A. Whaler, "Numerical Methods for Establishing Solutions to the Inverse Problem of Electromagnetic Induction," *J. Geophys. Res.*, **86**(1981), 9574–9584.

[174] B.N. Parlett, *The Symmetric Eigenvalue Problem*, Prentice-Hall, Englewood Cliffs, NJ, 1980.

[175] R.V. Patel and P. Misra, "Numerical Algorithm for Eigenvalue Assignment by State Feedback," *Proc. IEEE* , **72**(1984), 1755–1764.

[176] H. Perfect, " Methods of Constructing Certain Stochastic Matrices," *Duke Math. J.*, **20**(1953), 395–404.

[177] P.Hr. Petkov, N.D. Christov and M.M. Koustantinov, "A Computational Algorithm for Pole Assignment of Linear Single-input Systems," *IEEE Trans. Autom. Control*, **AC-29**(1984), 1045-1048.

[178] P.Hr. Petkov, M.M. Koustantinov and N.D. Christov, "Computational Algorithms for Linear Control Systems: a Brief Survey," *INT. J. Systems Sci.*, **16**(1985), 465–477.

[179] P.Hr. Petkov, N.D. Christov, and M.M. Koustantinov, "A Computational Algorithm for Pole Assignment of Linear Multi-input Systems," *IEEE Trans. Automat. Contr.*, **AC-31**(1986), 1044-1047.

[180] F. Rellich, *Perturbation Theory of Eigenvalue Problems*, Lecture Notes, Math Department, New York University, New York, 1953.

[181] J.R. Rice, "A Theory of Condition," *SIAM J. Numer. Anal.*, **3** (1966), 287–310.

[182] S. Richter and R. DeCarlo, "A Homotopy Method for Eigenvalue Assignment Using Decentralized State Feedback," *IEEE Trans. Automat. Contr.*, **AC-29**(1984), 148–158.

[183] H. Rutishauser, "On Jacobi Rotation Patterns," in *Experimental Arithmetic, High Speed Computing and Mathematics, Proc. Symp. Appl. Math. 15, Amer. Math. Soc.*, Providence, 1963, 219–239.

[184] F.L. Salzmann, "A Note on Eigenvalues of Nonnegative Matrices," *Lin. Alg. and Its Applic.*, **5**(1972), 329–338.

[185] S. Šarman, "Bemerkungen zu inversen Eigenwertproblemen," *Computing*, **4**(1969), 207–215.

[186] I. Schur, "On the Characteristic Roots of a Linear Substitution with an Application to the Theory of Integral Equations," *Math. Ann.*, **66**(1909), 488–510.

[187] B. Shafai and S.P. Bhattacharyya, "An Algorithm for Pole Assignment in High Oder Multivariable Systems," *IEEE Trans. Automat. Contr.*, **AC-33**(1988), 870–876.

[188] A. Shapiro, "On the Unsolvability of Inverse Eigenvalue Problems Almost Everywhere," *Lin. Alg. and Its Applic.*, **49**(1983), 27–31.

[189] Q.J. Shen, "An Algorithm of Jacobi Matrix Inverse Eigenvalue Problems," *Acta Scientiarum Naturalium Universitatis Nankaiensis*, No. 1, 1986, 12–17.

[190] Q.J. Shen, "A New Fast Algorithm of Reducing a Symmetric Bordered Diagonal Matrix to Tridiagonal Form," in *Proceedings of the 1991 Tianjin Conference on Computational Mathematics*, Tianjin, China, 521–523.

[191] M. Shub, "The Implicit Function Theorem Revisited," *IBM J. Res. Develop.*, **38**(1994), 259–264.

[192] R.A. Smith, "The Condition Numbers of the Matrix Eigenvalue Problem," *Numer. Math.*, **10**(1967), 232-240.

[193] G.W. Stewart, *Introduction to Matrix Computations*, Academic Press, New York, NY, 1973.

[194] G.W. Stewart, "The Economical Storage of Plane Rotations," *Numer. Math.*, **25**(1976), 137–138.

[195] G.W. Stewart and J.G. Sun, *Matrix Perturbation Theory*, Academic Press, New York, NY, 1990.

[196] H.R. Suleimanova, "Stochastic Matrices with Real Eigenvalues," *Dukl. Akad. Hauk SSSR*, **66**(1949), 343–345.

[197] J.G. Sun, "Eigenvalues and Eigenvectors of a Matrix Dependent on Several Parameters," *J. Comput. Math.*, **3**(1985), 351–364.

[198] J.G. Sun, "Some New Results on Algebraic Inverse Eigenvalue Problems," in *Proceedings of the First Chinese National Conference on Inverse Eigenvalue Problems*, 1986.

[199] J.G. Sun, "The Unsolvability of Multiplicative Inverse Eigenvalue Problems Almost Everywhere," *J. Comput. Math.*, **4**(1986), 227–244.

[200] J.G. Sun, "The Stability Analysis of the Solutions of Inverse Eigenvalue Problems," *J. Comput. Math.*, **4**(1986), 345–353.

[201] J.G. Sun, "On Numerical Methods for Robust Pole Assignment in Control System Design," *J. Comput. Math.*, **5**(1987), 119–134.

[202] J.G. Sun, "On Numerical Methods for Robust Pole Assignment in Control System Design (II)," *J. Comput. Math.*, **5**(1987), 352–363.

[203] J.G. Sun, "On the Sufficient Conditions for the Solubility of Algebraic Inverse Eigenvalue Problems," *Math. Numer. Sinica*, **9**(1987), 49–59.

[204] J.G. Sun, "Measures of Robustness in Control System Design," *Math. Numer. Sinica*, **9**(1987), 319–326.

[205] J.G. Sun, "A Note on Simple Non-zero Singular Values," *J. Comput. Math.*, **6**(1988), 258–266.

[206] J.G. Sun, "A Note on Pole Assignment," *Math. Numer. Sinica*, **10**(1988), 438–443.

[207] J.G. Sun, "Multiple Eigenvalue Sensitivity Analysis," *Lin. Alg. and Its Applic.*, **137/138**(1990), 183–211.

[208] J.G. Sun, "On Perturbation Bounds for the QR Factorization," *Lin. Aly. and Its Applic.*, **215**(1995), 95–112.

[209] J.G. Sun, "Perturbation Analysis of the Pole Assignment Problem," *SIAM J. Matrix Anal. Appl.*, **17**(1996), 313–331.

[210] J.G. Sun and Q. Ye, "The Unsolvability of Inverse Algebraic Eigenvalue Problems Almost Everywhere," *J. Comput. Math.*, **4**(1986), 212–226.

[211] D.G. Teng, "Inverse Eigenvalue Problem for Some Special Matrices," *Lin. Alg. and Its Applic.*, **139**(1990), 63–65.

[212] R.C. Thompson, "Counterexamples Concerning the Diagonal Elements of Normal Matrices," *Lin. Alg. and Its Applic.*, **49**(1983), 285–291.

[213] P. Van Dooren, "The Generalized Eigenstructure Problem in Linear System Theory," *IEEE Trans. Automat. Contr.*, **AC-26**(1981), 111–129.

[214] P. Van Dooren, "Deadbeat Control: a Special Inverse Eigenvalue Problem," *BIT*, **24**(1984), 681–699.

[215] A. Varga, "A Schur Method for Pole Assignment," *IEEE Trans. Automat. Contr.*, **AC-26**(1981), 517–519.

[216] R.S. Varga, *Matrix Iterative Analysis*, Prentica-Hall, Englewood Cliffs,NJ, 1962.

[217] J.Y. Wang and B.S. Garbow, "A Numerical Method for Solving Inverse Real Symmetric Eigenvalue Problems," *SIAM J. Sci. Stat. Comput.*, **4**(1983), 45–51.

[218] Z.K. Wang and T.A. Gao, *An Introduction to Homotopy Methods* (in Chinese), Chongqing Publishing House, Chongqing, China, 1990.

[219] D.S. Watkins, *Fundamentals of Matrix Computations*, John Wiley and Sons, New York, 1991.

[220] J.H. Wilkinson, *Rounding Errors in Algebraic Processes*, Prentice-Hall, Englewood Cliffs, NJ, 1963.

[221] J.H. Wilkinson, *The Algebraic Eigenvalue Problem*, Clarendon Press, Oxford, England, 1965.

[222] W.M. Wonham, "On Pole Assignment in Multi-input Controllable Linear Systems," *IEEE Trans. Automat. Control*, **AC-12**(1967), 660–665.

[223] W.M. Wonham, *Linear Multivariable Control: A Geometric Approach*, 2nd ed., Springer-Verlag, New York, 1979.

[224] S.F. Xu, "On the Qualitative Theory and Numerical Methods for the Classical Algebraic Inverse Eigenvalue Problems," Ph.D dissertation, Computing Center, Academia Sinica, Beijing, China, 1989.

[225] S.F. Xu, "The Solvability of Algebraic Inverse Eigenvalue Problems Almost Everywhere," J. Comput. Math., Supplementary Issue, 1992, 152–157.

[226] S.F. Xu, "On the Necessary Conditions for the Solubility of Algebraic Inverse Eigenvalue Problems," *J. Comput. Math.*, **10**(1992), 93–97.

[227] S.F. Xu, "On the Sufficient Conditions for the Solubility of Algebraic Inverse Eigenvalue Problems," *J. Comput. Math.*, **10**(1992), 171–180.

[228] S.F. Xu, "A Stability Analysis of the Solutions of General Algebraic Inverse Eigenvalue Problems," *Numer. Math. Sinica*, **14**(1992), 33–43, in Chinese; Translation in *Chinese J. Numer. Math. and Appl.*, **14**(1992), 13–26.

[229] S.F. Xu, "A General Formulation for Inverse Algebraic Eigenvalue Problem," *Numer. Math. Sinica*, **14**(1992), 498–505, in Chinese; translation in *Chinese J. Numer. Math. and Appl.*, **15**(1993), 84–93.

[230] S.F. Xu, "The Sensitivity Analysis of the Solutions of Additive Inverse Eigenvalue Problems," in *Post Doctoral Theses of China*, eds. G.Q. Wang et al., Peking University Press, Beijing, 1991, 51–61.

[231] S.F. Xu, "A Homotopy Algorithm for Solving Inverse Eigenvalue Problems," *Math. Numer. Sinica*, **14**(1992), 199–206.

[232] S.F. Xu, "A Homotopy Algorithm for Solving the Inverse Eigenvalue Problems for Complex Symmetric Matrices," *J. Comput. Math.*, **11**(1993), 7–19.

[233] S.F. Xu, "A Stability Analysis of the Jacobi Matrix Inverse Eigenvalue Problem," *BIT* **33**(1993), 695-702.

[234] S.F. Xu, "A Survey of Algebraic Inverse Eigenvalue Problems," in *Proceedings of Conference on Scientific and Engineering Computing for Young Chinese Scientists*, eds. J.Z. Cui, Z.C. Shi, and D.L. Wang, National Defence Industry Press, Beijing, 1994, 87–92.

[235] S.F. Xu, *Theory and Methods of Matrix Computations* (in Chinese), Peking University Press, Beijing, 1995.

[236] S.F. Xu, " On the Jacobi Matrix Inverse Eigenvalue Problem with Mixed Given Data," *SIAM J. Matrix Anal. Appl.*, **17**(1996), 632–639.

[237] S.F. Xu, "A Small Singular Value Method for Solving Inverse Eigenvalue Problems," *J. Comput. Math.*, **14**(1996), 23–31.

[238] Q. Ye, "A Class of Iterative Algorithm for Solving Inverse Eigenvalue Problems," *Math. Numer. Sinica*, **9**(1987), 144–153.

[239] Q. Ye, "The Unsolvability of Inverse Eigenvalue Problems for Hermitian Matrices Almost Everywhere," *Math. Numer. Sinica*, **9**(1987), 225–232.

[240] H.Y. Zha and Z.Y. Zhang, "A Note on Constructing a Symmetric Matrix with Specified Diagonal Entries and Eigenvalues," *BIT*, **35**(1995), 448–452.

[241] Z.Y. Zhang, "Inverse Eigenproblems for Tridiagonal Matrices," *Math. Numer. Sinica*, **13**(1991), 76–83.

[242] Z.Y. Zhang, "An Inverse Extreme Eigenpair Problem and Its Parallel Iterative Solution for Symmetric Tridiagonal Matrices," *SIAM J. Numer. Anal.*, **32**(1995), 1620–1634.

[243] S.Q. Zhou and H. Dai, *Algebraic Inverse Eigenvalue Problems* (in Chinese), Henan Science and Technology Press, Henan, China, 1991.

[244] B.R. Zhu, "The Perturbation Method for Inverse Eigenvalue Problems," *Acta Math. Appl. Sinica*, **10**(1987), 106–113.

[245] B.R. Zhu, K.R. Jacson and R.P.K. Chan, "Inverse Spectrum Problems for Block Jacobi Matrix," *J. Comput. Math.*, **11**(1993), 313–322.

Index